Applied Electrospray Mass Spectrometry

PRACTICAL SPECTROSCOPY
A SERIES

1. Infrared and Raman Spectroscopy (in three parts), *edited by Edward G. Brame, Jr., and Jeanette G. Grasselli*
2. X-Ray Spectrometry, *edited by H. K. Herglotz and L. S. Birks*
3. Mass Spectrometry (in two parts), *edited by Charles Merritt, Jr., and Charles N. McEwen*
4. Infrared and Raman Spectroscopy of Polymers, *H. W. Siesler and K. Holland-Moritz*
5. NMR Spectroscopy Techniques, *edited by Cecil Dybowski and Robert L. Lichter*
6. Infrared Microspectroscopy: Theory and Applications, *edited by Robert G. Messerschmidt and Matthew A. Harthcock*
7. Flow Injection Atomic Spectroscopy, *edited by Jose Luis Burguera*
8. Mass Spectrometry of Biological Materials, *edited by Charles N. McEwen and Barbara S. Larsen*
9. Field Desorption Mass Spectrometry, *László Prókai*
10. Chromatography/Fourier Transform Infrared Spectroscopy and Its Applications, *Robert White*
11. Modern NMR Techniques and Their Application in Chemistry, *edited by Alexander I. Popov and Klaas Hallenga*
12. Luminescence Techniques in Chemical and Biochemical Analysis, *edited by Willy R. G. Baeyens, Denis De Keukeleire, and Katherine Korkidis*
13. Handbook of Near-Infrared Analysis, *edited by Donald A. Burns and Emil W. Ciurczak*
14. Handbook of X-Ray Spectrometry: Methods and Techniques, *edited by René E. Van Grieken and Andrzej A. Markowicz*
15. Internal Reflection Spectroscopy: Theory and Applications, *edited by Francis M. Mirabella, Jr.*
16. Microscopic and Spectroscopic Imaging of the Chemical State, *edited by Michael D. Morris*
17. Mathematical Analysis of Spectral Orthogonality, *John H. Kalivas and Patrick M. Lang*
18. Laser Spectroscopy: Techniques and Applications, *E. Roland Menzel*
19. Practical Guide to Infrared Microspectroscopy, *edited by Howard J. Humecki*
20. Quantitative X-ray Spectrometry: Second Edition, *Ron Jenkins, R. W. Gould, and Dale Gedcke*
21. NMR Spectroscopy Techniques: Second Edition, Revised and Expanded, *edited by Martha D. Bruch*
22. Spectrophotometric Reactions, *Irena Nemcova, Ludmila Cermakova, and Jiri Gasparic*
23. Inorganic Mass Spectrometry: Fundamentals and Applications, *edited by Christopher M. Barshick, Douglas C. Duckworth, and David H. Smith*
24. Infrared and Raman Spectroscopy of Biological Materials, *edited by Hans-Ulrich Gremlich and Bing Yan*

25. Near-Infrared Applications in Biotechnology, *edited by Ramesh Raghavachari*
26. Ultrafast Infrared and Raman Spectroscopy, *edited by M. D. Fayer*
27. Handbook of Near-Infrared Analysis: Second Edition, Revised and Expanded, *edited by Donald A. Burns and Emil W. Ciurczak*
28. Handbook of Raman Spectroscopy: From the Research Laboratory to the Process Line, *edited by Ian R. Lewis and Howell G. M. Edwards*
29. Handbook of X-Ray Spectrometry: Second Edition, Revised and Expanded, *edited by René E. Van Grieken and Andrzej A. Markowicz*
30. Ultraviolet Spectroscopy and UV Lasers, *edited by Prabhakar Misra and Mark A. Dubinskii*
31. Pharmaceutical and Medical Applications of Near-Infrared Spectroscopy, *Emil W. Ciurczak and James K. Drennen III*
32. Applied Electrospray Mass Spectrometry, *edited by Birendra N. Pramanik, A. K. Ganguly, and Michael L. Gross*

ADDITIONAL VOLUMES IN PREPARATION

Applied Electrospray Mass Spectrometry

edited by

Birendra N. Pramanik
Schering-Plough Research Institute
Kenilworth, New Jersey

A. K. Ganguly
Stevens Institute of Technology
Hoboken, New Jersey

Michael L. Gross
Washington University
St. Louis, Missouri

CRC Press is an imprint of the
Taylor & Francis Group, an informa business

ISBN: 0-8247-0618-8

This book is printed on acid-free paper.

Headquarters
Marcel Dekker, Inc.
270 Madison Avenue, New York, NY 10016
tel: 212-696-9000; fax: 212-685-4540

The publisher offers discounts on this book when ordered in bulk quantities. For more information, write to Special Sales/Professional Marketing at the headquarters address above.

Copyright © 2002 by Marcel Dekker   All Rights Reserved.

Neither this book nor any part may be reproduced or transmitted in any form or by any means, electronic or mechanical, including photocopying, microfilming, and recording, or by any information storage and retrieval system, without permission in writing from the publisher.

Current printing (last digit):
10 9 8 7 6 5 4 3

PRINTED IN THE UNITED STATES OF AMERICA

Preface

In the last decade, electrospray ionization mass spectrometry (ESI–MS) has become the most universally applied mass spectrometric technique for nonvolatile, thermal-labile molecules ranging in size from small compounds (300–1000 Da) to large biomolecules with molecular weights in excess of 150 kDa. Electrospray ionization emerged from the pioneering experiments of J. Zeleny (1917) and later those of Malcolm Dole (late 1960s), who performed a series of experiments defining the parameters required for introducing large molecules as ions into the gas phase of a mass spectrometer. In 1984, John Fenn and his coworker M. Yamashita in the United States and M. L. Aleksandrov and his coworkers in the U.S.S.R. first reported the successful coupling of ESI with quadrupole and magnetic sector mass analyzers, respectively. Because ESI is relatively simple and works at atmospheric pressure, it can be easily coupled to both high-performance liquid chromatography, allowing the analysis of complex mixtures, and devices for automated sample analysis — satisfying the quest among mass spectrometrists to provide an interface to liquid separation methods. From the beginning, ESI showed signs that it would become the primary MS tool in pharmaceutical and biomedical research, especially in high-throughput screening of large chemical libraries, impurity analysis, pharmacokinetics and metabolite identification, and protein characterization.

This book brings together the work of contributors from academia and industry who have played important roles in the continuing development of this technique. Our goal is to present a compilation of the latest applications of ESI to practitioners of mass spectrometry, as well as to other scientists interested in applying this technique to their areas of research. The principal theme of the book is mass spectrometric analysis in the areas of pharmaceutical, peptide, and protein chemistry.

The book begins with a historical review of the ESI technique — its principles and instrumentation (Chapter 1), followed by a description of nanospray development and applications in macromolecules (Chapter 2). The remaining chapters cover applications of small molecules (Chapters 3, 4, and 5), peptides/proteins (Chapters 6, 7, and 8), noncovalent complexes (Chapter 9), special topics in protein folding/unfolding (Chapter 10), and clinical/biomedical applications (Chapter 11). Topics such as the structural elucidation of complex natural products (Chapter 3), analysis of libraries of compounds prepared using combinatorial chemistry (Chapter 4), and the application of ESI-MS in the field of pharmacokinetics and drug metabolism studies (Chapter 5) are of great importance to scientists in the pharmaceutical industries and are described in great detail. Chapters 6, 7, and 8 explore the critical role of the ESI technique in structural characterization of peptides/proteins, especially in the rapidly evolving area of proteomics and in identifying gene-expressed proteins in a high-throughput process (Chapter 8). Dynamic structures of proteins, such as noncovalent interactions and folding/unfolding, are further covered in Chapters 9 and 10. Finally, Chapter 11 deals with the application of microflow ESI–MS in problems in the field of clinical and biomedical research.

We acknowledge the special efforts of all the authors, who have made great contributions to this book. We also thank the Schering-Plough Research Institute, Washington University, and the National Center for Research Resources of the National Institutes of Health for their support. Special thanks go to the acquisitions and production editors at Marcel Dekker, Inc., for their assistance.

Birendra N. Pramanik
A. K. Ganguly
Michael L. Gross

Contents

Preface   iii
Contributors   vii

1. Electrospray Ionization Mass Spectrometry: History, Theory, and Instrumentation   1
   Robert B. Cody

2. Nanospray Electrospray Ionization Development: LC/MS, CE/MS Application   105
   Tom R. Covey and Devanand Pinto

3. Characterization of Pharmaceuticals and Natural Products by Electrospray Ionization Mass Spectrometry   149
   A. K. Ganguly, Birendra N. Pramanik, Guodong Chen, and Petia A. Shipkova

4	Electrospray Mass Spectrometry Applications in Combinatorial Chemistry *Mike S. Lee*	187
5	Electrospray Mass Spectrometry in Contemporary Drug Metabolism and Pharmacokinetics *Mark J. Cole, John S. Janiszewski, and Hassan G. Fouda*	211
6	Electrospray Mass Spectrometry of Peptides and Proteins: Methodologies and Applications *Joseph A. Loo and Greg W. Kilby*	251
7	Structural Analysis of Glycoproteins by Electrospray Ionization Mass Spectrometry *Anthony Tsarbopoulos, Ute Bahr, Birendra N. Pramanik, and Michael Karas*	283
8	Advanced Mass Spectrometric Approaches for Rapid and Quantitative Proteomics *Richard D. Smith, Gordon A. Anderson, Thomas P. Conrads, Christophe D. Masselon, Mary S. Lipton, Ljiljana Paša Tolić, and Timothy D. Veenstra*	307
9	Detection of Noncovalent Complexes by Electrospray Ionization Mass Spectrometry *A. K. Ganguly, Birendra N. Pramanik, Guodong Chen, and Anthony Tsarbopoulos*	361
10	Application of Hydrogen Exchange and Electrospray Ionization Mass Spectrometry in Studies of Protein Structure and Dynamics *Yinsheng Wang and Michael L. Gross*	389
11	Microelectrospray Analysis Combined with Microdialysis Sampling of Neuropeptides and Drugs *Per E. Andrén and Richard M. Caprioli*	411

Index   431

Contributors

Gordon A. Anderson Pacific Northwest National Laboratory, Richland, Washington

Per E. Andrén Uppsala University, Uppsala, Sweden

Ute Bahr J.W.-Goethe University of Frankfurt, Frankfurt, Germany

Richard M. Caprioli Department of Biochemistry, Vanderbilt University, Nashville, Tennessee

Guodong Chen Schering-Plough Research Institute, Kenilworth, New Jersey

Robert B. Cody JEOL USA, Inc., Peabody, Massachusetts

Mark J. Cole Pfizer Inc., Groton, Connecticut

Thomas P. Conrads National Cancer Institute, Frederick, Maryland

Contributors

Tom R. Covey MDS Sciex, Concord, Ontario, Canada

Hassan G. Fouda Pfizer Inc., Groton, Connecticut

A. K. Ganguly Department of Chemistry, Stevens Institute of Technology, Hoboken, New Jersey

Michael L. Gross Department of Chemistry, Washington University, St. Louis, Missouri

John S. Janiszewski Pfizer Inc., Groton, Connecticut

Michael Karas J.W.-Goethe University of Frankfurt, Frankfurt, Germany

Greg W. Kilby Pfizer Global Research and Development, Ann Arbor, Michigan

Mike S. Lee Milestone Development Services, Newtown, Pennsylvania

Mary S. Lipton Pacific Northwest National Laboratory, Richland, Washington

Joseph A. Loo Pfizer Global Research and Development, Ann Arbor, Michigan

Christophe D. Masselon Pacific Northwest National Laboratory, Richland, Washington

Devanand Pinto National Research Council of Canada, Institute of Marine Biosciences, Halifax, Nova Scotia, Canada

Birendra N. Pramanik Schering-Plough Research Institute, Kenilworth, New Jersey

Petia A. Shipkova Schering-Plough Research Institute, Kenilworth, New Jersey

Richard D. Smith Pacific Northwest National Laboratory, Richland, Washington

Ljiljana Paša Tolić Pacific Northwest National Laboratory, Richland, Washington

Anthony Tsarbopoulos Bioanalytical Department, GAIA Research Center, The Goulandris Natural History Museum, Kifissia, Greece

Timothy D. Veenstra National Cancer Institute, Frederick, Maryland

Yinsheng Wang Department of Chemistry, Washington University, St. Louis, Missouri

1
Electrospray Ionization Mass Spectrometry

History, Theory, and Instrumentation

Robert B. Cody
JEOL USA, Peabody, Massachusetts

I. INTRODUCTION: THE SPRAY REVOLUTION

A. Mass Spectrometry Before ESI

It can be argued that certain technical developments have dramatically changed the entire field of mass spectrometry by making it possible to ionize and analyze new classes of compounds. For example, in 1966, the development of chemical ionization (CI) by Munson and Field (1) made "soft ionization" practical and provided a means to obtain molecular weight information from volatile organic compounds that undergo extensive fragmentation under electron ionization (EI) conditions. Fast atom bombardment (FAB), introduced in 1981 (2), provided a rapid and convenient method for analyzing polar and thermally labile compounds, including peptides and small proteins.

In the 1980s, two particularly challenging applications remained for practical mass spectrometry: (1) the analysis of high molecular weight compounds and (2) the development of an efficient interface between liquid chromatography and mass spectrometry. Electrospray ionization (ESI) (3–10) offered a solution that addressed both needs. As a result, electrospray ionization and the closely related technique of atmospheric pressure chemical ionization (APCI) (11,12) dramatically changed the technology and applications of mass spectrometry.

The range of applications of electrospray ionization is truly incredible. ESI has been applied to the analysis of an enormous variety of compound classes, including synthetic organic compounds (13,14), pharmaceutical compounds and their metabolites (15–17), natural products (18), illicit drugs (19,20), proteins (21–30), carbohydrates (31–34), nucleotides and DNA (35–42), lipids (43–49), polymers (50,51), inorganic and organometallic compounds (52–63), fullerenes (64–66), surfactants (67,68), and even self-assembled monolayers (69a,69b) and micelles (69c). ESI has also made it possible to combine sample introduction methods such as liquid chromatography (70), capillary electrophoresis and capillary electrochromatography (71), supercritical fluid chromatography (72), and gel permeation chromatography (73) as well as others (74–78) with mass spectrometry. Following the demonstration that intact viral material could remain viable following electrospray ionization and passage through a mass spectrometer (79), preparative ESI mass spectrometry was proposed (80). Recent developments in high throughput mass spectrometry have relied heavily on electrospray ionization techniques. The impact of electrospray ionization on the field of mass spectrometry can best be put into context by a brief review of other approaches for dealing with the challenges of high mass analysis and liquid chromatography coupled with mass spectrometry (LC/MS).

1. *The Search for a Practical LC/MS Interface*

An excellent historical review of mass spectrometer interfaces for both gas chromatography and liquid chromatography was recently presented by Abian (81). A brief overview of the LC/MS interface is presented here. The earliest efforts to develop an LC/MS interface were based upon direct liquid introduction (DLI) and were reported in Russian by Tal'roze and coworkers (82,83). This interface was based upon electron ionization (EI) and was limited to flow rates of about 1 μL/min. A few years later, McLafferty and coworkers showed that an organic solvent could act as a chemical ionization reagent (84), paving the way for a more practical DLI LC/MS interface (85). The use of the solvent as a chemical ionization reagent later became an important feature of atmospheric pressure chemical ionization sources. Moving wire (86) and moving belt (87) interfaces were developed, and they used mechanical means for removing the solvent and transporting the analyte into the mass spectrometer ion source. Despite early popularity, mechanical complexity and solvent limitations led to the decline of this approach.

The thermospray interface developed by Vestal and coworkers (88,89) became one of the most popular LC/MS interfaces in the 1980s, owing to its ability to handle high liquid flow rates and to provide a means for analyzing thermally labile compounds. Despite its wide popularity, the effectiveness of the technique and the optimum operating conditions were somewhat sample-dependent, and thermospray has been largely replaced by API sources in recent years.

The monodisperse aerosol generating interface for chromatography (MAGIC) particle beam interface introduced by Willoughby and Browner in 1984 (90) was used with EI, CI, and FAB ion sources. Particle beam LC/MS interfaces remain useful for cases where one needs library-searchable EI mass spectra for compounds separated by liquid chromatography.

Continuous-flow FAB, first introduced as "frit FAB" by Ito et al. (91) and developed by Caprioli (92), was used for a wide variety of applications including in vivo analyses and peptide mapping. Although it greatly reduced the problems of sample suppression and chemical background that are observed with "static" FAB, the technique is still subject to a relatively high chemical background and is most useful for analytes with molecular masses of less than ~2000 u.

One LC/MS interface deserves special mention: atmospheric pressure chemical ionization (APCI) (11,12). APCI is a unique approach that avoids the difficulties of introducing large amounts of solvent directly into the vacuum system by forming the ions at atmospheric pressure. APCI and ESI share a common vacuum interface, and the two ion sources are often sold together. Because APCI is closely associated with electrospray ionization, APCI will be discussed in more detail later in this chapter.

2. The Challenge of High-Mass Analysis

The analysis of large biomolecules was a highly sought after and particularly challenging goal. Working with Fourier transform ion cyclotron resonance (FTICR) mass spectrometry in the early 1980s, I can recall requests to show "mass spectra of anything above m/z 1000." In 1983, successful attempts to analyze "middle molecules" (those with masses between 1000 and 10,000 u) by mass spectrometry were referred to as "heroic efforts" (93). By 1984, FAB mass spectra of bovine insulin had become a common sight at mass spectrometry conferences. However, the practical upper mass limit for FAB and the closely related technique of liquid matrix secondary ion mass spectrometry (LSIMS) was found to be just above 10,000 u for organic molecules. Field desorption (FD) (94) demonstrated some successes for polymers with molecular masses exceeding 10,000 u, but FD was not particularly useful for large biomolecules. Plasma desorption mass spectrometry (PDMS), introduced in 1976 (95), remained for many years the only reliable mass spectrometric technique for analyzing proteins with molecular masses greater than 10,000 u.

Matrix-assisted laser desorption ionization (MALDI) (96–98) became a viable mass spectrometric technique in the early 1990s, within the same time period in which ESI was beginning to receive notice. It seemed remarkable that after many years of searching for methods to analyze large molecules by mass spectrometry, mass spectrometrists were given not one but two new and complementary techniques: MALDI and ESI.

B. The Development of Electrospray Ionization

1. Electrospray Before and Beyond Mass Spectrometry

The fact that a strong electric field can produce a spray of fine droplets from a liquid surface was well known at the time of the invention of ESI. Studies of the electrospray phenomenon go back more than 80 years to the original research by Zeleny (99–101) and later studies (102–104,108,109). Electrostatic spraying has been used to generate fine droplets for a variety of applications ranging from paint spraying (105) and electrostatic painting (106, 107), to crop spraying (111), electrostatic emulsification (110,111), fuel atomization (112–115), and the preparation of polymer coatings on electrodes (116). Electrospray methods have also been used for sample preparation for β-counting experiments (117), plasma desorption mass spectrometry (PDMS) (118), and static secondary ion mass spectrometry SIMS (119), and even for DNA analysis by an electrospray scanning mobility particle sizer (120). The benefit of using electrospray sample deposition techniques for preparing samples for desorption ionization is that the sample is deposited as a fine microcrystalline surface. This produces an evenly distributed sample target and reduces sample–sample interactions. Fig. 1 shows an electron microscopic image ($10,000 \times$ magnification) of the surface of a static-SIMS target onto which vitamin B_{12} had been electrosprayed.

2. The 1960s: Early Developments [121–127]

A few researchers recognized in the late 1960s that it might be possible to introduce a beam of macromolecules into a mass spectrometer source from an

ESI-MS: History, Theory, Instrumentation

aerosol. A remarkable unpublished master's thesis in physics (121) was submitted to the University of Wisconsin in 1967 by Erwin Neher, under the direction of W. W. Beeman. In this thesis, Neher considered various methods for generating aerosols of macromolecules as a step toward developing a biopolymer molecular beam source for mass spectrometry. Among the methods discussed were thermal evaporation, ultrasonics, electrostatic separation, and solution spraying (including electrospray). A spray apparatus using a gas–liquid–glass-ball aerosol generator called a Vaponefrin nebulizer was used to generate molecular beams for catalase, lysozyme, and R-17 viral particles. An electrostatic mobility analyzer was used to obtain the size distributions of the charged particles, which could be interpreted as monomers, dimers, and trimers. Neher did not continue this line of research, but he went on to win the Nobel Prize in Physiology with Bert Sakmann in 1991 for research into the functions of single-ion channels in cells.

The earliest successful efforts to combine an electrospray ion source with mass spectrometry were published in 1968 in the pioneering work by Malcolm

Figure 1 Electron microscopic photograph of a SIMS target onto which vitamin B_{12} has been electrosprayed. Magnification 10,000 ×. (Photograph courtesy of J. Amster, University of Georgia.)

Dole and coworkers (122,123). Working with the Bendix Corporation, the early manufacturer of time-of-flight mass spectrometers, Dole and coworkers demonstrated that polystyrene solutions with an average molecular mass of 51,000 u or larger could be electrosprayed from a solution of benzene and acetone. They also showed that the desolvated ions could be analyzed to determine kinetic energy differences due to different masses or charge states. Dole's instrumentation introduced the concept of heat transfer to a concurrent flow of an inert gas (nitrogen) to accomplish desolvation. The desolvated ions passed through an orifice into the vacuum region and were accelerated and analyzed by using a repeller grid with a potential that could be adjusted to reject ions with different kinetic energies. Dole and coworkers concluded that polystyrene with an average molecular mass of 51,000 u existed primarily as a dimer, whereas polystyrene with an average molecular mass of 411,000 u formed triply charged monomers.

In the early 1960s, Sugden and coworkers (124,125) developed an atmospheric pressure sampling system to permit mass spectrometric investigation of the chemistry of flames, and Shahin (126,127) described declustering reactions resulting from CID in a study on corona discharges in air. This led the way for subsequent developments in atmospheric pressure ion sources.

3. The 1970s: The Development of APCI

Between Dole's landmark paper and the 1980s, systematic research into electrospray as an ion source for mass spectrometry was relatively dormant. Following work with electrohydrodynamic ionization (discussed in the following paragraph), Simons et al. (128) suggested the use of electrospray ionization for LC/MS in 1975. Electrospray deposition methods were used as a sample preparation method for other soft ionization techniques (117–119). Because electrospray can generate an aerosol with finely divided droplets, sample surfaces prepared by electrospray deposition were characterized by evenly distributed microcrystalline deposits that can be more easily desorbed by techniques such as desorption chemical ionization (DCI), plasma desorption, and static SIMS. Despite the relative lack of activity directly involving electrospray, research in several closely related techniques paved the way for the development of ESI.

The technique of electrohydrodynamic ionization (EHD) (129–131), reported by Colby and Evans in 1973 (129), involves the use of electric fields to remove ions directly from solution in a matrix such as glycerol within the vacuum system of the mass spectrometer. Although this is similar in many respects to ESI, EHD is carried out in vacuum. In EHD, the sample is placed in a capillary and a high electric potential is applied. The sample emerges from

ESI-MS: History, Theory, Instrumentation

the capillary in a conical shape, and the field at the tip of the cone is sufficiently high to desorb ions directly from the solution.

Many developments in atmospheric pressure chemical ionization (APCI) that occurred in the mid-1970s had a direct influence on the atmospheric pressure ionization interface, which is common to both ESI and APCI. Following early work on the mass spectrometric analysis of flames (124,125), Horning and coworkers (132–134) developed atmospheric pressure chemical ionization with both ^{63}Ni and corona discharge ion sources, demonstrated the use of APCI as an LC/MS interface, and introduced the term "atmospheric pressure ionization (API)." Kambara and Kanomata (135,136) developed what is now referred to as "in-source dissociation," making use of collisions in the intermediate pressure region of the API interface to break up solvent clusters. French and coworkers (137,138) combined in-source dissociation with the use of a "curtain gas" to reduce solvent clustering and minimize solvent contamination of the mass spectrometer.

4. The 1980s: Thermospray, APCI, and Electrospray [139–158]

Thermospray ionization (TSP), developed by Vestal and coworkers (88,89) in the early 1980s, became the LC/MS technique of choice for a large number of applications and remained in use for more than a decade. Thermospray ionization involves the introduction of a solution through a precisely controlled heated capillary tube directly into a reduced pressure spray chamber. Ions produced in the spray chamber are introduced into the mass spectrometer through a conical sampling aperture. The thermospray ionization process was discovered when it was found that ions were still being detected even after the electron filament in the thermospray interface had burned out. The ion formation process consists of a combination of (1) desolvation of preformed analyte and buffer ions and (2) gas-phase chemical ionization reactions occurring between the desolvated ions and vaporized neutrals. Thermospray ionization, therefore, combines features of both ESI and APCI. The principal differences between TSP and ESI are

1. In thermospray ionization the solvent is thermally vaporized directly into a reduced-pressure spray chamber. In ESI (and also APCI) the solvent is sprayed at atmospheric pressure.
2. In ESI, the solvent is sprayed at a high electric potential, which not only affects the spraying process but also assists in the separation of positive and negative ions at the needle tip. The high electric field is important in producing the multiply charged ions that are characteristic of ESI mass spectra. A high electric potential is not applied to the TSP capillary.

The critical work that was responsible for the development and recognition of electrospray ionization as an ion source and the LC/MS interface was carried out in the research group of John Fenn at Yale's Department of Engineering and Applied Science. Following his early research into free-jet expansions, Fenn revisited and extended Dole's work. Dole introduced a counterflow of bath gas to improve desolvation and reduce contamination entering the vacuum region and began to simplify the problem of understanding the electrospray process by examining small molecules instead of larger polymers. Yamashita and Fenn described their early results in papers published in 1984 (139,140), and in 1985, Whitehouse, Dreyer, Yamashita, and Fenn improved upon the ESI source (which they referred to as "ESPI") and described its use as an LC/MS interface (141).

At the same time that Fenn's research group was developing the electrospray ion source, the research group of Aleksandrov and coworkers at the University of Leningrad independently developed an electrospray ion source that they called "extraction of ions from solution at atmospheric pressure" or EIS AP. These results were published in Russian chemical journals in 1984 (142,143) and 1985 (144) and were summarized in English in 1986 (145). Whereas Fenn's group developed an ESI source for a quadrupole mass spectrometer, the ESI source of Aleksandrov et al. was installed on a magnetic sector mass spectrometer.

Several key features of the new ionization method became apparent, including high sensitivity and the capability of inducing "in-source" fragmentation. Both Aleksandrov and Fenn observed multiply charged ions resulting from the electrospray ionization of small peptides. However, the implications of multiple charging were not fully appreciated until after Fenn's group investigated multiple charging in the ESI mass spectra of polyethylene glycols (146,147) and reported these results at the 1987 ASMS meeting (146). It became apparent that multiple charging could make it possible to determine the mass of large molecules without requiring a mass spectrometer with an extended mass-to-charge ratio (m/z) range. The practical application of this concept became possible when Fenn and coworkers recognized that different charge states could be interpreted as independent measurements of molecular weight and that an averaging method based upon the solution of simultaneous equations could provide accurate molecular weight estimates for large molecules (148,149).

The earliest ESI sources were limited to solvent flow rates of a few microliters per minute and were fairly restrictive in terms of solution composition. A major advance in the practical utility of electrospray ionization came with the application of pneumatically assisted electrospray by Bruins et al. in 1987 (150). This was called ion spray or IonSpray (the latter being trade name used by SCIEX). Pneumatically assisted electrospray (150–154), originally reported by

Dole and coworkers (123), made it possible to use a wider range of solvent flow rates and compositions, electrolyte concentrations, and electric potentials (151). The ion spray interface could handle solvent flow rates of several hundred microliters per minute, a much more practical range for LC/MS applications.

Although ion spray provided a practical solution to the requirement of higher solvent flow rates for the LC/MS interface, a different approach was needed to handle the very low liquid flow rates associated with capillary electrophoresis (CE). Olivares, Smith, and coworkers (155,156) made use of small-diameter spray needles and introduced a coaxial sheath-flow "makeup" solvent in a CE/ESI/MS interface. Lee and coworkers (157) reported a CE/MS interface based upon an ion spray source. Another significant event in ESI history occurred in 1989 when McLafferty's research group at Cornell, working together with Hunt and Shabanowitz at the University of Virginia, reported the combination of an electrospray ion source with Fourier transform ion cyclotron resonance (FTICR, also FTMS) mass spectrometry (158).

5. The 1990s: Electrospray Becomes Commonplace

The 1990s were a period of tremendous growth and development in both the instrumentation and applications of electrospray ionization mass spectrometry. The fundamental mechanisms of ion formation in ESI were studied by many groups, with major contributions by both Kebarle and Fenn. In 1990, van Berkel et al. (159) reported the combination of ESI with an ion trap mass spectrometer (ITMS). An ESI source for magnetic sector mass spectrometers was reported by Larsen and McEwen (160) in 1991 with the roughing pumps floated at high voltage to avoid a corona discharge between the ESI source and the pumps. Successful adaptations of ESI sources for other magnetic sector mass spectrometers were soon reported by other laboratories (161–163). Early ESI/TOF systems were reported in 1991 by Sin et al. (164) and also by Boyle and Whitehouse (165). The combination of ESI with orthogonal extraction time-of-flight mass spectrometry (166) was first reported in 1991 by Dodonov et al. (167) at the 12th International Conference on Mass Spectrometry in Amsterdam, and other reports soon followed (168,169). The unique capabilities of combining ESI with FTICR for high resolution and high mass analysis were reported by many research groups, with a large number of publications coming from the research groups of McLafferty at Cornell and Smith at the Pacific Northwest National Laboratories. Commercial versions of ESI sources (commonly sold with interchangeable APCI sources) became widely available for all the major types of mass spectrometers, ranging from low cost, fully automated benchtop models to large, high performance instruments. A more detailed discussion of various mass analyzers follows in a separate section.

Changes to the ESI and API source designs have led to better performance for both low flow and high flow applications, and sensitivity and reliability have been greatly improved. These improvements in reliability and ease of operation have led to the recent trend toward automation and high throughput mass spectrometry.

6. Current Trends: Automation and High Throughput Mass Spectrometry [170–174]

Most vendors now supply ESI and APCI mass spectrometer systems that can be operated as "walk-up" or "open access" systems. These systems allow a system administrator to assign accounts on a mass spectrometer system so that chemists who are not expert in mass spectrometry can analyze their own samples. Increased access has resulted in a tremendous growth in mass spectrometry, especially for well-characterized applications such as protein and peptide analysis (see Chapter 6). This has generally been a benefit for both mass spectrometrists and the chemists who rely on these analyses. However, not all applications are well behaved, and it is important to recognize that unexpected artifacts or contaminants can be misunderstood without sufficient knowledge of the chemistry or instrumentation.

Making it easier to analyze samples has led to a dramatic increase in the number of samples that can be analyzed. The ultimate extrapolation of this trend is the development of high throughput mass spectrometry and microchip-based methods, which have evolved in response to the need to characterize the results of combinatorial chemistry and the increasing number of samples to be analyzed by the pharmaceutical industry (170–174). High throughput mass spectrometry will be discussed in more detail in this book in the chapters on combinatorial chemistry (Chapter 4) and pharmacokinetics and drug metabolism (Chapter 5).

II. PRINCIPLES OF OPERATION

A. Components of an ESI/APCI Source

The ESI "source" actually consists of two separate but interdependent components. The atmospheric pressure region includes the ESI spray needle and auxiliary hardware, and the vacuum interface provides a means for ion transport into the mass spectrometer (Fig. 2). In commercial designs, the ESI source is generally interchangeable with an APCI source.

ESI-MS: History, Theory, Instrumentation

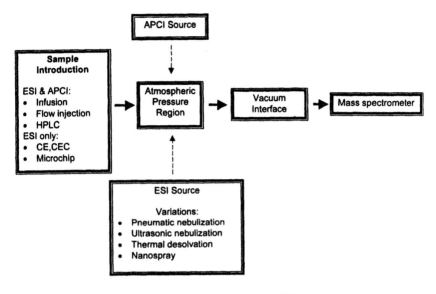

Figure 2 Block diagram showing components of an API system.

The atmospheric pressure region for an ESI source typically consists of

An electrospray needle at high electric potential
A means for establishing an electric potential difference between the spray needle and the orifice into the high vacuum region
An optional desolvation apparatus (usually thermal or pneumatic)

The atmospheric pressure region for an APCI source typically consists of

A spray needle
A thermal desolvation chamber
A corona discharge needle

The vacuum interface typically consists of

An orifice or capillary through which spray is introduced
A set of skimmers and pumping stages
An RF ion guide

A functional description of each of these components will be given in the following sections.

B. Electrospray Ionization: "Desolvation, Not Desorption"

The early work by Aleksandrov and coworkers referred to the process as "extraction of ions from solution at atmospheric pressure" (144). Electrospray ionization is often referred to as a desolvation process, in contrast to "desorption" methods such as FAB, SIMS, and MALDI. Although ESI does remove ions from solution, there is plenty of evidence that the ions observed in ESI mass spectra are not the same as the ions that exist in the solution phase. This subject will be discussed later. Perhaps a better way to classify ion production methods is to follow Fenn's suggestion and define the categories as "field desorption" techniques and "energy-sudden" techniques (5). ESI belongs to the "field desorption" category.

Electrospray ionization and atmospheric pressure chemical ionization share a common vacuum interface through which the ions are introduced into the high vacuum region of the mass spectrometer ion source. This interface is often called the atmospheric pressure ionization (API) interface. Although there are numerous variations on the interface design, modern API interfaces share several common design features. The API interface hardware can be divided into two regions: (1) the atmospheric pressure region and (2) the vacuum interface. The atmospheric pressure regions for ESI and APCI are separate but interchangeable (Fig. 3).

The basic electrospray process is simple to summarize. An aerosol spray consisting of fine droplets is created when a high electric potential is applied to a needle containing a solution with a polar solvent. The spray process can be pneumatically or ultrasonically assisted. A drying bath gas or thermal

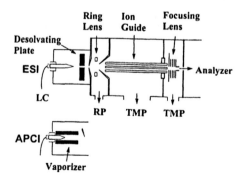

Figure 3 Schematic diagram of JEOL ESI source and vacuum interface showing optional APCI source.

desolvation method is used to eliminate clustering as the droplets are cooled by supersonic expansion. The droplets are introduced into the high vacuum region through an orifice or skimmer. The vacuum interface consists of pumping stages and ion optics designed to maximize ion transmission. Collision-induced dissociation (CID) in the vacuum interface aids in breaking up solvent clusters and provides a means for generating fragment ions, which are often structurally significant.

From an instrumental point of view, it is convenient to break up the discussion of operating principles into three sections:

1. The electrospray process
2. The vacuum interface
3. Variations and auxiliary techniques

C. The ESI Process: Events Occurring at the Spray Needle Under "Pure" ESI Conditions (175–187)

The electrospray process is electrophoretic in nature (187). That is, the applied electric field results in the separation of positive and negative charges existing in the solution (Fig. 4). For example, in positive ion mode, the electrospray needle has a relatively high positive potential relative to the vacuum orifice. Anionic species are attracted to the needle tip, whereas cations predominate at the meniscus surface. The positive charges at the surface repel each other, and the liquid surface expands away from the needle tip. When the electrostatic force and the surface tension are balanced, the cone-shaped liquid surface has a

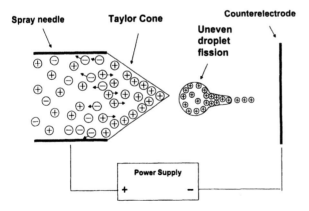

Figure 4 Charge separation and droplet fission in an ESI source.

half-angle of 49.3° at the apex. This is referred to as a "Taylor cone" following the work of G. Taylor (175). As the droplet grows small, the potential increases, the excess positive charge overcomes the surface tension, and droplet formation occurs from the tip of the Taylor cone.

Given the electrophoretic mechanism for electrospray, it would seem that a polar and conductive solvent like water would be ideal. However, water's high surface tension can cause a problem. The electric potential needed to spray pure water can be high enough to cause an electric discharge. This is especially problematic when the spray needle is held at a high negative potential (negative ion mode), and electron emission can occur. Although an electric discharge is the basis for APCI, an electric discharge in the ESI mode can suppress the electrospray ionization process and can even damage components of the source. A sheath-flow gas (such as O_2 or SF_6) that acts as an electron capture agent can be used to suppress the electric discharge in negative ion mode (176,177).

An equation derived from the work of D.P.H. Smith (178) can be used to explain several aspects of the electrospray process. The potential V_{on} (kV) required for the onset of electrospray is related to the radius r (μm) of the electrospray needle, the surface tension of the solvent, γ (N/m), and the distance d (mm), between the needle tip and the counter electrode (the vacuum orifice):

$$V_{on} \approx 0.2 \sqrt{r\gamma} \ln\left(\frac{4000\,d}{r}\right) \tag{1}$$

With methanol as the solvent ($\gamma = 0.0226$ N/m), a spray needle radius of 50 μm, and a needle–counter electrode distance of 5 mm, the onset potential is 1.27 kV. Changing the solvent to water ($\gamma = 0.073$ N/m) increases the onset potential to 2.29 kV.

One solution to the problem of electric discharge is to reduce the needle diameter (179). In the pure water example above, changing the needle diameter from 50 μm to 10 μm decreases the onset potential from 2.29 kV to 1.3 kV. A reduction in the potential required to initiate a spray is one of several benefits of nanospray techniques.

It is interesting to note that an electric potential does not necessarily have to be applied to the spray needle. In the JEOL design (Fig. 3), the potential is not applied to the needle itself but to the surrounding metal tube that supplies the nebulizing gas. The Analytica of Branford design (Fig. 5) uses a metal-tipped glass capillary, which allows the needle to be grounded, and a high negative potential is applied to the capillary entrance. This avoids the problem of current leakage back through the conductive solvent into the LC or syringe

ESI-MS: History, Theory, Instrumentation

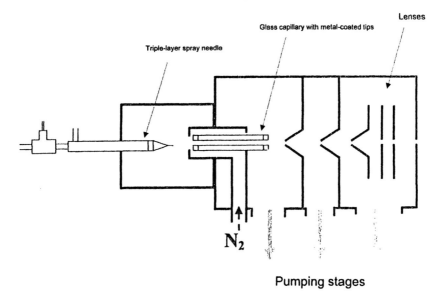

Figure 5 Schematic diagram of an early Whitehouse-type ESI source of the type manufactured by Analytica of Branford.

pump but does not eliminate the possibility of discharge from the needle tip. Recently, Wang and Hackett (180) showed that charge separation induced by using a cylindrical capacitor can produce an aerosol spray with no voltage applied to the needle tip. Hirabayshi and coworkers (181) demonstrated ion formation by spraying solutions with pneumatic nebulization using a gas velocity of Mach 1. No electric potential was applied to the spray needle; friction electrification is proposed as a charging mechanism.

1. ESI as an Electrochemical Process (187–203)

It is important to understand that electrochemical reactions can occur during the electrospray process. Electrochemical oxidation and reduction can lead to unexpected results, such as oxidation of proteins. However, electrochemistry can also be exploited to permit the analysis of compound classes that would otherwise be difficult to analyze by ESI-MS.

Electrochemical contributions to the closely related process of electrohydrodynamic ionization (EHD) were recognized very early by Colby and Evans (129) and Cook (130). Kebarle and coworkers (187) recognized that charge balance considerations must be invoked to explain the electrospray ionization mechanism and that the electrospray ion source can be viewed as a special kind of electrochemical cell. The ESI source differs from a traditional electrochemical cell because charge transport in ESI does not occur through a solvent but as a stream of charged droplets moving at atmospheric pressure between the spray needle and the counter electrode. It is interesting to note that the ESI current has been measured and found to have a frequency of about 300 kHz due to the rate of droplet formation (188). In the positive ion mode, oxidation occurs at the needle tip and reduction occurs at the counter electrode. This was demonstrated by measuring Zn^{2+} ions in the ESI-MS spectrum of a solution sprayed from a zinc spray needle. Although zinc is more easily oxidized than iron, Fe^{2+} ions can be observed when a stainless steel spray needle is used (187).

Van Berkel et al. (189) reported the unexpected observation of molecular radical cations for the ESI mass spectra of porphyrins. This led to extensive investigations (190) into methods for taking advantage of oxidation and reduction reactions in ESI-MS to permit the analysis of species that are not easily ionized by proton (or cation) addition or abstraction. Examples of such "difficult" compounds include porphyrins (189), carotenoids such as beta carotene (199), polynuclear aromatic hydrocarbons (PAHs) (192,200), and electrochemically ionizable derivatives of alcohol and sterols (191).

Zhou and van Berkel (195) showed that the ESI source acts as a controlled current electrolytic (CCE) cell where the rate of charged droplet formation is the current-limiting step. The reactions that are observed depend on the current available. In the positive ion mode, the most easily oxidized species react first, and the more difficult to ionize species are observed only if there is sufficient current. The best results are obtained if one eliminates easily oxidized species; this can involve eliminating trace contaminants or changing the solvent or even the spray needle. A platinum spray needle is less easily oxidized than a stainless steel spray needle and may give better results for some compound classes. Adding an electrolyte (such as lithium triflate) that increases the colution conductivity but does not tend to suppress the analyte ion current can also increase the current. A compact in-line microelectrochemical probe for ESI was also developed by Cole and coworkers (200).

D. Events Occurring as the Droplets Dry

The aerosol droplets carry an excess positive charge when the spray needle is at a positive potential. As the solvent evaporates and the droplets shrink, the ratio of surface charge to surface area increases until the charge repulsion

ESI-MS: History, Theory, Instrumentation

overcomes the surface tension and the droplet breaks apart. The point at which the charge repulsion is equal to the surface tension is the Rayleigh stability limit (204), defined by the equation

$$q_{Ry} = 8\pi(\varepsilon_0 \gamma R^3)^{1/2} \tag{2}$$

where q is the droplet charge, R is the droplet radius, ε_0 is the permittivity of vacuum, and γ is the surface tension.

A droplet relieves "Coulombic stress" by emitting a stream of smaller droplets (Fig. 4). The droplet then continues to lose solvent by evaporation until the Rayleigh stability limit is again reached and another stream of droplets is emitted (205–209). Because the larger droplets break up into streams of smaller droplets, this process has been termed "uneven fission" or "droplet-jet fission" by Kebarle and coworkers (207,208).

There are two theories that describe the ultimate fate of the drying droplets. The "charged residue mechanism" (CRM), proposed by Dole and coworkers and elaborated upon by Röllgen and coworkers (210), states that the ions detected in ESI-MS are the charged species that remain after all of the solvent has evaporated from a droplet. The "ion evaporation" model proposed by Iribarne and Thomson (211,212) states that as a droplet reaches a radius of less than about 10 nm, direct emission of solvated ions can occur from the droplet. Neither the ion evaporation theory nor the charged residue theory has achieved universal acceptance (213–215), and it has been suggested that the mechanism for formation of ions from small molecules may differ from the mechanism for formation of large multiply charged ions (208).

The response in ESI shows a dependence on analyte concentration, which has several practical consequences. A linear dependence of response vs. concentration is observed only up to a maximum concentration of about 10^{-5} M. Above this concentration, the response curve flattens. This is attributed to the fact that at a certain concentration the droplet surface is completely saturated, and increasing the concentration will not increase the number of surface charges available for ion formation (216). The dynamic range of ESI is limited on the low end by the sensitivity, which depends on the efficiency of both the ion source and the analyzer and the ability to discriminate against chemical background. This varies widely with instrumentation and application, but typical dynamic ranges for ESI applications are on the order of three to four orders of magnitude.

The concentration dependence of ESI is illustrated in Fig. 6, which shows that for an ESI source that can handle high flow rates, the signal intensity does not vary as the solution flow rate is changed from 100 µL/min to 1 mL/min. Although much more material is actually injected at the higher flow

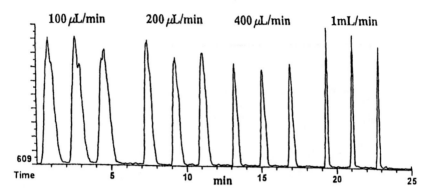

Figure 6 Signal as a function of flow rate in ESI. The ESI response is concentration-dependent.

rates, the signal reponse is not increased. This is attributed to a decrease in droplet charging efficiency and higher droplet diameters at higher flow rates. The problem is accentuated for electrochemically generated ions, where the sensitivity can decrease as flow rate increases. This is attributed to the fact that there is less time for an electrochemical reaction to occur at the needle tip if the solution flow rate is increased.

It is worth noting that ESI sensitivity is frequently discussed in terms of the amount of "sample consumed." This is really more a measurement of sensitivity (signal for a given amount of sample introduced) than a statement of practically achievable detection limits (the smallest amount of sample that can be detected with a given signal-to-noise ratio). Even if the "amount consumed" is reported to be very low, this does not necessarily imply that a low detection limit can be achieved under practical conditions. The amount consumed can be very low for a very high concentration of analyte measured over a very short time. Practical detection limits depend on low easily one can handle small amounts of sample. The best practical detection limits will be achieved if the sample handling and separation methods can increase the sample concentration during the ESI measurement. In combination with ESI, separation methods that operate at lower flow rates, such as microcapillary LC or capillary electrophoresis, will have the best detection limits in terms of total sample consumption. Carrying this concept to extremes, Feng and Agnes (217) reported that single electrostatically levitated droplets can be a source of ions.

Ion evaporation and the charged-residue mechanism are not the end of the story. The end result of either mechanism is a solvated ion. Furthermore,

ESI-MS: History, Theory, Instrumentation

free-jet expansion in the transition region between atmospheric pressure and vacuum can lead to cooling and resolvation. The nature of the ions that are actually observed in ESI-MS depends strongly on what happens in the next stage as the ions are transmitted through the atmospheric pressure ionization interface into the high vacuum region of the mass spectrometer. This process is described in the following section.

E. The Vacuum Interface

The minimum interface between the atmospheric pressure region and the vacuum region consists of a small aperture or a capillary tube that acts as a conductance limit. Skimmers and pumping stages are used to reduce the pressure in stages, and ion optic elements are used to confine the ion beam and improve ion transmission into the mass spectrometer (216–221). The vacuum interface also serves an important function in breaking up solvent clusters, and it provides a means to induce fragmentation by collision-induced dissociation (CID).

The end result of either ion evaporation or the charged-residue mechanism is a solvated ion. Furthermore, Fenn and coworkers recognized that free-jet expansion into the vacuum region can cool the spray, and ions can recondense with solvent molecules in the vacuum interface. A counterflow of drying nitrogen was used to block neutral solvent molecules from entering the vacuum interface. In addition, CID inside the vacuum interface plays a critical role in completing the desolvation of solvated ions. Control of the dynamics and gas composition in the vacuum interface region has been used to study solvation effects and gas-phase reactions (222–228).

1. Collision-Induced Dissociation in the Vacuum Interface: "In-Source CID"

The potentials inside the vacuum interface can be adjusted so that solvated ions are accelerated to higher kinetic energies in a region of relatively high pressure where they can undergo CID. Under mild conditions, solvent clusters are broken up, completing the desolvation process. If the kinetic energy is increased, ion fragmentation can be induced. This has been referred to by a variety of terms including "nozzle/skimmer CID," "in-source CID," and "up-front CID." This differs from true MS/MS experiments because there is no precursor ion selection and fragments can result from the dissociation of all precursor ions, including solvent and background species.

The in-source CID process is very similar to that observed for triple-quadrupole MS/MS systems in that it usually involves multiple collisions and

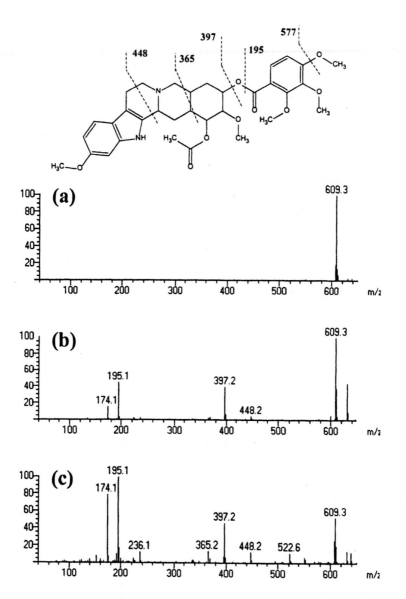

Figure 7 In-source CID. Effect of changing vacuum interface potentials on the ESI mass spectrum of reserpine. The potential difference between orifice 1 and the ring lens is increased from (a) to (b) to (c).

relatively low collision energies (e.g., 10–100 eV collisions). The observed fragmentation depends strongly on the collision energy, which in turn depends on the potentials in the vacuum interface. An example is given in Fig. 7, which shows the ESI mass spectra of reserpine taken at three different sets of vacuum interface potentials. Breakdown curves can be obtained by plotting the relative abundance of fragment ions as the vacuum interface potentials are varied (229). Breakdown curves have been used to determine the optimum conditions for detecting a characteristic fragment ion (230) and in an attempt to standardize fragmentation conditions to allow library searching for drugs in serum (231).

The choice of vacuum interface potentials is critical for success in the ESI experiment. Mild conditions (low potential difference and low collision energies) permit the observation of solvent adducts and clusters, and stronger conditions (moderate potential differences and moderate collision energies) are used to observe singly and multiply charged pseudomolecular ions. The optimum conditions for analyzing small molecules can be different from the optimum conditions for analyzing larger molecules such as proteins. Even stronger conditions result in fragmentation. Several researchers have used a technique in which the vacuum interface potentials are switched during scanning between mild conditions and conditions favoring fragmentation (232–234). This has been useful for identifying certain compound classes such as phosphoproteins. It should be noted that the optimum conditions for analyzing small molecules differ from the optimum conditions for analyzing larger molecules such as proteins. The best strategy is to optimize the ESI source and vacuum interface conditions by using a standard sample that is similar in nature to the species to be analyzed.

F. Variations and Auxiliary Techniques

1. ESI Ion Source Designs

a. Pneumatic Nebulization and Desolvation. Although electrostatic spraying works well for nanospray applications, electrospray is relatively difficult to achieve and maintain for high flow applications. To initiate a spray requires very well defined electric fields, and factors such as applied electric potential, needle diameter, and position are critical. Pneumatic nebulization, also known as "ion spray," was mentioned previously. This method uses a concurrent flow of a nebulizing gas (dry nitrogen) to break up the liquid droplets and form an aerosol. A spray occurs even for relatively high flow rates (a few hundred microliters per minute) in the absence of an electric field. Of course, the applied electric field is still required to induce charge separation. Pneumatic nebulization can be combined with thermal or pneumatic nebulization to produce a more robust ESI source.

A heated stream of gas can be applied as either a counterflow or a crossflow to aid in droplet drying. The counterflow of drying gas is a feature of the original Fenn and Whitehouse ESI source, whereas the cross-flow of drying gas was introduced later by Covey et al. (235) and referred to as Turbo IonSpray, a Sciex trademark.

b. Thermal Desolvation and Dissociation (236–240). Adding heat can facilitate the droplet drying process (240). The early ESI source designs such as the Analytica of Branford source made use of a counterflow of heated nitrogen. A heated metal capillary interface introduced by Chait and coworkers in 1990 (236) provided a simple and effective alternative to the glass capillary and curtain gas interfaces. Rockwood et al. (237,239) showed that the heated capillary could be used to induce thermal decomposition, and Allen and Vestal (238) described an ESI source based solely on thermal desolvation with no gas counterflow.

Modern variations on the thermal desolvation theme include the MicroMass "pepper pot" and JEOL "desolvating plate" designs. These designs provide an indirect path through a desolvating chamber to eliminate the problem of neutral solvent droplets that can enter the vacuum interface in "line-of-sight" designs. If the temperature of the desolvating chamber is kept to a reasonable value (e.g., 200°C for the JEOL design), the heat achieves desolvation without inducing decomposition of the analyte. Therefore, many "fragile" species such as thermally labile analytes and noncovalently bound complexes can readily be analyzed with an ESI source that makes use of thermal desolvation.

Species that cannot tolerate high temperatures, such as some unstable organometallic complexes, can be analyzed by the "cold ESI" technique developed by Yamaguchi and coworkers (61). This technique makes use of an electrospray ionization source maintained at ambient or subambient temperatures and a liquid nitrogen bath to cool the nebulizing gas. The method was shown to be effective for Grignard reagents (61a) and platinum cage complexes (61b).

c. Ultrasonic Nebulization. Ultrasonics are another method for aerosol generation, and ultrasonic nebulization is the basis for a particle beam LC/MS interface developed by Ligon and coworkers (241). Banks et al. (242) have also applied ultrasonic nebulization to ESI-MS. This combination, termed Ultraspray, was trademarked and commercialized by Analytica of Branford and showed some performance advantages over pneumatic nebulization (243).

ESI-MS: History, Theory, Instrumentation 23

However, the added cost and complexity of the ultrasonic spray needle and power supply, together with the difficulty in replacing the needle in the event of clogging, have limited the popularity of this approach.

d. Other Source Designs (Perpendicular Spray, etc.). Numerous variations on the ESI source design have been developed in an attempt to reduce ion source contamination and improve the ability of ESI to work with nonvolatile buffers. The concept of using a counterflow of drying gas or a curtain gas flow to reduce the flow of neutrals into the vacuum interface has already been discussed. Another approach is to use a mechanical barrier to block the direct flow of neutrals. This is the basis of the MicroMass "pepper pot" interface and the JEOL interface, both of which provide a complex flow path through the thermal desolvation region. The JEOL interface also adds an off-axis RF ion guide in the vacuum interface to further reduce neutral flow into the mass spectrometer.

The "off-axis" spray concept, a characteristic feature of the original Sciex ESI source, was widely adopted in many high flow ESI source designs. The pneumatically assisted aerosol is sprayed at about a 45° angle away from the orifice into the vacuum interface. Charged droplets are extracted from the aerosol by the presence of the electric field, which also acts to induce charge separation and consequent droplet fission. Carrying the concept of an indirect spray further, a perpendicular spray was developed by Hewlett-Packard Analytical (now Agilent). Even more extreme examples of indirect spraying include the MicroMass Z-Spray ESI source and the ThermoQuest α-Q-α source (which also includes a concurrent "wash" spray nozzle).

Although techniques such as these reduce the introduction of contamination into the mass spectrometer, a complete mechanical solution to the problem of nonvolatile buffers does not yet exist. The problem lies in the deleterious effects of excess salts and other contaminants on the ion formation process itself. Keeping the salts and nonvolatile buffers out of the mass spectrometer prevents clogging and helps to keep the ESI source and vacuum interface clean, but it does not address problems such as peak analyte signal suppression and peak broadening due to excessive adduct formation.

e. Nanospray (244–282). Although high flow rates are convenient for use with conventional HPLC and flow injection methods, there are certain benefits to using extremely low flow rates for other applications. Microspray and nanospray sources operate in the low microliter per minute to manoliter per minute flow ranges. In brief, nanospray involves using a low flow rate and a small needle diameter. "Pure" electrospray techniques are used, and the spray is introduced directly into the vacuum interface without pneumatic, ultrasonic, or themal nebulization. Because the spray is generated by strictly electrostatic means, the needle diameter, position, and applied potential are critical. It can be more difficult to initiate a stable nanospray than to obtain an assisted spray at higher flow rates. However, once a spray is initiated, a signal can be maintained for hours from microliter-level sample loadings.

One of the challenges in developing an effective nanospray source is to provide a means to apply the electric potential to the solution exiting a small-diameter needle. Several techniques (244–250) have been described for coating the needle tip with a conductive material such as gold. This is an effective approach, but the conductive coating can wear off after prolonged use, and research has been directed at developing a sturdy, conductive coating that can provide a nanospray over a prolonged period without wear. Alternatives for low flow applications include the use of sheath-flow liquids (156,251). This is very easy to implement, but it does result in dilution of the sample with a consequent decrease in detection limits. Several researches (252–254) have applied the potential to a fine wire inserted into the capillary tip. This avoids the problems of wear and sample dilution, but it can require some skill to implement.

Nanospray ESI applications include interfacing with capillary electrophoresis (CE) and microcapillary liquid chromatography (μLC). These low flow separation techniques use very small sample quantities, and they offer the possibility of fast separations and high separation efficiencies. Nanospray methods allow the analyst to extend the time available to analyze a small sample volume. This provides a convenient means for making high resolution measurements, signal averaging, or performing a variety of experiments with a single sample loading.

Other benefits of nanospray methods arise from the reduction in onset potential that comes with decreasing the needle diameter. This facilitates the use of aqueous solutions and reduces the likelihood of a corona discharge. A further benefit of nanospray is a reduction in analyte suppression due to salt contamination. Juraschek et al. (255) recently showed that nanospray can tolerate salt concentrations about an order of magnitude higher than can be handled with an IonSpray source. This is attributed to the formation of droplets about an order of magnitude smaller in nanospray than in conventional ESI sources and reduced competition between salt and analyte ions for the droplet surface as the ratio of surface area to droplet volume decreases.

Nanospray methods are discussed in greater detail in Chapter 2.

2. Other Techniques

a. Sheath-Flow Interface. The addition of a secondary liquid by using a sheath-flow interface was introduced by Smith and coworkers (156) in 1988.

ESI-MS: History, Theory, Instrumentation 25

This provides a makeup solvent to increase the flow rate for low flow techniques such as capillary electrophoresis (for which it was initially developed). The sheath liquid can also be used to modify the solvent composition and provide a means for applying the electric potential required to induce charge separation. The principal disadvantage to the sheath-flow approach is that it results in dilution of the sample, which increases detection limits. However, the sheath-flow approach is effective and easily implemented. It was shown that the pH of the sheath-flow liquid can affect the spectra observed in CE/MS (279), and a method for optimizing the needle and sheath position during operation was reported (251).

b. Lab-on-a-Chip. A logical extension of the nanospray technique is to combine the nanospray ESI source with microchip technology. Several researchers have combined lab-on-a-chip technology (such as microscale capillary electrophoresis) with microfabricated ESI sources (283–285). Microchip technology combines the benefits of parallel analysis, small size, low sample consumption, and a potentially low manufacturing cost. All of these characteristics are attractive for high throughput analysis. The technology requires a small sample volume and a fast and sensitive analysis method. Nanospray ESI meets these requirements, especially when combined with a fast and sensitive analyzer such as a time-of-flight mass spectrometer. Microchip ESI is discussed in more detail in Chapter 11.

c. High Speed Separations and Multiplex Sources (286–288). To increase the number of samples that can be analyzed in a given time, one can increase the analysis speed for sequential analysis. This has been the goal of research involving high speed separations combined with high speed mass spectrometry. High speed separations can be accomplished by capillary electrophoresis or high speed liquid chromatography. For quantitative analysis, it is usually sufficient to monitor specific ions (selected ion monitoring) or reactions (selected reaction monitoring), and complete chromatographic separations are not always necessary. Excellent results were demonstrated by Romanyshyn et al. (287) by using high mobile-phase flow rates and short LC columns. Other applications may require data acquisition. Time-of-flight (TOF) mass spectrometry is well suited for high speed analysis, and an overview of TOF mass spectrometers will be presented in the discussion on mass analyzers for ESI-MS.

An alternative is to analyze several samples at the same time. Parallel analysis has already been mentioned in relation to microchip ESI devices, which reduce the size and cost of the individual ESI sources. However, the mass spectrometer is still the largest and most expensive component of a parallel analysis system. One solution has been developed by Kassel et al. (286a) and also by

Bateman et al. (286b). In this approach, multiple spray nozzles are used to multiplex several sample streams that share a single mass spectrometer analyzer. Each mass spectrum can be correlated with the stream that was being sampled at the time it was acquired. This works well as long as the analyzer (the ESI-MS system) is much faster than the sample introduction and separation system.

In all of the preceding methods, each mass spectrum contains information about only one analyte. An alternative approach has been proposed by Cody (286c). Instead of measuring n samples one at a time, one can measure $n/2$ samples simultaneously in n linearly independent combinations. This provides a set of n equations in n unknowns, and the mass spectrum for each analyte can be obtained by using Hadamard transform techniques. This provides a signal-to-noise (S/N) improvement that increases with n:

$$S/N = \frac{\sqrt{n}}{2} \qquad (3)$$

For large n, it reduces the time required to achieve a given signal-to-noise ratio to $4/n$. This approach has been demonstrated in my laboratory for ESI-MS analysis of three- and seven-component mixtures of small drug molecules. An added benefit is that background contaminants present in all samples can be identified by calculating the geometric mean of the n mass spectra, and the calculated chemical background can be subtracted to provide a further enhancement in the signal-to-noise ratio for the analyte species.

High throughput is discussed further in Chapter 5.

6. Solvents and Buffers

1. Solvents for ESI

Almost all of the early protein ESI work in our laboratory made use of a solvent consisting of a 1:1 mixture of water and methanol with the addition of 2% glacial acetic acid. This solvent mixture could be electrosprayed without pneumatic or ultrasonic assistance, and the acetic acid facilitated protein solubility and provided a proton source for cation formation. However, this solvent mixture caused protein denaturation and was of limited use for many other compound classes also. Pneumatically assisted ESI increased the available range of solvent compositions and made it possible to match the solvent composition to the analytical requirements. Although a wide range of solvents can be used with ESI, there are some limitations.

Because ESI is electrophoretic, the solvent must provide a conductive solution. Spraying "pure" water is no problem as long as pneumatic or ultrasonic nebulization or nanospray methods are used. However, the relatively high surface tension makes unassisted ESI of water difficult for reasons described previously.

Solvents that are commonly used for reverse-phase HPLC, such as water, methanol, and acetonitrile, are compatible with ESI. Other useful solvents include dichloromethane and dichloromethane–methanol mixtures, dimethyl sulfoxide (DMSO), higher alcohols such as isopropanol and butanol, tetrahydrofuran (THF), acetone, and dimethyl formamide (DMF). Solvents that do not work well for ESI are hydrocarbons such as hexane, aromatics such as benzene, and other nonpolar solvents such as carbon tetrachloride. Toluene is not particularly well suited for use as an ESI solvent, but it has been used in some fullerene studies (64,65).

2. Solvent Purity and Sample Contaminants (289–297)

Solvent purity is an important consideration. The ESI source is not a very "bright" (sensitive) ion source in the sense that it produces a large number of ions from a given analyte concentration. However, the chemical background is relatively low compared to other kinds of ion sources such as fast atom bombardment. If the chemical background can be kept low, ESI-MS is capable of extremely low detection limits. For this reason, ESI-MS is very sensitive to the presence of contaminants, and solvents should be free of salts and compounds that might contribute to the chemical background or suppress the analyte (295–297). For example, we observed that some brands of commercially available "HPLC grade" water, although free of UV absorbers that might interfere with a conventional HPLC detector, contained large amounts of salts such as NaCl that caused a severe deterioration of the quality of ESI mass spectra of noncovalently bound protein adducts. The salt contamination can be avoided by using a deionizer or by purchasing water that is specified to have a conductivity of ≥ 18 MΩ/cm^3.

It is nearly impossible, and generally undesirable, to remove all traces of salts from the solvents and solvent delivery system. Small amounts of electrolytes are required for the electrospray ionization process to work. Furthermore, traces of cations are needed to produce detectable ions such as $[M+Na]^+$ from compounds such as polyethers or oligosaccharides. The common reference standards poly(ethylene oxide) (also known as polyethylene glycol or PEG) and poly(propylene oxide) (also known as polypropylene glycol or PPG) are almost always observed as $[M+Na]^+$ ions even when no sodium is added. The sodium is probably introduced to the solvents through contact with glassware or glass containers or transfer lines to the ESI source.

Salt contamination is especially troublesome for the analysis of oligonucleotides, and to a lesser extent carbohydrates, where cation attachment can reduce the analyte signal and cause peak broadening due to the presence of multiple adducts. Special care must be paid to sample cleanup for these analyses. Methods include precipitation (289,290), additives (291), microdialysis, and ion-exchange chromatography (292,293).

Other kinds of impurities are present in organic solvents. One of the most commonly observed contaminants in mass spectrometry is the plasticizer bis(ethylhexyl) phthalate (also called dioctyl phthalate), which appears as an $[M+H]^+$ species at m/z 391 or the $[M+Na]^+$ species at m/z 413. This compound is so ubiquitous that the peaks at m/z 391 or m/z 413 are often used as reference peaks for tuning the mass spectrometer and ESI source.

Impurities can also be present in the sample. Alkali metal cations such as sodium and potassium can be particularly troublesome for ESI-MS analysis of certain compound classes, especially oligonucleotides. Sulfate and phosphate impurities in protein samples can lead to adducts with a mass of 98 u; this can be eliminated by precipitating out the sulfate or phosphate prior to ESI-MS analysis (289). Common surfactants such as sodium dodecyl sulfate (SDS) or Triton X in protein samples can completely mask the presence of a protein signal. The use of a sample cleanup method such as microdialysis or of a separation method such as LC or CE can often eliminate or reduce the deleterious effect of impurities in a sample. Small volumes of protein solutions can also be purified for ESI-MS analysis by a simple procedure involving a small amount of C18 LC column packing material in a micropipette tip (294).

3. Buffers and Additives

Buffers and additives are used in HPLC to control the pH and ensure reliable separations. Although many recent ESI sources have been designed with indirect spraying mechanisms in an attempt to be more tolerant of the presence of contaminants and nonvolatile buffers, no LC/MS interface is truly compatible with nonvolatile buffers. Therefore, phosphate or sulfate buffers should be avoided for ESI-MS because these buffers can cause ion source contamination and interfere with ion production from the analyte (298). Volatile buffers and additives such as acetic acid, formic acid, ammonium acetate, ammonium hydroxide, and trifluoroacetic acid (TFA) are commonly used to control the pH for LC/MS analyses. Typical concentrations are in the 0.1–1% range.

The strong ion-pairing agent trifluoroacetic acid is known to reduce the analyte signal in ESI because analyte cations pair with trifluoroacetate anions, resulting in charge neutralization. This signal suppression can be reduced by postcolumn addition of a weaker acid with a higher boiling point than that of TFA (299). A postcolumn addition of 50% propanoic acid in isopropanol is effective in improving sensitivity for separations performed with a solvent containing 0.1% TFA.

The problem of excesss alkali metal cations was discussed in the preceding section. Although it is commonly necessary to remove excess Na^+ and K^+, it is occasionally necessary to add a cation source to promote the formation of species such as $[M + Na]^+$ or $[M + NH_4]^+$. Additives such as 10–50 μM sodium, potassium, or ammonium acetate can be useful in this case. Ammonium acetate is also useful a useful additive in APCI because it can help to direct the preferential formation of $[M + NH_4]^+$, making it easier to interpret the APCI mass spectrum. Additives such as lithium triflate enhance electrochemical reactions in the ESI source.

H. Kinds of Ions Observed in ESI-MS

1. Singly Charged Ions

The ions observed for ESI of small molecules are similar to those observed for other common "soft" ionization techniques such as FAB. Positive ion ESI mass spectra of small polar or basic molecules are dominated by proton attachment to produce the $[M + H]^+$ species or cation attachment to produce $[M + Na]^+$, $[M + K]^+$, or $[M + NH_4]^+$. Preformed ions such as quaternary ammonium salts are observed as M^+. Negative ion mass spectra show $[M - H]^-$ species for compounds that can lose a proton. Under the right conditions, electrochemical reactions can also produce molecular radical ions from compounds such as PAHs or porphyrins.

The tendency to form dimers such as $[2M + H]^+$ or $[2M + Na]^+$ increases with analyte concentration. This is a common problem with all soft ionization methods, and it can cause problems in interpreting the data. Reducing the analyte concentration can reduce dimer formation. Steffanson et al. (300) also showed that the tendency to produce multimers could be reduced by adding primary amines. Dimers resulting from primary amine attachment are less stable than those formed as proton- or sodium-bound dimers.

Electrospray ionization tends to produce ions with relatively low internal energies (301), and therefore fragment ions are relatively uncommon in ESI mass spectra. However, the API source relies on collision-induced dissociation to break apart solvent clusters, and if the collision energy is high enough, fragment ions can be observed. The compound naproxen has been suggested as a "thermometer" molecule to test the "softness" of an ESI source (216). One should be able to adjust the API interface potentials to produce the $[M - H]^-$ species at m/z 229 without losing CO_2 to produce a fragment at m/z 185.

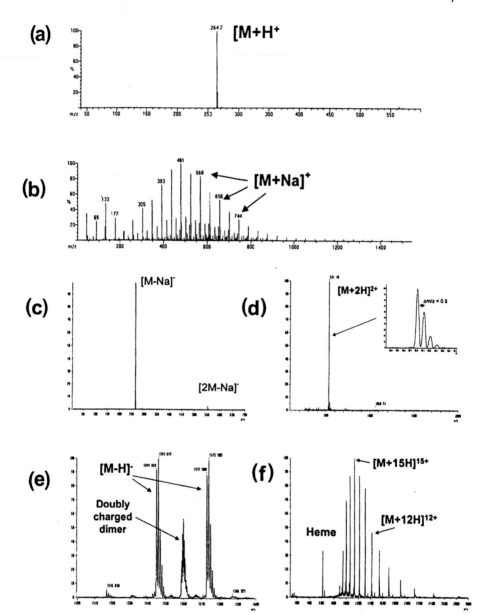

Figure 8 Examples of ESI mass spectra for (a) nortriptylene (ESI+), (b) polyethylene glycol (ESI+), (c) sodium dodecyl sulfate (ESI−), (d) bradykinin (ESI+), (e) ganglioside (G_{Mla}: ESI−), (f) myoglobin in 1:1 water–methanol (ESI+).

2. Multiply Charged Ions [302–314]

One of the most interesting aspects of ESI is the tendency to form multiply charged ions such as $[M + nH]^{n+}$, $[M + nNa]^{n+}$, or $[M - nH]^{n-}$. The degree of charging tends to increase with increasing molecular mass. Small peptides such as leucine enkephalin produce almost exclusively singly charged ions, $[M + H]^+$, whereas slightly larger peptides such as the cyclic peptide Gramicidin S produce both an abundant $[M + 2H]^{2+}$ and a less abundant $[M + H]^+$. Proteins such as myoglobin or lysozyme produce a distribution of charge states (Fig. 8).

The observed distribution of charges depends on both the solution-phase conditions and the ESI source conditions (302–314). The factors that affect charge-state distributions have been discussed in detail by Wang and Cole (302), and only a few examples will be mentioned here. Changing the protein conformation by changing the pH, solvent composition, or adding denaturants such as dithiothreitol (to reduce double bonds) can change the number of exposed basic sites, and this can affect the number of protons that can be attached (303). The bath gas composition can be modified to induce proton transfer reactions and control the observed charge distribution (225). The fact that the observed charge state can be influenced by controlling the vacuum interface potentials has been used to favor singly charged species and simplify ESI mass spectra of small polymers (50,161).

I. Noncovalent Adducts

Noncovalently bound adducts can also be observed in ESI mass spectra, and this has become an important area of study (315–331). A well-known example (319c) is the observation of noncovalently bound heme adducts in the ESI mass spectrum of myoglobin solutions at pH 7. Cyclodextrin inclusion complexes have also been observed by ESI-MS (325,326). However, one report suggested that gas-phase reactions can lead to the same observations and that caution should be exercised in interpreting these data as evidence of solution-phase inclusion complexes (327).

Noncovalent complexes is discussed in greater detail in Chapter 9.

J. Relationship Between Ions Observed in ESI Mass Spectra and Ions in the Solution Phase

The description of ESI as a desolvation process might suggest that the ions observed in the ESI mass spectrum are a direct reflection of the ions that exist in the solution phase. It is true that the nature of the solution has an influence on what is observed in the ESI mass spectrum (332). For example, the charge distributions of protein ESI mass spectra can be influenced by pH changes or by solution additives that denature the protein (change its conformation). Denaturation leads to more opportunities for charging and a shift to lower m/z. However, there are many other factors in the ESI process itself that can have a stronger influence on what is observed in the mass spectrum. Two examples that were already discussed are the electrochemical processes that can occur as a result of the ESI process and the effect of surface activity on ESI sensitivity.

The desolvation process itself can also have a pronounced effect. A familiar analogy from condensed-phase chemistry is crystallization of sodium chloride from solution. Solvated sodium and chlorine ions in an aqueous solution are physically different from sodium chloride crystals formed as the water evaporates. Likewise, the loss of solvent from the drying droplets can change the environment experienced by the ions.

An often-cited example of the difference between solution-phase chemistry and the ions observed in ESI mass spectra are the reports by Loo et al. (333) and Kelly et al. (334), who reported on ESI-MS of proteins in solutions at high pH. Although the proteins exist in these basic solutions as deprotonated negatively charged species, multiply protonated species are observed in the positive ion ESI mass spectra. Likewise, the relative abundances of singly charged ions and doubly charged ions observed in the ESI mass spectra of small peptides do not change significantly with pH (335). In contrast, the relative ratio of doubly charged ions to singly charged ions varies over seven to eight orders of magnitude with pH in solution. Effects such as these are attributed to differences in the local pH and the solution composition at the surface of the drying droplet compared to the bulk solution.

The atmospheric pressure interface relies on collision-induced dissociation to remove solvent molecules from solvated ions. However, species that are stable in solution may not be stable in the gas phase, and vice versa. As an example, Blades and Kebarle (336) reported that unsolvated SO_4^{2-} cannot be observed in ESI mass spectra, whereas $SO_4^{2-}(H_2O)_n$ can. This is attributed to electron detachment from SO_4^{2-} in the gas phase to produce SO_4^-. Furthermore, proton transfer reactions in the CID step are observed when one attempts to produce a hydrate with less than three waters:

$$SO_4^{2-}(H_2O)_3 \rightarrow HOSO_3 + (OH)^-(H_2O)_2 \qquad (4)$$

Under the right conditions, even very low collision energies can produce extensive fragmentation for certain ionic species (337), and one should always be aware that fragmentation and rearrangement processes can happen in the vacuum interface.

Another example of how interface conditions can change the observed mass spectra is found in the ESI mass spectra of polymers (161). At normal (moderate) vacuum interface potentials, a mixture of $[M + H]^+$ and $[M + Na]^+$ can usually be observed for polyethers such as poly(ethylene oxide) or poly(propylene oxide). Multiply charged species such as $[M + 2H]^{2+}$ and $[M + 2Na]^{2+}$ can also be observed as the average molecular weight increases.

If the vacuum interface potentials are increased, the $[M + H]^+$ ions tend to diminish while the relative abundance of the $[M + Na]^+$ ions increases. This can be attributed to preferential dissociation of the protonated species, whereas the sodiated species tend to survive the harsher collision conditions. Furthermore, the relative abundance of the multiply charged species also diminishes to produce a simpler mass spectrum dominated by singly charged sodium adducts. This has been attributed to charge-stripping reactions (161).

K. Comparison with Atmospheric Pressure Chemical Ionization

Because APCI sources are easily interchangeable with ESI sources, it is worthwhile to mention briefly the APCI source and how the two sources complement each other.

1. Overview of APCI

Atmospheric Pressure Chemical Ionization is based upon gas-phase chemical ionization reactions between the thermally vaporized analyte and reagent ions generated from the solvent mixture by a corona discharge. Because the pressure is high, the chemical ionization reactions leading to the formation of reagent ions can be complex. The resulting spectra depend strongly on the nature of the solvent and additives. A mixture of adducts such as $[M + H]^+$, $[M + Na]^+$, $[M + K]^+$, and $[M + NH_4]^+$ can often be observed in APCI. This can make interpretation difficult, and it can be useful to add a buffer or dopant such as ammonium acetate to favor the production of a specific adduct species such as $[M + NH_4]^+$. Because APCI is a chemical ionization process, fragmentation is much more common than in ESI.

The APCI solution passes through a hot vaporizing chamber at much higher temperatures than those used for thermal desolvation in ESI. For example, the JEOL ESI source uses a typical ESI desolvation temperature of ~200°C, whereas the JEOL APCI source uses a vaporizing chamber temperature of up to 500°C. Therefore, ESI may be better suited for the analysis of thermally labile species.

Also, a much higher minimum flow rate is required for APCI than for ESI. Typical flow rates for APCI vary from a minimum of 100–200 µL/min up to a maximum of 1–2 mL/min. In contrast, ESI flow rates are on the order of 1 µL/min to ~1 mL/min, and nanospray techniques permit even lower flow rates. Both ESI and APCI can be easily coupled with HPLC operated at typical microbore and analytical column flow rates, but ESI is much better suited for low flow applications such as microcolumn HPLC and capillary electrophoresis (CE).

2. APCI or ESI: Choosing the Appropriate Source

In addition to physical considerations such as temperature and flow rate, the choice between APCI and ESI is also determined by the analyte chemistry. ESI is favored for basic or polar species that are likely to be charged in solution. Basic compounds can easily attach protons, whereas acidic species can lose protons. Precharged species such as quaternary ammonium salts or sulfates are also readily detected by ESI, and cation attachment can also be observed for compounds such as polyethers and carbohydrates, which contain heteroatoms. Multiply charged species such as dications or dianions or high molecular weight compounds are candidates for ESI.

Atmospheric pressure chemical ionization is usually the first choice for smaller, less polar compounds such as steroids or carotenoids, which are uncharged in solution and have relatively low proton affinities. The optimum polarity (positive ion or negative ion mode) is also compound-dependent. Many neutral compounds that can be analyzed by APCI can also be analyzed by electrochemical ESI with comparable sensitivity. The choice between ESI+, ESI−, APCI+, and APCI− is not always obvious, and it is best to evaluate all four methods when developing an analytical method for a particular compound class.

Normal-phase liquid chromatography is more easily interfaced with an APCI source than an ESI source. Solvents such as hexane that are used for normal-phase LC do not favor ion formation in solution and are not suitable for the charge-separation mechanism of ESI. However, these nonpolar solvents and their typical solutes tend to be sufficiently volatile to be analyzed by APCI, and hydrocarbon solvents form reagent ions that are effective proton transfer agents. Postcolumn addition of a suitable solvent may be considered if it is necessary to interface normal-phase LC with an ESI source.

To further complicate the choice, the reader is reminded that other sample introduction methods (such as GC/MS) and ionization methods (such as MALDI) may be just as good as, or better than, ESI or APCI for a given analysis. The detection limits for electron ionization mass spectrometry or electron capture negative ion mass spectrometry can be in the low femtogram range. In a recent study, we evaluated both GC/MS and LC/MS methods for analyzing compounds suspected to have endocrine-disrupting activity (338). Both approaches were successful and gave comparable detection limits. A benefit of the LC/MS approach is that sample derivatization is not required.

III. MASS ANALYZERS

The operating principles of the various mass analyzers have been discussed at length in numerous other books and reviews, and a detailed discussion of each analyzer is beyond the scope of this chapter. The following discussion concerns only those features that are relevant to each mass analyzer's use with electrospray ionization. It will be seen that each mass analyzer has its own unique benefits and disadvantages and no single mass analyzer is ideal for all applications. As always, one must match the solution to the problem.

A. Mass Spectrometers with Continuous Ion Formation

1. Quadrupole Mass Spectrometers

Quadrupole mass spectrometers (339) have been used more widely for ESI and APCI applications than any other kind of mass analyzer. The principal advantage of single-quadrupole mass analyzers is that they are relatively reliable and low cost systems that offer good quantitative and qualitative analytical capabilities. Quadrupole mass spectrometers are generally limited to producing low resolution mass spectra, although exact mass measurements can be obtained with a quadrupole mass spectrometer provided that the sample is relatively pure and chemical background impurities do not cause unresolved interferences (340).

Quantitative analysis with a single-quadrupole mass spectrometer relies on selected ion monitoring (SIM). The detection limit for a SIM analysis is usually determined by how well one can discriminate between the target analyte and other components in the sample, including background ions. There are two ways to improve analyte selectivity: high resolution selected ion monitoring (HRSIM) and selected reaction monitoring (SRM), in which one monitors specific MS/MS reactions. Neither of these experiments is possible with a single-quadrupole mass spectrometer.

Although true tandem mass spectrometry (MS/MS) is not available with single-quadrupole mass spectrometers, in-source CID can provide some MS/MS functionality by providing a method for inducing fragmentation (341).

In-source CID does not provide any precursor ion selectivity, and product ions will be observed for background contaminants and coeluting species as well as for the analyte.

If one needs MS/MS capabilities for structural analysis or for improving target compound selectivity by selected reaction monitoring, a triple-stage quadrupole mass spectrometer (TQMS) must be used. In its simplest form, a triple-stage quadrupole mass spectrometers uses a first-stage quadrupole (MS-1) for precursor ion selection, an RF-only quadrupole as a collision region, and a second-stage (MS-2) quadrupole for product ion analysis. By far the majority of mass spectrometers in use for quantitative analysis by ESI-MS have been TQMS systems.

A TQMS instrument is capable of all of the analytically useful MS/MS scans, including product ion scans (the most common MS/MS scan type), precursor ion scans, and constant-neutral-loss scans. The collision-induced dissociation process (CID) is a low energy process, with typical collision energies in the 10–100 eV range (laboratory frame). Because low energy CID is sensitive to operating conditions, it has been difficult to achieve reproducible product ion spectra between different models and manufacturers. However, tandem mass spectra obtained with a triple-quadrupole mass spectrometer tend to be more informative and less susceptible to rearrangements and artifacts than the tandem mass spectra obtained with ion trap mass spectrometers.

It is interesting to note that some of the limitations customarily associated with quadrupole mass spectrometers such as low mass-to-charge ratio range and low resolving power have been overcome in special-purpose, research grade quadrupole mass spectrometers (345,346).

2. Magnetic Sector Mass Spectrometers

Although ESI sources were first implemented on both quadrupole mass spectrometers and magnetic sector mass spectrometers, lower ion source potentials made the initial implementation easier for quadrupole systems. Interfacing the ESI source to the kilovolt-range ion source of a magnetic sector mass spectrometer is a minor engineering problem that has been completely solved for modern magnetic sector mass spectrometers. One way to avoid a corona discharge between the first pumping stage in the API source and the first-stage pumps is to enclose the roughing pumps in an isolation box and float them to the ion source potential.

Great improvements have been made in all types of mass spectrometer analyzers, and magnetic sector instruments are no exception. A few decades ago, magnetic sector mass spectrometers were large and relatively difficult to

ESI-MS: History, Theory, Instrumentation

maintain and operate. However, some recent models can fit on a laboratory benchtop, and most are fully automated.

Magnetic sector mass spectrometers are capable of high resolving power. Therefore, they are used for both exact mass measurements and high resolution selected ion monitoring. The maximum resolving power that can be obtained with a double-focusing magnetic sector mass spectrometer varies from 5000 for a benchtop system to over 100,000 for a large instrument operated in the electron ionization (EI) mode. The resolving power is determined by the slit widths, so an increase in resolving power is accompanied by a loss of sensitivity. A resolving power of 10,000 can be obtained on a routine basis with a magnetic sector system operated in ESI mode. This is generally more than sufficient for exact-mass measurements for small molecules. As an example, 10 repeated measurements of the $[M + H]^+$ ion of reserpine were made on one of our magnetic sector mass spectrometers operated at a resolving power of 10,000 (10% valley definition). The mean error was 0.00003 u (0.05 ppm) with a standard deviation of 0.00068 u (1.1 ppm). Shipkova et al. (34) used an HPLC instrument coupled with a magnetic sector mass spectrometer to analyze impurities and degradation products in the oligosaccharide antibiotic everninomycin. Takahashi et al. (354) demonstrated exact mass measurements by using a novel multiple sprayer nanoelectrospray source and a magnetic sector mass spectrometer. The resolution of individual isotopes has been demonstrated in our laboratory (161) and others (162) for proteins such as lysozyme (m/z 14,305.2) and myoglobin (m/z 16,950.7). Exact mass measurements of fragment ions generated by in-source CID were reported by Starrett and DiDonato (347). Selected-ion monitoring at high resolving power has been demonstrated as a method for reducing background interferences and improving target compound selectivity for quantitative assays of bile acid glucuronides (230) and environmental contaminants suspected to have endocrine-disrupting activity (338).

Magnetic sector mass spectrometers are scanning instruments. Sensitivity is determined by how much of the ion beam can be sampled at a given time. A slower scan speed can improve the sensitivity by allowing more time to integrate the ion current at each mass-to-charge ratio. However, this is often impractical for time-varying experiments such as LC/MS. An alternative approach is to make use of a focal plane detector (348–350). It was already mentioned that decreasing the slit widths can result in a loss of sensitivity. Eliminating the collector slit to detect a wider range of mass-to-charge ratios simultaneously provides a multichannel advantage and improved sensitivity. Detection limits can be improved whenever the signal-to-noise ratio is limited by ion current and not chemical background. We showed that the detection limits for ESI-MS can be improved by a factor of 50 when a photodiode array detector is substituted for a conventional point detector on a magnetic sector mass spectrometer (348). Scanning array detectors such as position- and time-resolved ion counting (PATRIC) detectors (349) and charge-coupled device (CCD) array detectors (350) have been used with good success for ESI-MS experiments with magnetic sector mass spectrometers.

Tandem mass spectrometry capabilities are possible with linked-scanning methods, which also provide kilovolt collision energies for high energy CID experiments that tend to be information-rich and relatively insensitive to variations in experimental conditions (Fig. 9). Linked-scan MS/MS can be used for product ion scans, precursor ion scans, neutral-loss scans, and selected reaction monitoring. The principal limitation for linked-scan MS/MS experiments is that one can have either good product ion resolving power with limited precursor selectivity (for collisions in the first field-free region, FFR1) or good precursor selectivity and poor resolving power for product ions (for collisions occurring in the second field-free region, FFR2). In practice, this is not a severe limitation because unit resolution is not generally required for precursor selection in

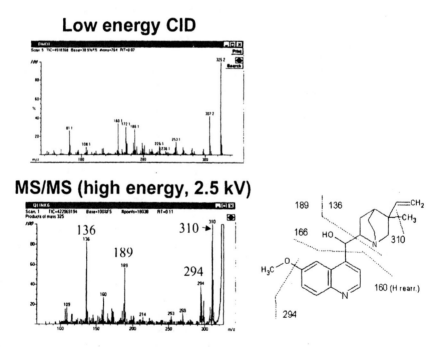

Figure 9 Comparison of in-source CID with high energy CID for quinine.

LC/MS/MS experiments, and the presence of isotopic fragments can sometimes be helpful in determining the charge state of a fragment ion. A tandem (351–354) or hybrid mass spectrometer is required to have good resolving power for both precursor selection and product detection.

B. Pulsed Mass Spectrometers

Sector and quadrupole mass spectrometers operate by continuous ion formation with scanning detection. Another category of mass spectrometers requires pulsed or gated-ion formation or ion injection. This category includes the trapped-ion mass spectrometers (Fourier transform ion cyclotron resonance and quadrupole ion traps) and the time-of-flight mass spectrometers. Scanning mass spectrometers detect only the small portion of the ions formed that are allowed to reach the detector at any instant during the scan. This is relatively inefficient. In contrast, all three types of pulsed mass spectrometers can detect all of the ions present in the ion source at the start of an event cycle. This leads to high detection efficiency. The multichannel detection capabilities of pulsed mass spectrometers are often referred to as "simultaneous" ion detection. Strictly speaking, the only true simultaneous ion detection is accomplished with Fourier transform ion cyclotron resonance (FTICR) mass spectrometry. Although ion detection efficiency is not 100% in pulsed mass spectrometers because of duty cycle limitations, ion accumulation techniques and efficient duty cycles can lead to excellent detection limits.

The pulsed nature of these mass spectrometers requires that the continuous ion beam produced in the ESI source be gated. This can be done by pulsed injection of ions into a trapped-ion mass spectrometer or orthogonal acceleration of ions into a time-of-flight mass spectrometer.

1. Trapped-ion Mass Spectrometers

a. FTICR. Fourier transform ion cyclotron resonance mass spectrometry (FTICR or FTMS) has provided some of the most creative and spectacular research related to electrospray ionization, and the number of publications on ESI/FTICR fundamentals and applications is worthy of a separate review. Several summaries have been published describing of the state of the art in FTICR analysis of large biomolecules. First reported in 1989 in a collaborative effort between researchers in McLafferty's group at Cornell and Hunt's group at the University of Virginia (158), the ESI/FTICR combination is a particularly good match between the strengths of the source and the analyzer (355).

FTICR is capable of extremely high resolving power and excellent mass accuracy, and it can provide isotopic resolution for multiply charged molecular and fragment ions from large proteins. The multiple charging resulting from ESI also helps to overcome a fundamental problem in FTICR.

Unlike quadrupole mass spectrometers, which have found widespread application in pharmacokinetic studies and small-molecule analysis, the majority of published applications of ESI/FTICR mass spectrometry have been research-oriented applications where high resolving power and/or tandem mass spectrometry is needed to analyze intact high molecular mass species. Instrumentation for this application requires superconducting magnets and pumping systems capable of producing ultrahigh vacuum conditions; therefore this instrumentation tends to be relatively large and expensive.

The signal intensity in FTICR is related to the number of ion charges detected as image charges on the detector plates. The longer the ion image currents can be detected, the better the resolving power. For good sensitivity and resolving power, one desires a large number of charges detected for a long time. As a trapped-ion mass spectrometer, an FTICR instrument is subject to problems arising from space-charge effects (electrostatic repulsion between ions). As ions are trapped and detected for a longer time, space-charge repulsion becomes more important. Space-charge effects can lead to loss of coherence (preferential loss of the signal from less abundant ions), loss of ions, peak distortion, frequency shifts, and a variety of other symptoms.

Electrospray ionization offers a solution to this problem. Multiple charging provides more charges and therefore produces a stronger image-current signal and better sensitivity. Ion–ion repulsion is dramatically reduced when many charges reside on a single ion. Fewer ions are required to produce a signal. As an extreme example, Chen et al. (401,402) detected very large multiply charged single ions by FTICR.

The very high resolution that can be achieved by using FTICR makes it the method of choice when high resolving power must be obtained for large molecules. For comparison, a magnetic sector mass spectrometer can resolve isotopic peaks for medium-sized proteins such as lysozyme (14,305 u) and myoglobin (16,951 u), whereas FTICR has demonstrated unit resolution for a protein having a molecular mass of approximately 112,000 u (356). Excellent mass accuracy can be obtained by FTICR, and sometimes this can be done without using an internal reference compound. Because space-charge effects can cause shifts in the measured frequencies, however, changes in the number of ions must be taken into account for exact mass measurements (406a,406b).

Although virtually all mass analyzers can be used to observe fragmentation of larger molecules ionized by ESI, the isotopic resolution in FTICR provides a means for determining the charge states of fragment ions, which makes it possible to interpret readily data from MS/MS experiments involving larger biomolecules. FTICR also offers a wide range of ion activation methods including numerous variations on collisional activation, photodissociation, and blackbody infrared dissociation. Collisional activation in FTICR (357) generally consists of variations on low energy, multicollision techniques. The most effective of these methods for dissociating large molecules is sustained off-resonance irradiation (SORI) (359–364), a technique closely related to the collisional activation technique used in quadrupole ion traps. The maximum achievable collision energy in FTICR decreases with increasing mass-to-charge ratio and increases with increasing magnetic field. High energy CID is possible for small m/z ions, but it is not generally practical for a full mass range. McLafferty and coworkers (408) reported an intriguing alternative to CID that is based upon electron capture by large, multiply charged ions.

Multistage mass spectrometry experiments in FTICR and quadrupole ion trap mass spectrometers are referred to as "tandem-in-time" as opposed to the "tandem-in-space" experiments performed with magnetic sector and quadrupole mass spectrometers. MS/MS experiments in FTICR (and also in ion traps) consist of a series of sequential manipulations of the trapped ions. These manipulations include ion ejection (used as a means to select precursor ions), activation by methods such as those described in the preceding paragraph, ion–molecule reactions, ion–electron recombination reactions, and ion detection (357,358). An interesting application to ESI-MS involves the use of infrared multiphoton photodissociation (365) and SORI (364) to remove adducts from large biomolecules. This "in-trap" sample cleanup procedure does not address the effect of contamination during the droplet evaporation process, but it can result in significant improvements in signal-to-noise ratio in the measured mass spectrum. The capability for performing sequential MS/MS stages [MS^n (366)] has been carried to elegant extremes by using FTICR techniques.

Fourier transform ion cyclotron resonance detects the image currents produced on detector plates by ions circulating in the magnetic field. This is inherently less sensitive than using an ion multiplier, but because FTICR is a nondestructive technique it is possible to improve the signal-to-noise ratio by ion remeasurement (367–369). This is done by repeatedly accelerating a set of trapped ions and measuring the image currents of the coherent ion packets, cooling the ions, and then measuring them again. This is a powerful technique for working with small ion populations, and the remeasurement technique can be done at high resolving power.

A high vacuum (typically 10^{-7} torr for low resolving power or better for high resolving power) is required for FTICR to minimize ion–molecule collisions, and it works best with a superconducting magnet with a very high magnetic field. The ESI source is external to the superconducting magnet, and the vacuum interface, ion transfer optics, and ion-trapping mechanism are important aspects of instrument design and operation. FTICR is an extremely powerful mass analyzer for ESI and MALDI, but the complexity of the hardware and dynamic nature of the ion manipulation mechanics require a good understanding on the part of the operator if one is to achieve good results.

b. Quadrupole ion trap. The quadrupole ion trap is a smaller and less expensive relative of the FTICR mass spectrometer. The FTICR mass spectrometer can be described as a "static" ion trap because ions are confined by a combination of a static magnetic field and small DC potentials applied to the trapping plates. In contrast, the quadrupole ion trap is a "dynamic" ion trap that relies on high amplitude RF electric fields to confine the ions. Quadrupole ion traps are also capable of MS^n experiments involving sequential manipulations of trapped ions. The combination of small size and low cost together with MS^n capabilities has led to widespread use of ion traps as analyzers for GC/MS and LC/MS applications (411).

Unlike FTICR, quadrupole ion traps are normally operated as scanning mass spectrometers. Ions formed externally in an ESI source are injected into the ion trap and accumulated (421). The trapped ions can be manipulated by using techniques analogous to those used in FTICR. However, ion detection is accomplished by applying a ramp that causes sequential ejection of ions through apertures in the end-cap electrode into the ion multiplier. The combination of ion accumulation with ion multiplier detection provides excellent sensitivity.

The principal mass spectrometric factors affecting detection limits are the efficiencies of the ion injection and ion trapping steps and the overall duty cycle. Ion traps operated with electron ionization (EI) sources can be operated in a configuration in which ion formation occurs inside the trap (internal ionization); this is highly efficient because virtually all of the ions formed can be trapped and detected. For ESI applications, ions are formed in an external ion source and injected into the ion trap; ion loss can occur in the injection and trapping stages.

The number of ions that can be trapped places a relatively severe limit on the dynamic range of an ion trap, that is, the range between the smallest number of ions that can be detected and the largest number of ions that can be effectively contained within the trap. Space-charge effects (ion–ion repulsion) can cause frequency shifts as the number of trapped ions increases. Fortunately, the number of trapped ions can be limited by controlling the number of ions injected into the ion trap from the external ESI source. Ion ejection can also be used to remove unwanted ions from the trap, with consequent improvements in performance.

Ion traps tend to be operated as low-to-medium resolving power mass spectrometers. In fact, ultrahigh resolving power mass spectra have been obtained by using research grade ion traps (412), but true high resolving power capabilities are not available in commercial traps. "Zoom scans" can be readily used to obtain resolving power sufficient to determine the charge state for multiply charged peptides, but that which can be obtained with commercially available ion traps tends to be much less than that obtained by FTICR. Furthermore, high resolving power does not always imply mass accuracy (413). Shifts in the measured frequencies of trapped ions can occur due to space-charge effects and other electric field effects. In FTICR, these frequency shifts are limited by the presence of the powerful static magnetic field, making exact mass measurements possible. Frequency shifts are problematic for exact mass measurements in ion traps, and the exact mass experiment is not routinely possible with ion traps.

Collisional activation is accomplished in ion traps by applying an excitation waveform that gently accelerates and decelerates the ions in the presence of a collision gas. This is very similar to the SORI experiment in FTICR. Slow, gentle collisional heating of the ions tends to lead to highly efficient, low energy collision-induced dissociation. The low collision energies and relatively long activation times favor rearrangement reactions over simple bond cleavage. Product ion mass spectra taken with an ion trap tend to be dominated by a few abundant fragments, and multistage mass spectrometry (MS^n) is often required to obtain a complete sequence. Because of the tendency toward rearrangement (which increases with each added MS/MS stage) and the possibility of ion–molecule reactions during long trapping times, artifacts can occur (414,415) and caution is needed in interpreting product ion mass spectra for complete unknowns. Protein database searching reduces this problem for proteomic applications because it is often sufficient to know the peptide molecular masses and some sequence from the presence of a few key fragment ions to identify a protein.

Multiply charged ions from large biomolecules [such as ionized duplex DNA (416) and noncovalently bound heme–myoglobin complexes (417)] are stable in an ion trap for relatively long periods, permitting the study of ion–molecule reactions [such as gas-phase H/D exchange (418)] and ion–ion reactions (419). Both cations and anions can be easily trapped in a quadrupole ion trap. This is not generally possible with FTICR because the polarity of the trapping plate potentials determines the polarity of the ions that can be stored.

McLuckey et al. (426) showed that proton transfer reactions can be used to improve the effective mass resolution in an ion trap, and the same group showed that cation–anion recombination reactions can be used to reduce the observed charge state for multiply charged ions (425).

Quantitative analysis is not a strength of ion trap mass spectrometers, because the dynamic range of the analyzer is not high. Further, the dynamic range and the numerous ion–molecule and ion–ion interactions that can occur during the trapping period may adversely affect quantitation. However, the combination of low cost, relatively high sensitivity, and MS/MS capabilities makes ion traps well suited for target compound screening and qualitative analysis of proteins and peptides.

It should be noted that the quadrupole ion trap is not the only kind of ion trap. For example, Benner (420) reported on a unique gated electrostatic trap to measure the charge and mass-to-charge ratio of large ions generated by ESI.

2. Time-of-Flight Mass Spectrometers (433–441)

Time-of-flight (TOF) mass spectrometers are experiencing a period of rapid growth. Fast electronics and powerful laboratory computers have brought about a renewed interest in a technology that was relatively uncommon a decade ago. Newer TOF systems are capable of very high speed, high sensitivity, and high resolution, all attractive capabilities for an ESI-MS system (433).

All commercially available ESI/TOF systems at the time of this writing use orthogonal acceleration ion introduction. Orthogonal acceleration (OA) (166) accepts a continuous ion beam (such as one from an ESI source) and accelerates a gated bunch of ions into the TOF analyzer in a direction perpendicular to the axis of introduction. This is an effective way to deal with differences in ion energy and position as ions are introduced into the pulsed TOF mass analyzer from a continuous ion source. A reflectron TOF mass analyzer provides improved resolution, and commercial ESI/OA-TOF systems can demonstrate a resolving power of 5000 [full width at half-maximum (FWHM) definition] or better. An attractive feature of most ESI/TOF systems is that they always operate at near their maximum resolving power, and there is little or no benefit to be gained by decreasing the resolution.

Time-of-flight mass spectrometers are also capable of exact mass measurements. A relatively simple calibration equation combined with the use of a lock mass to correct for long-term drift offers an attractive approach to dealing with the need for an internal standard for exact mass measurements.

Time-of-flight mass spectrometers are also the fastest of all mass analyzers, and some commercial ESI/TOF systems can measure 100–200 summed mass spectra per second. For comparison, all of the other mass analyzers (quadrupole, magnetic sector, FTICR, and quadrupole ion trap) can operate at a maximum practical data acquisition rate of about 5–10 spectra per second. To obtain a mass spectrum with a TOF system and a time-to-digital converter, a large number of individual measurements are summed together to produce the observed mass spectrum. Although all ESI/TOF systems use fast acquisition rates, the figure of merit for high speed data acquisition is the number of summed spectra with adequate signal-to-noise ratios that can be acquired per second. This figure varies from 5–10 spectra per second for the MicroMass and Perseptive systems up to a maximum rate of 100 spectra per second in the Leco and Analytica systems.

It must be noted that ESI/TOF systems are not without limitations. A time-to-digital converter (TDC) detector is used on virtually all ESI/TOF systems. The TDC is an ion-counting detector that records the time at which each ion strikes the detector. A large number of individual acquisitions are summed together to produce the observed mass spectrum; this is a fast, efficient, and low noise approach to TOF detection. However, TDCs have a finite recovery time (typically a few nanoseconds), which means that the ion current must be sufficiently low that only one ion strikes the detector within their recovery period. Multiple ion events are counted as a single event, and this limits the dynamic range of the ESI/TOF system. Peaks measured for high incident ion currents can be distorted. This means that exact mass measurements must be obtained under conditions that produce a low ion current (dilute samples with low chemical background). Quantitative analysis is also adversely affected by limited detector dynamic range.

One approach to dealing with the limitations of a single TDC-based detector with a single time-to-digital converter is to use multiple detectors with multiple time-to-digital converters (438). This requires that the ion beam be defocused so that as ion current increases, individual ions have a higher probability of striking physically different detectors. This reduces the likelihood of two or more ions striking the detector during the TDC recovery period. The benefits are improved signal-to-noise ratio (S/N theoretically proportional to the number of detectors) and an improved ability to operate under high-speed conditions. The multianode detector approach is the basis of the Ionwerks detector (four or eight anodes) and the LECO ESI/TOF system (36 anodes).

The LECO system (164b,164c) is a refinement of the early ESI/TOF work by Sin et al. Brigham Young University (164a). This system is unique in that it is the only commercial ESI/TOF system that does not use a reflectron TOF analyzer. Instead, a parabolic-field flight tube [called an "inverted Perfectron" by Rockwood (439)] corrects for the different spatial origins of ions introduced by the orthogonal ESI source. A resolving power of 5000

(FWHM definition) is possible under certain circumstances by using single-anode detection, demonstrating that a reflectron is not necessarily required to achieve high resolving power.

An alternative solution is to avoid the TDC altogether. The Analytica ESI/TOF system follows the work of Boyle and Whitehouse and uses a high speed analog-to-digital converter (ADC) with an integrating transient recorder. This avoids the ion-counting limitations of the TDC.

In addition to detector considerations, the sensitivity of a time-of-flight mass spectrometer depends on the size of the ion packet that can be transferred from the ion source to the TOF analyzer. In MALDI/TOF systems, essentially all of the ions produced by each laser pulse are analyzed. Adapting a continuous ion source such as an ESI source poses problems. Most ESI/TOF systems reduce the problem by taking a "slice" of the ion beam and extracting it orthogonally. Another approach was taken by Quian and Lubman (440), who developed a hybrid quadrupole ion trap/TOF system in which ions can be accumulated in the trap and transferred to the TOF for analysis. Zare and coworkers (441) developed a Hadamard transform time-of-flight mass spectrometer to address the issue of duty cycle. The Hadamard transform method can improve the duty cycle to nearly 50%.

Faster data acquisition and higher resolving power lead to another problem: enormous data sets. Even a relatively short analysis acquired as profile data at a rate of 100 spectra per second can present difficulties for real-time display and lead to data sets that are tens or hundreds of megabytes in size. Although a completely satisfactory solution does not exist at the time of this writing, approaches to dealing with this problem involve rapid peak detection and high speed data compression algorithms. Increasing data volume is a general problem that must be solved for high throughput mass spectrometry. In some respects, this is an extension of the problem that was first encountered when GC/MS became practical. Fortunately, rapid advances in data storage as well as data processing hardware and software should lead to a solution.

C. Hybrid and Tandem Systems

Hybrid and tandem mass spectrometers are MS/MS systems that combine a first mass spectrometer (MS-I) for precursorion selection and a second (MS-II) for production analysis. A triple-stage quadrupole mass spectrometer is one kind of tandem mass spectrometer, and a four-sector mass spectrometer is another kind. Multisector tandem mass spectrometers offer exceptionl resolving power and high collision energies (351–354), but large size, complex operation, and high cost have limited their availability for the average researcher.

Hybrid systems combine two different mass analyzers. Examples include the sector/quadrupole hybrid, sector/TOF hybrids, a sector/trap hybrid, an ion trap/TOF hybrid (440, 442–446), and the quadrupole mass spectrometer/quadrupole ion trap hybrid (QTOF)]. Sector/quadrupole hybrids were the first commercially available hybrid mass spectrometers, followed by sector/OA-TOF and sector/trap systems and, more recently, the quadrupole/OA-TOF hybrid (447–451). The ion trap/TOF hybrid has been the subject of numerous publications by Lubman and coworkers. This combination shows very promising characteristics, but it is not commercially available at this time.

Many other combinations of mass analyzers and activation methods can be envisioned. In collaboration with Wysocki and Nikolaev, we recently developed a sector/in-line TOF hybrid that allows a choice of high energy CID or surface-induced dissociation (SID) for ion activation (452). The very high efficiency of SID and the wide range of internal energies that can be achieved make SID attractive for MS/MS applications. Furthermore, SID does not require a collision gas, so one avoids the problems associated with introducing a collision gas into the high vacuum of the analyzer region. The SID/TOF combination should be adaptable to other MS/MS systems.

Of all the hybrid systems, the QTOF instruments have enjoyed the greatest commercial success. The advantage of the QTOF is it combines MS/MS capabilities with high sensitivity and capabilities for performing accurate mass measurements on both precursor and productions. Disadvantages include a relatively high cost and dynamic range limitations associated with time-to-digital converter detectors.

IV. MASS SPECTROMETER CONSIDERATIONS
A. Calibration (453–465)

Reference compounds are required to calibrate the mass scale of any mass spectrometer, and it is important to find reference compounds that are compatible with a particular ion source. A variety of reference compounds that have been used to calibrate mass spectrometers for electrospray ionization mass spectrometry are described in this section.

A mass spectrometer can be calibrated for ESI by using a different ion source (such as an FAB ion source) and then switching to an ESI source (161). This is effective for low resolution ESI analysis, and it is convenient for calibrating magnetic sector mass spectrometers for linked-scan MS/MS experiments, which require reference peaks at very low m/z values. However, this approach requires an additional ion source and is not practical for exact mass measurements or for continuous operation in ESI mode. Multiply charged species produced by ESI of proteins have been used as reference standards for calibration (161,370,453). However, unless one uses a high resolution mass spectrometer such as a magnetic sector or FTICR mass spectrometer, it is not possible to resolve individual isotope peaks, and one must calibrate by using the average molecular mass for the unresolved isotope clusters. This in unsuitable for exact mass measurements of small molecules. The danger in using large molecules as reference standards is illustrated by the fact that equine myoglobin was used as a reference standard by many groups until it was pointed out by Zaia et al. (454) that the amino acid sequence used to calculate the molecular mass of myoglobin was incorrect. The correct molecular mass of equine myoglobin was actually 1 u higher than the value used as a reference standard.

Cesium iodide and alkali metal salt clusters (464) cover a wide m/z range and are commonly used to calibrate mass spectrometers for fast atom bombardment. Electrospray ionization of cesium iodide solutions produces singly and doubly charged species from m/z 133 up to m/z 3510 or higher (464a). The persistence of cesium iodide in the ESI source and the possibility of sample suppression or unwanted cation attachment have limited the use of CsI solutions for ESI-MS calibration, although Hop (464a) showed good results by calibrating with clusters produced with dilute solutions of CsI and tetrabutylammonium acetate.

Polymers such as poly(ethylene oxide) or poly(propylene oxide) are widely used for ESI-MS calibration (161). Cation attachment is the dominant ion-formation mechanism for polyethers, and sodium attachment is frequently observed due to traces of sodium in solvents and glassware. The positive ion ESI mass spectra of these polymers are characterized by abundant $[M + n\text{Na}]^{n+}$ and some $[M + n\text{H}]^{n+}$ species. When using poly(ethylene oxide) reference standards in our laboratory, we have found that it is useful to increase the vacuum interface potentials to promote charge stripping and produce small fragments such as $(C_2H_4O)H^+$ (m/z 45.03404), $(C_2H_4O)_2H^+$ (m/z 89.06025), and $(C_2H_4O)_3H^+$ (m/z 133.08647). We have also added trace amounts of amines and quaternary ammonium salts to the reference mixture to produce distinct reference ions at low m/z (455b). Other cation sources such as ammonium acetate can also be added to the solvent system. Yamaguchi et al. (463) demonstrated the use of macrocyclic polyethers and crown ethers as reference standards for exact mass measuements.

Nonderivatized polyethers are less than ideal reference compounds for some applications. The polyethers are sometimes difficult to flush out of the ion source. Complex mass spectra resulting from the presence of several different cation sources and different charge states can sometimes be difficult to interpret. Negative ion ESI mass spectra show relatively weak [M − H]⁻ peaks that can be observed only with difficulty; thus, polyethers are not useful calibration compounds for negative ion ESI analysis.

Derivatized polyethers have been investigated for both positive ion and negative ion calibration. Following Hemling's use of polyether sulfates for FAB calibration, we found that polyether sulfates can be used for negative ion ESI calibration (161). Although polyether sulfates are not commercially available, they are easily synthesized. Commercial dishwashing soaps also contain compounds such as lauryl sulfate ethoxylates that are convenient reference compounds for negative ion ESI. Later work in our laboratory (455a) led to the observation that dilute solutions of polyether amines and quaternary ammonium salts can be used as positive ion reference standards. These commercially available (456) compounds do not exhibit significant sodium or potassium adducts, and they are more easily flushed out of the mass spectrometer ion source than are nonderivatized polyethers. In addition, doubly charged polyether diamines (such as the Huntsman Corporation's Jeffamine D230) can produce reference peaks at low m/z values (455a).

Moini (459b) described the use of perfluoroalkyl triazines (457) such as Lancaster's Ultramark 1621 (458) [a common FAB reference standard (459)] as an exact mass reference standard for ESI. Although effective, this standard is very "sticky," and it is very difficult to remove from the ion source. Moini et al. (460) described several other reference standards that have desirable characteristics including sodium acetate and sodium trifluoracetate clusters, which produce useful reference peaks for both positive and negative-ion ESI (460).

Konig and Fales (461) recently described the use of cesium salts of monobutyl phthalate and several perfluorinated acids to generate cluster ions up to m/z 10,000.

Water clusters have been successfully used as reference standards by numerous groups (216,462). Water clusters provide closely spaced reference peaks, and their use as a reference avoids entirely the problem of source contamination. The only real disadvantage is that ESI source and vacuum interface conditions required to observe water clusters are necessarily different from those required to observe desolvated analyte ions. This is a limitation when an internal reference standard is needed for exact mass measurements.

Fujiwara et al. (465) developed fluorinated derivatives of glyphosate and aminomethylphosphonic acid that can be used as reference compounds for negative ion ESI as well as positive ion chemical ionization with ammonia reagent gas. These reference standards were synthesized as a set of individual compounds that can give singly charged reference ions over the m/z range 140–772. Because these are individual compounds rather than a single compound that gives a distribution of oligomers or cluster ions, individual reference masses can be selected to bracket the mass of the analyte.

B. Resolving Power

The interpretation of ESI mass spectra of small molecules is made easier if sufficient resolving power is available (161,339,466) that the number of charges can be determined from the spacing between isotope peaks (see the discussion in Section V). If resolving power is defined as $M/\Delta M$, then the resolving power needed to separate the isotopes of a multiply charged species for a compound is the same as the resolving power needed to separate the isotope of the singly charged species. A resolving power of 2000 is needed to resolve the isotopes for the $[M + H]^+$ for a compound with a molecular mass of 2000 u, and the same resolving power is also sufficient to separate the isotopes of $[M + 2H]^{2+}$ or $[M + 3H]^{3+}$ for the same compound.

It is worth noting that several definitions of resolving power are in use. The oldest and strictest definition is that used in describing magnetic sector mass spectrometers. Resolving power ($M/\Delta M$) is defined as the separating ability necessary to produce a 10% valley between two peaks. This is generally equivalent to defining ΔM as the width of a single peak at the 5% height level. A resolving power of 2000 means that m/z 1999 and m/z 2000 will be separated with a 10% valley between the two peaks, and m/z 100 will be separated from m/z 100.05 to the same extent.

Other mass analyzers use different definitions. Quadrupole mass analyzers use a "unit resolution" definition that states that two successive integer masses will be resolved up to a given mass. For example, unit resolution up to m/z 2000 means that an ion of m/z 99 will be just resolved from that of m/z 100, and an ion of m/z 1999 will be just resolved from that of m/z 2000.

A full width at half-maximum (FWHM) definition is presently favored for ion traps, ion cyclotron resonance, and time-of-flight mass spectrometers. This defines ΔM as the width of a single peak at 50% height. In FTICR, this is because the peak tends to be Lorentzian in shape and has excessive broadening of the skirts compared to a Gaussian peak.

In comparing different mass analyzers, one should be aware that the different resolution definitions do not mean the same thing. A resolving power of 10,000 using the 10% valley definition indicates narrower peaks (better separating ability) than a resolving power of 10,000 using the FWHM definition.

C. Mass Accuracy

For small molecules, one usually desires to know the mass of the most abundant isotope peak. Exact mass measurements can provide valuable elemental composition information. Larger molecules present a different problem. It is often sufficient to measure the average molecular mass of a macromolecule using the centroid of an unresolved isotope cluster. For example, the analytical requirement might be to determine whether an amino acid substitution has occurred in a protein sequence or to examine noncovalently bound complexes. These problems can readily be handled by using average molecular masses. However, there are limitations to the accuracy of average molecular mass measurements for large biomolecules, and inhererent properties of the isotopic distributions can lead to underestimation of the average molecular mass (466). The highest accuracy will be obtained under conditions that permit isotopic resolution, and even those data may be difficult to interpret owing to isotope dilution (93).

V. DATA PROCESSING FOR ESI-MS

One of the unique features of ESI is the fact that multiply charged ions can be easily detected. The degree of charging tends to increase as the molecular weight of the analyte increases. This can be beneficial because large molecules can be detected at mass-to-charge (m/z) ratios that are compatible with the m/z ranges for commonly available mass spectrometers. A disadvantage of using multiply charged ions is that one must interpret the mass spectra to determine the molecular weights of the analytes. Mann et al. (149) were the first to describe methods for determining the molecular weight of an analyte based upon multiply charged ions.

A. Simultaneous Equations

1. Two Successive Charge States Known

Mann et al.'s "averaging algorithm" (149) is based upon solving two simultaneous equations in two unknowns. Suppose that we have an unknown with molecular mass M that can undergo proton attachment leading to two different charge states. The mass-to-charge ratio for charge state m_1 can be represented as

$$m_1 = \frac{M + nH}{n} \tag{5}$$

where n is the number of charges and H is the mass of a proton (1.007825 u).
The mass-to-charge ratio for charge state m_2 can be represented as

$$m^2 = \frac{M + (n+1)H}{n+1} \tag{6}$$

Note that m_2 is the higher charge state and m_1 is the higher mass-to-charge ratio.
If we solve for M in the expression for m_1,

$$M = m_1 n - nH \tag{7}$$

Substituting for M in the expression for m_2,

$$\frac{m_1 n - nH + nH + H}{n+1} = m^2 \tag{8}$$

Solving for n, we find

$$n = \frac{m_2 - H}{m_1 - m_2} \tag{9}$$

Once n is known, M can be found from expression (5). The molecular mass measurements of ions having a different charge are independent determinations, and they can be averaged together to improve accuracy.

2. One Charge State and Two Adducts Known

The molecular weight of an analyte can be determined from a single charge state if two adduct species can be detected. Suppose that we have both $[M+H]^+$ and $[M+Na]^+$. The two adduct peaks will differ by approximately 22 (the mass difference between H and Na). If the species are doubly charged, $[M+2H]^{2+}$ and $[M+H+Na]^{2+}$, then the peaks will differ by 11 (half of 22 u). A similar approach was taken by Senko et al. (469) and by Cunniff and Vouros (470), who deliberately added species to permit charge determination from the difference in m/z ratios of two adduct types.

B. Resolved Isotopic Peaks

If the mass analyzer provides enough resolving power to separate isotopic peaks, the charge state of a given species can be easily determined. Singly charged ions have isotope peaks that have mass-to-charge ratios that differ by approximately 1. Doubly charged ions will have isotope peaks with mass-

ESI-MS: History, Theory, Instrumentation

to-charge ratios that differ by 0.5, and triply charged ions will have isotope peaks with mass-to-charge ratios that differ by 0.33. For example, the positive ion electrospray mass spectrum of the cyclic peptide antibiotic Gramicidin S is dominated by an isotopic cluster having peaks at m/z 571.4, 571.9, 572.4, and so on. These peaks are separated by a mass-to-charge ratio difference of 0.5, indicating doubly charged species ($[M + 2H]^{2+}$). A much smaller isotope cluster shows peaks at m/z 1141.7, 1142.7, 1143.7, etc., indicating singly charged species ($[M + H]^{+}$).

A more dramatic example is shown in Fig. 10 for the nanospray analysis of the protein ubiquitin (average molecular mass 8564.9 u) by using a high performance magnetic sector mass spectrometer set to a resolving power of 8,000 (10% valley definition). The charge states for the $[M + 9H]^{9+}$ and $[M + 10H]^{10+}$ species can be easily determined from the isotope spacing. For example, the peaks for the $[M + 10H]^{10+}$ species have mass-to-charge ratios that differ by 0.1.

The assignment of charge state based upon resolved isotope peaks can be automated. Senko et al. (467) and Horn et al. (471) described algorithms based upon Patterson functions and Fourier transforms that are powerful enough to use for high resolving power mass spectra with low signal-to-noise ratios.

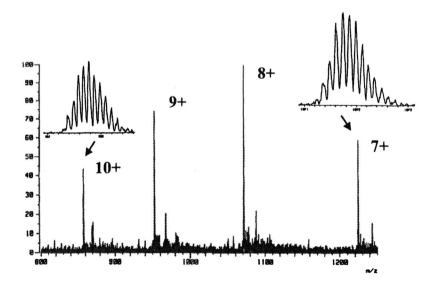

Figure 10 Multiply charged ions in nanospray mass spectrum of ubiquitin.

These algorithms have been incorporated into a program developed for fully automated interpretation of high-resolution ESI mass spectra of large molecules.

C. Transform ("Deconvolution") Methods

1. The Mann "Deconvolution" Algorithm

Mann et al. (149) described another data interpretation method that they called the "deconvolution" algorithm, which has been shown by Reinhold and Reinhold (472) to be a special case of a least-squares-based algorithm. Following the publication by Mann and coworkers, the term "deconvolution" became widely used to refer to mathematical methods for determining the molecular weight from the mass spectra of multiply charged species. In my opinion, this is unfortunate, because it may lead to confusion with the more common use of the term "deconvolution" to describe the inverse of the convolution of two functions. The Mann algorithm and related algorithms can be considered to be *transforms* from the multiply charged (m/z) domain to the uncharged (mass) domain.

The Mann transform is relatively simple to describe.

1. The user specifies the ion polarity (positive or negative), the expected charge carrier (proton, sodium, etc.), and a molecular weight range within which the analyte molecular weight is expected to fall.
2. The algorithm breaks the user-defined molecular mass range up into a number of individual data points. Each data point is treated as an estimate of the analyte molecular weight that will correspond to a x-axis data point in the transformed mass spectrum.
3. For each data point the algorithm calculates the expected mass-to-charge ratios for an analyte having the estimated molecular weight.
4. The abundances are summed for each data point in the mass spectrum (multiply charged domain) that fall within a given error tolerance of the calculated mass-to-charge ratios. This sum corresponds to the y-axis abundance for the data point in the transformed (uncharged domain) molecular mass spectrum.
5. The molecular weight having the highest abundance in the transformed (uncharged domain) spectrum is assigned as the molecular weight of the analyte.

This algorithm is easily implemented, but it has the disadvantage that artifact peaks can appear in the transformed spectrum because the predicted mass-to-charge ratio for an incorrectly estimated molecular mass can correspond to an actual mass-to-charge ratio for an actual component with a different molecular

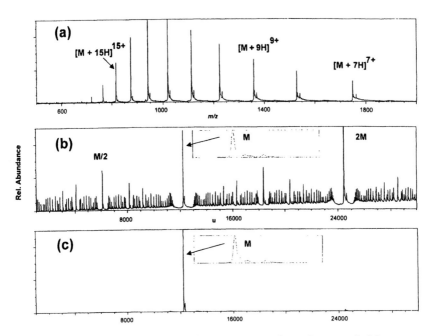

Figure 11 "Deconvolution." (a) ESI mass spectrum of cytochrome c; (b) Mann transform; (c) PCM/SHARF transform.

mass. For example, all of the m/z values for a compound of mass M (Fig. 11a) will be summed (together with background noise) for a molecular mass estimate of $2M$ (see Fig. 11b).

2. Other Transforms

Other algorithms have been devised that improve upon various aspects of the Mann transform. Labowski et al. (473) extended the Mann transform to a three-dimensional implementation that dealt with adduct mixtures, making it easier to assess molecular mass estimates and identify calibration errors. However, this algorithm requires a large number of user inputs, is calculationally intensive, and requires some interpretation of the three-dimensional output.

Reinhold and Reinhold (472) used an entropy-based algorithm that plots the error between the predicted and observed mass spectrum for each estimated molecular mass. This algorithm discriminates against the artifacts produced by the Mann transform, especially when a singly charged impurity is present in the mass spectrum of a multiply charged component.

Hagen and Monnig (474a) described a multiplicative correlation algorithm that multiplies the abundances of data points corresponding to predicted mass-to-charge ratios for a given molecular mass estimate. This strongly enhances the relative abundance of a component whose peaks are distributed over several different charge states, but the peak heights in the resulting transform no longer linearly depend on the relative abundances of components present in the sample.

Kato and Ishihara (475) described an algorithm called the partial correlation method (PCM) that dramatically reduces artifact peaks that appear in the mass domain spectrum at higher mass than the analyte mass (Fig. 11c). A postprocessing of the transformed data with a subharmonic artifacts removal filter (SHARF) can remove remaining artifacts (appearing at masses lower than the analyte mass) to produce a relatively clean transformed spectrum. These algorithms are based upon functions that calculate a weak or zero abundance in the mass domain spectrum if a peak due to one of the expected mass-to-charge ratios is absent from the m/z domain spectrum. For example, the PCM algorithm defines a function for each estimated molecular mass (such as the harmonic mean or geometric mean) that will be zero if one of the peaks due to a predicted mass-to-charge ratio has zero abundance in the original m/z domain spectrum. This eliminates artifacts appearing in the transformed spectrum at an incorrect mass greater than the actual analyte mass.

Zhang and Marshall's Zscore algorithm (476) uses an adjustable scoring scheme to assign the charge state to each multiply charged species that produces a peak with an abundance greater than a predefined threshold. Peaks due to other charge states are located, and their abundances are summed to generate the abundance for the estimated molecular mass. The peaks due to other charge states are then marked as "processed" and removed from the spectrum. The algorithm can be applied to the automatic processing of mass spectra taken at low or high resolving power as well as from peak tables.

3. Maximum Entropy Method

The maximum entropy method (MEM) (477–484) is a mathematical technique for estimating spectra or images from data sets that have been distorted by a "damage" function such a noise or blurring. In essence, the method works by applying a damage model to the estimated spectrum or image to produce a trial spectrum (a model) that can be compared to the measured data. The model is iteratively refined until the trial spectrum and measured spectrum are in good agreement. The method acts as a kind of "Occam's razor" algorithm that produces the simplest model that can explain the observed data.

Maximum entropy works best when one has prior information about the damage function. For example, the observed peaks in low resolution electrospray ionization mass spectra of large molecules are broadened by a known instrumental broadening function (the mass spectrometer resolving power) and the isotopic distribution of the analyte peaks. One can estimate how the isotopic distribution varies with molecular mass for typical members of certain compound classes such as proteins. By using the instrumental and isotopic broadening functions to develop a damage model, the maximum entropy method can eliminate the isotopic natural abundance distribution and achieve an effective resolving power enhancement for ESI mass spectra of large molecules. Ferrige et al. (479,480) first described the application of the method, and Zhang et al. (482) provided a detailed description of the algorithm and its application. An example of the application of MEM is shown in Fig. 12.

4. Chromatographic Enhancement

Electrospray ionization produces a relatively low chemical background. However, the high sensitivity of the method makes it possible to detect very small amounts of analyte. Furthermore, ESI is also very sensitive to the presence of solvent impurities. This means that it is often difficult to identify eluting components based upon a total ion current chromatogram. Several approaches have been developed for component detection that can be useful for separation methods combined with ESI mass spectrometry.

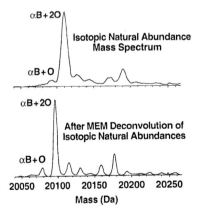

Figure 12 Illustration of maximum entropy method. (From Ref. 482.)

Windig et al. (485) described a "component detection algorithm" (CODA) that combines mass chromatograms with low noise and background to produce a reduced-noise total ion current chromatogram (Fig. 13). Low noise mass chromatograms are determined by calculating a similarity index that compares the mass chromatogram with its smoothed and mean-subtracted equivalent. The CODA algorithm is available with some commercial data systems and can be used to identify components in LC/MS or GC/MS analyses with relatively high background, including exact-mass analyses with the continuous introduction of a reference standard. Muddiman et al. (485) demonstrated the application of sequential paired covariance to detect components in capillary electrophoresis combined with ESI mass spectrometry. Bertrand and coworkers (486a,486b) developed a component detection program called "TICstrip" that identifies background peaks based upon their relatively high

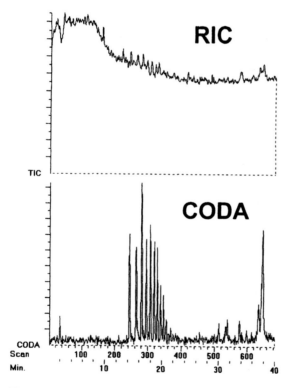

Figure 13 Example of CODA for LC/MS chromatographic enhancement.

frequency of occurrence in repetitive scanning; this algorithm is also highly effective in finding components in the presence of high background. Finally, the Automated Mass Spectral Deconvolution and Identification System (AMDIS) developed by the U.S. National Institute of Standards and Technology (NIST) (Gaithersburg, MD) for chemical weapons treaty verification can also be applied to LC/MS and related analyses. This software can produce a relatively "clean" mass spectrum even in the presence of high chemical background and coeluting interferences.

VI. CONCLUDING REMARKS

Electrospray ionization mass spectrometry has developed rapidly over the past decade to become much more than "just another LC/MS interface." The range of applications for ESI-MS is extremely wide and growing, and the ability to produce highly charged ions has made it possible to apply mass spectrometry to new classes of compounds and to new kinds of gas-phase ion chemistry. A better understanding of the mechanisms involved in ion production in the ESI source has led to significant improvements in instrumentation. If the present trends continue, we can expect to see ESI-MS instrumentation become smaller, faster, more sensitive, more accurate, and better integrated with sample handling and separations methods. The mass spectrometry laboratories of tomorrow will bear little resemblance to those of today.

REFERENCES

1. MSB Munson, FH Field. J Am Chem Soc 88:2621, 1966.
2. (a) M Barber, RS Bordoli, GJ Elliott, RD Sedgwick, AN Tyler. J Chem Soc Chem Commun 1981:325. (b) MBarber, RS Bordoli, GJ Elliott, RD Sedgwick, AN Tyler. Anal Chem 54:645A, 1982.
3. RB Cole, ed. Electrospray Ionization Mass Spectrometry: Fundamentals, Instrumentation, and Applications. New York: Wiley, 1997.
4. JB Fenn, M Mann, CK Meng, SF Wong. Science 246:64–71, 1989.
5. JB Fenn, M Mann, CK Meng, SF Wong. Mass Spectrom Rev 9:37, 1990.
6. EC Huang, T Wachs, JJ Conboy, JD Henion. Anal Chem 62:713A–725A, 1990.
7. (a) J Am Soc Mass Spectrom 4 (7):523–610. Electrospray. Theory, Instrumentation, Method. (Special Focus Issue). (b) J Am Soc Mass Spectrom 4(8):615–684, 1993. Electrospray. Applications to Peptide and Protein Chemistry. (Special Focus Issue).
8. (a) J Mass Spectrom 35:761–817 2000. (b) JF de la Mora, GJ Van Berkel, CG Enke, RB Cole, M Martinez-Sanchez, JB Fenn. J Mass Spectrom 35:939–952, 2000.

9. SJ Gaskell. Electrospray: Principles and practice. J Mass Spectrom 32:677–688, 1997.
10. AP Bruins. Mechanistic aspects of electrospray ionization. J Chromatogr A 794:345–357, 1998.
11. Anon. The API Book. Sciex, Mississauga, ON, Canada, 1990.
12. DM Garcia, SK Huang, WF Stansbury. Characterization of the atmospheric pressure chemical ionization liquid chromatography mass spectrometry interface. J Am Soc Mass Spectrom 7:59–65, 1996.
13. G Perkins, F Pullen, C Thompson. Automated high resolution mass spectrometry for the synthetic chemist. J Am Soc Mass Spectrom 10:546–551, 1999.
14. KL Duffin, JD Henion, JJ Shieh. Electrospray and tandem mass spectrometric characterization of acylglycerol mixtures that are dissolved in nonpolar solvents. Anal Chem 63:1781–1788, 1991.
15. GK Poon. Drug metabolism and pharmacokinetics. In: RB Cole, ed. Electrospray Ionization Mass Spectrometry. New York: Wiley, 1997, Chap. 14.
16. YV Lyubarskaya, SA Carr, D Dunnington, WP Prichett, SM Fischer, ER Appelbaum, CS Jones, BL Karger. Screening for high-affinity ligands to the Src SH2 domain using capillary isoelectric focusing-electrospray ionization ion trap mass spectrometry. Anal Chem 70:4761–4770, 1998.
17. H Keski-Hynnilä, L Luukkanen, J Taskinen, R Kostiainen. Mass spectrometric and tandem mass spectrometric behaviour of nitrocatechol glucuronides: A comparison of atmospheric pressure chemical ionization and electrospray ionization. J Am Soc Mass Spectrom 10:537–545, 1999.
18. RB van Breemen, C-R Huang, Z-Z Lu, R Rimando, HS Fong, JF Fitszloff. Electrospray liquid chromatography/mass spectrometry of ginsenosides. Anal Chem 67:3985–3989, 1995.
19. C Wu, WF Siems, HH Hill Jr. Secondary electrospray ionization ion mobility/mass spectrometry of illicit drugs. Anal Chem 72:396–403, 2000.
20. C Dass, JJ Kusmierz, DM Desiderio, SA Jarvis, BN Green. Electrospray mass spectrometry for the analysis of opioid peptides and for the quantification of endogenous methionine enkephalin and β-endorphin. J Am Soc Mass Spectrom 2:149–156, 1991.
21. JA Loo, R Ogorzalek Loo. Electrospray ionization mass spectra of peptides and proteins. In: RB Cole, ed. Electrospray Ionization Mass Spectrometry. New York: Wiley, 1997, Chap 11.
22. IA Kaltashov, C Fenselau. Stability of secondary structural elements in a solvent-free environment: The alpha helix. Proteins 27:165–170, 1997.
23. (a) A Li, C Fenselau, IA Kaltashov. Stability of secondary structural elements in a solvent-free environment. II. The beta-pleated sheets. Proteins Suppl 2:22–27, 1998. (b) SJ Eyles, T Dresch, LM Gierasch, IA Kastashov. Unfolding dynamics of a beta-sheet protein studied by mass spectrometry. J Mass Spectrom 34:1289–1295, 1999.
24. (a) H Niwa, S Inouye, T Hirano, T Matsuno, S Kojima, M Kubota, M Ohashi, FI Tsuji. Chemical nature of the light emitter of the Aequorea green fluorescent

protein. Proc Natl Acad Sci USA 93:13617–13622, 1996. (b) T Shimizu, T Takao, K Hozumi, K Nunomura, S Ohta, Y Shimonishi, S Ikegami. Structure of a covalently cross-linked form of core histones present in the starfish sperm. Biochemistry 36:12071–12079, 1997. (c) T Takao, T Shimizu, S Ikegamim, Y Shimonishi. High-sensitivity mass spectrometry for analysis of post-translational modifications. J Protein Chem 16:409–413, 1997.

25. SL Cohen, JC Padovan, BT Chait. Mass spectrometric analysis of mercury incorporation into proteins for X-ray diffraction phase determination. Anal Chem 72:574–579, 2000.
26. F He, CL Hendrickson, AG Marshall. Unequivocal determination of metal atom oxidation state in naked heme proteins: Fe(III) myoglobin, Fe(III) cytochrome c, Fe(III) cytochrome b5 and Fe(III) cytochrome b5 L47R. J Am Soc Mass Spectrom 11:120–126, 2000.
27. JCY Le Blanc, KWM Siu, R Guevremont. Electrospray ionization mass spectrometric study of protein-ketone equilibria in solution. Anal Chem 66:3289–3296, 1994.
28. B Xiang, J Ferretti, HM Fales. Use of mass spectrometry to ensure purity of recombinant proteins: A cautionary tale. Anal Chem 70:2188–2190, 1998.
29. KP Braun, RB Cody Jr, DR Jones, CM Peterson. Structural assignment for a stable acetaldehyde-lysine adduct. J Biol Chem 270:11263–11266, 1995.
30. M Busman, KL Schey, JE Oatis Jr, DP Knapp. Identification of phosphorylation sites in phosphopeptides by positive and negative-mode electrospray ionization-tandem mass spectrometry. J Am Soc Mass Spectrom 7:243–249, 1996.
31. Y Ohashi. Electrospray ionization mass spectrometry of carbohydrates and lipids. In: RB Cole, ed. Electrospray Ionization Mass Spectrometry. New York: Wiley, 1997, Chap. 13.
32. J Charlwood, H Birrell, ESP Bouvier, Langridge, P Camilleri. Analysis of oligosaccharides by microbore high-performance liquid chromatography. Anal Chem 72:1469–1474, 2000.
33. K Yoshino, T Takao, H Murata, Y Shimonishi. Use of the Derivatizing agent 4-aminobenzoic acid 2-(diethylamino)ethyl ester for high-sensitivity detection of oligosaccharides by electrospray ionization mass spectrometry. Anal Chem 67:4028–4031, 1995.
34. PA Shipkova, L Heinmark, PL Bartner, G Chen, BN Pramanik, AK Ganguly, RB Cody, A Kusai. J Mass Spectrom 35:1252–1258, 2000.
35. PF Crain. Electrospray ionization mass spectrometry of nucleic acids and their constituents. In: RB Cole, ed. Electrospray Ionization Mass Spectrometry. New York: Wiley, 1997, Chap. 12.
36. J Ding, RJ Anderegg. Specific and nonspecific dimer formation in the electrospray ionization mass spectrometry of oligonucleotides. J Am Soc Mass Spectrom 6:159–164, 1995.
37. PA Limbach, PF Crain, JA McCloskey. Molecular mass measurement of intact ribonucleic acids via electrospray ionization quadrupole mass spectrometry. J Am Soc Mass Spectrom 6:27–39, 1995.

38. J Ni, SC Pomerantz, J Rozenski, Y Zhang, JA McCloskey. Interpretation of oligonucleotide mass spectra for determination of sequence using electrospray ionization and tandem mass spectrometry. Anal Chem 68:1989–1999, 1996.
39. DLD Deforce, FPK Ryniers, EG Van Den Eeckhout, F Lemière, EL Esmans. Analysis of DNA adducts in DNA hydrolysates by capillary zone electrophoresis and capillary zone electrophoresis-mass spectrometry. Anal Chem 68:3575–3584, 1996.
40. DC Muddiman, X Cheng, HR Udseth, RD Smith. Charge-state reduction with improved signal intensity of oligonucleotides in electrospray ionization mass spectrometry. J Am Soc Mass Spectrom 7:697–706, 1996.
41. N Yu, SY Chiang, VE Walker, JA Swenberg. Quantitative analysis of 1,3-butadiene-induced DNA adducts in vivo and in vitro using liquid chromatography electrospray ionization tandem mass spectrometry. J Mass Spectrom 33:363–376, 1998.
42. (a) Y Naito, K Ishikawa, Y Koga, T Tsuneyoshi, H Terunuma, R Arakawa. Molecular mass measurement of polymerase chain reaction products amplified from human blood DNA by electrospray ionization mass spectrometry. Rapid Commun Mass Spectrom 9:1484–1486, 1995. (b) DS Wunschel, L Paša Tolić, B Feng, RD Smith. Electrospray ionization Fourier transform ion cyclotron resonance analysis of large polymerase chain reaction products. J Am Soc Mass Spectrom 11:333–337, 2000.
43. FF Hsu, J Turk. Distinction among isomeric unsaturated fatty acids as lithiated adducts by electrospray ionization mass spectrometry using low-energy collisionally activated dissociation on a triple-stage quadrupole instrument. J Am Soc Mass Spectrom 10:600–612, 1999.
44. FF Hsu, J Turk. Structural characterization of triacylglycerols as lithiated adducts by electrospray ionization mass spectrometry using low-energy collisionally activated dissociation on a triple-stage quadrupole instrument. J Am Soc Mass Spectrom 10:587–599, 1999.
45. DK MacMillian, RC Murphy. Analysis of lipid hydroperoxides and long-chain conjugated keto acids by negative-ion electrospray mass spectrometry. J Am Soc Mass Spectrom 6:1190–1201, 1995.
46. K Metzger, PA Rehberger, G Erben, WD Lehmann. Identification and quantification of lipid sulfate esters by electrospray ionization MS/MS techniques: Cholesterol sulfate. Anal Chem 67:4178–4183, 1995.
47. X Han, RW Gross. Structural determination of picomole amounts of phospholipids via electrospray ionization tandem mass spectrometry. J Am Soc Mass Spectrom 6:1202–1210, 1995.
48. HY Kim, TCL Wang, YC Ma. Liquid chromatography/mass spectrometry of phospholipids using electrospray ionization. Anal Chem 66:3977–3982, 1994.
49. FF Hsu, J Turk. Structural determination of sphingomyelin by tandem mass spectrometry with electrospray ionization. J Am Soc Mass Spectrom 11:437–449, 2000.

50. CN McEwen, WJ Simonsick Jr, BS Larsen, K Ute, K Hatada. The fundamentals of applying electrospray ionization mass spectrometry to low mass poly (methyl methacrylate) polymers. J Am Soc Mass Spectrom 6:906–911, 1995.
51. S Koster, MC Duursma, JJ Boon, RMA Heeren. Endgroup determination of synthetic polymers by electrospray ionization Fourier transform ion cyclotron resonance mass spectrometry. J Am Soc Mass Spectrom 11:536–543, 2000.
52. CL Gatlin, F Turecek. Electrospray ionization of inorganic and organometallic complexes. In: RB Cole, ed. Electrospray Ionization Mass Spectrometry. New York: Wiley, 1997, Chap 15.
53. II Stewart, G Horlick. Developments in the electrospray mass spectrometry of inorganic species. Trans Anal Chem 15:80–90, 1996.
54. JJ Gaumet, GF Strouse. Electrospray mass spectrometry of semiconductor nanoclusters: Comparative analysis of positive and negative ion mode. J Am Soc Mass Spectrom 11:338–344, 2000.
55. ME Ketterer, JP Guzowski. Isotope ratio measurements with elemental-mode electrospray mass spectrometry. Anal Chem 68:883–887, 1996.
56. (a) RLC Lau, J Jiang, DKP Ng, TW Chan. Fourier transform ion cyclotron resonance studies of lanthanide(III) porphyrin-phthalocyanine heteroleptic sandwich complexes by using electrospray ionization. J Am Soc Mass Spectrom 8:161–169, 1997. (b) II Stewart, G Horlick. Electrospray ionization mass spectra of lanthanides. Anal Chem 66:3983–3993, 1994.
57. J Shen, J Brodbelt. Formation of double charged transition metal-polyetherpyridyl mixed complexes by electrospray ionization. J Mass Spectrom 34:137–146, 1999.
58. CL Gatlin, F Turecek, T Valsar. Determination of soluble Cu(I) and Cu(II) species in jet fuel by electrospray ionization mass spectrometry. Anal Chem 66:3950–3958, 1994.
59. L Charles, D Pépin. Electrospray ion chromatography-tandem mass spectrometry of oxyhalides at sub-ppb levels. Anal Chem 70:353–359, 1998.
60. CECA Hop, JT Brady, R Bakhtiar. Transformation of neutral rhenium compounds during electrospray ionization mass spectrometry. J Am Soc Mass Spectrom 8:191–194, 1997.
61. (a) K Yamaguchi, S Sakamoto, H Tsuruta, T Imamoto. Structure analysis of unstable organometallic compounds using electrospray mass spectrometry. Presented at 46th Annual Conf on Mass Spectrometry (Japan), Takasaki, Japan, 1998. (b) S Sakamoto, K Yamaguchi, unpublished data.
62. (a) N Kubota, T Fukuo, R Arakawa. Mass spectrometric analysis of self-assembled 1,19-ferrocenedicarboxylic acid. J Am Soc Mass Spectrom 10:557–560, 1999. (b) R Arakawa, T Matsuo, K Nozaki, T Ohno, M Haga. Analysis of multiply charged ions of ruthenium(II) tetranuclear complexes by electrospray ionization mass spectrometry. M Inorg Chem 34:2464–2467, 1995. (c) R Arakawa. J Mass Spectrom Soc Jpn 46:219–227, 1998. (In Japanese) (d) M Okamoto, H Doe, K Mizuno, T Fukuo, R Arakawa. J Am Soc Mass Spectrom 9:966–969, 1998. (e) R Arakawa, J Lu, K Mizuno, H Inoue, H Doe, T Matsuo.

On-line electrospray mass analysis of photoallylation reactions of dicyanobenzenes by allylic silanes by photoinduced electron transfer. Int J Mass Spectrum Ion Proc 160:371–376, 1997. (f) R Arakawa, S Mimura, G Matsubayani, T Matsuo. Photolysis of (diamine)bis(2,29-bipyridine)ruthenium complexes using on-line electrospray mass spectrometry. Inorg Chem 35:5725–5729, 1996.
63. SA Pergantis, S Wangkam, KA Francesconi, JE Thomas-Oates. Identification of arsenosugars at the picogram level using nanoslectrospray quadrupole time-of-flight mass spectrometry. Anal Chem 72:357–366, 2000.
64. K Hiraoka, I Kudaka, S Fujimaki, H Shinohara. Observation of the fullerene anions C60 and C70 by electrospray ionization. Rapid Commun Mass Spectrom 6:254–256, 1991.
65. S Fujimaki, I Kudaka, T Sato, K Hiraoka, H Shinohara, Y Saito, K Nojima. Application of electrospray ionization to the observation of higher fullerene anions. Rapid Commun Mass Spectrom 7:1077–1081, 1993.
66. GJ Van Berkel, KG Asano. Chemical derivatization for electrospray ionization mass spectrometry. 2. Aromatic and highly conjugated molecules. Anal Chem 66:2096–2102, 1994.
67. KB Sherrard, PJ Marriott, MJ McCormick, R Colton, G Smith. Electrospray ionization mass spectrometric analysis and photocatalytic degradation of polyethoxylate surfactants used in wool scouring. Anal Chem 66:3394–3399, 1994.
68. BN Jewett, L Ramaley, JCT Kwak. Atmospheric pressure ionization mass spectrometry techniques for the analysis of Alkyl ethoxysulfate mixtures. J Am Soc Mass Spectrom 10:529–536, 1999.
69. (a) TD McCarley, RL McCarley. Toward the analysis of electrochemically modified self-assembled monolayers. Electrospray ionization mass spectrometry of organothiolates. Anal Chem 69:130–136, 1997. (b) KC Russell, E Leize, A Van Dorsselaer, JM Lehn. Investigation of self-assembled supramolecular species in solution by IL-ESMS, a new mass spectrometric technique. Angew Chem Int Ed Engl 34:209–213, 1995. (c) D Nohara, T Ohkoshi, T Sakai. The possibility of direct measurement of micelle weight by electrospray ionization mass spectrometry. Rapid Commun Mass Spectrom 12:1933–1935, 1998.
70. RD Voyksner. Combining liquid chromatography with electrospray mass spectrometry. In: RB Cole, ed. Electrospray Ionization Mass Spectrometry. New York: Wiley, 1997, Chap 9.
71. JC Severs, RD Smith. Capillary electrophoresis-electrospray ionization mass spectrometry. In: RB Cole, ed. Electrospray Ionization Mass Spectrometry. New York: Wiley, 1997, Chap. 10.
72. TR Baker, JD Pinkston. Development and application of packed-column supercritical fluid chromatography/pneumatically assisted electrospray mass spectrometry. J Am Chem Soc 9:498–509, 1998.
73. (a) L Prokai, WJ Simonsick Jr. Electrospray ionization mass spectrometry coupled with size-exclusion chromatography. Rapid Commun Mass Spectrom 7:853–856, 1993. (b) MWF Nielen. Characterization of synthetic polymers by

size-exclusion chromatography/electrospray ionization mass spectrometry. Rapid Commun Mass Spectrom 10:1652–1660, 1996.
74. H Iwahashi, CE Parker, RP Mason, KB Tomer. Combined liquid chromatography/electron paramagnetic resonance spectrometry/electrospray ionization mass spectrometry for radical identification. Anal Chem 64:2244–2252, 1992.
75. DA Barnett, B Ells, R Guevremont, R Purves. Separation of leucine and isoleucine by electrospray ionization-high field asymmetric waveform ion mobility spectrometry-mass spectrometry. J Am Soc Mass Spectrom 10:1279–1284, 1999.
76. (a) Y Liu, SJ Valentine, AE Counterman, CS Hoaglund, DE Clemmer. Injected-ion mobility analysis of biomolecules. Anal Chem 69:728A–735A, 1997. (b) SC Henderson, SJ Valentine, AE Counterman, DE Clemmer. Anal Chem 71:291–301, 1999. (c) SJ Valentine, AE Counterman, DE Clemmer. A database of 660 peptide ion cross sections: Use of intrinsic size parameters for bona fide predictions of cross sections. J Am Soc Mass Spectrom 10:1188–1211, 1999.
77. DI Papac, KL Schey, DR Knapp. Combination electrospray-liquid secondary ion mass spectrometry ion source. Anal Chem 63:1658–1660, 1991.
78. MM Siegel, K Tabei, F Lambert, L Candela, B Zoltan. Evaluation of a dual electrospray ionization/atmospheric pressure chemical ionization source at low flow rates (~50 ml/min) for the analysis of both highly and weakly polar compounds. J Am Chem Soc 9:1196–1203, 1998.
79. G Siuzdak, B Bothner, M Yeager, C Brugidou, CM Fauquet, K Hoey, CM Chang. Mass spectrometry and viral analysis. Chem Biol 3:45–48, 1996.
80. G Siuzdak, T Hollenbeck, B Bothner. Preparative mass spectrometry with electrospray ionization. J Mass Spectrom 34:1087–1088, 1999.
81. J Abian. The coupling of gas and liquid chromatography with mass spectrometry. J Mass Spectrom, 34:157–168, 1999.
82. VL Tal'roze, GV Karpov, IG Gorodetskii, VE Skurat. Zh Fiz Chim 42:1658, 1968.
83. VL Tal'roze, VE Skurat, GV Karpov. Zh Fiz Chim 43:241, 1969.
84. MA Baldwin, FW McLafferty. Org Mass Spectrom 7:1111, 1973.
85. P Arpino, MA Baldwin, FW McLafferty. Biomed Mass Spectrom 1:80, 1974.
86. RPW Scott, CG Scott, M Munroe, J Hess. J Chromatogr 99:394, 1974.
87. WH McFadden, HL Schwartz, S Evans. J Chromatogr 122:389, 1976.
88. CR Blakly, JJ Carmody, ML Vestal. Adv Mass Spectrom 8B:1616–1623, 1980.
89. CR Blakly, ML Vestal. Anal Chem 55:750, 1983.
90. RC Willoughby, RF Browner. Monodisperse aerosol generation interface for coupling liquid chromatography with mass spectroscopy. Anal Chem 56:2626–2631, 1984.
91. Y Ito, T Takeuchi, D Ishi, M Goto. J Chromatogr 346:161, 1985.
92. RM Caprioli, ed. Continuous Flow Fast Atom Bombardment Mass Spectrometry. New York: Wiley, 1990.
93. J Yergey, D Heller, G Hansen, RJ Cotter, C Fenselau. Isotope dilution in mass spectra of large molecules. Anal Chem 55:353–356, 1983.

94. L Prókai. Field Desorption Mass Spectrometry. New York: Marcel Dekker, 1990.
95. RD MacFarlane, DF Torgerson. Science 191:920, 1976.
96. K Tanaka, H Waki, Y Ido, S Akita, Y Yoshida. Rapid Commun Mass Spectrom 2 151–153, 1988.
97. M Karas, F Hillenkamp. Anal Chem 60:2299–2301, 1988.
98. M Karas, D Bachmann, U Bahr, F Hillenkamp. Int J Mass Spectrom Ion Process 8:53–68, 1987.
99. J Zeleny. Phys Rev 3:69, 1914.
100. J Zeleny. Phys Rev 10:1, 1917.
101. J Zeleny. Phys Rev 16:102, 1920.
102. WA Macky. Proc Ron Soc Lond A 133:565, 1931.
103. B Vonnegut, RL Neubauer. J Colloid Sci 7:616, 1952.
104. J Taylor. Proc Roy Soc Lond 280:383, 1964.
105. W Kleber. Tech Dig 9:550, 1967.
106. A Van den Abbeele. Gatuano 37:743, 1968.
107. AA Elmoursi. IEEE Trans Ind Appl 25:234, 1989.
108. TK Burayev, IP Vereshchagin. Fluid Mech-Sov Res 2:56, 1972.
109. VI Kozhenkov, NA Fuks. Russ Chem Rev 45:1179, 1976.
110. JF Hughes, JMC Roberts. Int J Cosmet Sci 6:103, 1984.
111. AG Bailey. Atomisation Spray Technol 2:95, 1986.
112. CP Mao, V Oechsle, N Chigier. J Fluids Eng 109:64, 1987.
113. AG Bailey. Electrostatic Spraying of Liquids. New York: Wiley, 1988.
114. AJ Yule, SM Aval. Fuel 68:1558, 1989.
115. RE Childs, NN Mansout. J Propuls Power 5:641, 1989.
116. B Hoyer, G Sørenson, N Jensen, DB Nielsen, B Larsen. Electrostatic spraying: A novel technique for preparation of polymer coatings on electrodes. Anal Chem 68:3840–3844, 1996.
117. E Bruninx, G Rudstam. Nucl Instrum Methods 13:131, 1961.
118. CJ McNeal, RD MacFarlane, EL Thurston. Anal Chem 51:2036, 1979.
119. RC Murphy, KL Clay, WR Mathews. Anal Chem 54:336, 1982.
120. S Mouradian, JW Skogen, FD Dorman, F Zarrin, SL Kaufmann, M Smith. DNA analysis using an electrospray scanning mobility particle sizer. Anal Chem 69:919–925, 1997.
121. E Neher. MSc (Physics) Thesis, University of Wisconsin 1967.
122. M Dole, LL Mack, RL Hines, RC Mobley, LD Ferguson, MA Alice. J Chem Phys 49:2240, 1968.
123. LL Mach, P Kralik, A Rheude, M Dole. J Chem Phys 52:4977, 1970.
124. PF Knewsubb, TM Sugden. Mass spectrometric studies of ionization in flames. I. The spectrometer and its application to ionization in hydrogen flames. Proc Roy Soc Lond A255:520–537, 1960.
125. AN Hayhurst, TM Sugden. Mass spectrometry of flames. Proc Roy Soc Lond A293:36–50, 1966.
126. MM Shahin. Mass spectrometric studies of corona discharges in air at atmospheric pressures. J Chem Phys 45:2600–2605, 1966.

127. MM Shahin. Use of corona discharges for the study of ion-molecule reactions. J Chem Phys 457:4392, 1968.
128. DS Simons, BN Colby, CA Evans. Int J Mass Spectrom Ion Phys 15, 1974.
129. BN Colby, CA Evans. Anal Chem 45:1884, 1973.
130. KD Cook. Mass Spectrom Rev 5:467, 1986.
131. T Dülcks, FW Röllgen. J Mass Spectrom 30:324–332, 1995.
132. EC Horning, ME Horning, DI Carroll, I Dzidic, RN Stilwell. New picogram detection system based on a mass spectrometer with an external ionization source at atmospheric pressure. Anal Chem 45:936–943, 1973.
133. EC Horning, ME Horning, DI Carroll, I Dzidic, RN Stilwell. Chemical ionization mass spectrometry. Adv Biochem Psychopharmacol 7:15–31, 1973.
134. EC Horning, DI Carroll, I Dzidic, KD Haegele, ME Horning, RN Stilwell. Atmospheric pressure ionization (API) mass spectrometry. Solvent-mediated ionization of samples introduced in solution and in a liquid chromatograph effluent stream. J Chromatogr Sci 12:725–729, 1974.
135. H Kambara, I Kanomata. Ionization characteristics of atmospheric pressure: Ionization by corona discharge. Mass Spectros 24:229–236, 1976.
136. H Kambara, I Kanomata. Collisional dissociation in atmospheric pressure ionization mass spectrometry. Mass Spectrosc 24:271–282, 1976.
137. JA Buckley, JB French, NM Reid. A multipurpose trace atmospheric gas analyzer. Can Aeronaut Space J 20:231–233, 1974.
138. JB French, NM Reid, CC Poon, J Buckley. Presented at the 25 Annual Conference on Mass Spectrometry and Allied Topics, Washington, DC May 1977. A solution to ion clustering interference in API mass spectrometry.
139. M Yamashita, JB Fenn. Electrospray ion source. Another variation on the free-jet theme. J Phys Chem 88:4451–4459, 1984.
140. M Yamashita, JB Fenn. Negative ion production with the electrospray source. J Phys Chem 88:4671, 1984.
141. CM Whitehouse, RN Dreyer, M Yamashita, JB Fenn. Electrospray interface for liquid chromatographs and mass spectrometers. Anal Chem 57:675, 1985.
142. ML Aleksandrov, LN Gall, VN Krasnov, VI Nikolaev, VA Pavlenko, VA Shkurov. Dokl Akad Nauk SSSR 277:379–383, 1984.
143. ML Aleksandrov, LN Gall, VN Krasnov, VI Nikolaev, VA Pavlenko, VA Shkurov, GI Baram, MA Gracher, VD Knorre, YS Kusner. Bioorg Khim 10:710, 1984.
144. ML Aleksandrov, LN Gall, NV Krasnov, VI Nikolaev, VA Shkurov. Mass spectrometric analysis of thermally unstable compounds of low volatility by the extraction of ions from solution at atmospheric pressure. Zh Anal Chem 40:1227–1236, 1985.
145. ML Aleksandrov, LN Gall, NV Krasnov, VI Nikolaev, VA Shkuzov, GI Baram, MA Grachev, YS Kusner. Adv Mass Spectrom, 10:605, 1986.
146. SF Wong, CK Meng, JB Fenn. Presented at the 35 ASMS Conference on Mass Spectrometry and Allied Topics, May 24–29, 1987, Denver, CO. Multiple charging in electrospray ionization of polyethylene glycols.

147. SF Wong, CK Meng, JB Fenn. Multiple charging in electrospray ionization of polyethylene glycols. J Phys Chem 92:546, 1988.
148. M Mann, CK Meng, SF Wong, JB Fenn. Presented at the 37 ASMS Conference on Mass Spectrometry and Allied Topics, May 21–26, 1989, Miami Beach, FL. Some Mass Spectrometric Implications of Multiply Charged Ions.
149. M Mann, CK Meng, JB Fenn. Interpreting mass spectra of multiply charged ions. Anal Chem 54:1702–1708, 1989.
150. AP Bruins, T Covey, JD Henion. Ionspray interface for combined liquid chromatography/atmospheric pressure ionization mass spectrometry. Anal Chem 59:2642–2646, 1987.
151. ME Ikonomou, AT Blades, P Kebarle. Electrospray-ion spray: A comparison of mechanisms and performance. Anal Chem 63:1989–1998, 1991.
152. TR Covey, RF Bonner, BI Shushan, J Henion. Determination of protein, oligonucleotide and peptide molecular weights by ionspray mass spectrometry. Rapid Commun Mass Spectrom 2:249–256, 1988.
153. EC Huang, JD Henion. Packed-capillary liquid chromatography/ion-spray tandem mass spectrometry determination of biomolecules. Anal Chem 63:732–739, 1991.
154. KL Duffin, T Wachs, JD Henion. Atmospheric pressure ion-sampling system for liquid chromatography/mass spectrometry analyses on a benchtop mass spectrometer. Anal Chem 64:61–68, 1992.
155. JA Olivares, NT Nguyen, CR Yonker, RD Smith. On-line mass Spectrometric detection for capillary zone electrophoresis. Anal Chem 59:1230–1232, 1987.
156. RD Smith, JA Olivares, NT Nguyen, HR Udseth. Capillary zone electrophoresis–mass spectrometry using an electrospray ionization interface. Anal Chem 60:436–441, 1988.
157. ED Lee, W Muck, JD Henion, TR Covey. On-line capillary zone electrophoresis-ion spray tandem mass spectrometry for the determination of dynorphins. J Chromatogr 458:313–321, 1988.
158. KD Henry, ER Williams, BH Wang, FW McLafferty, J Shabanowitz DF Hunt. Proc Nat Acd Sci USA 86:9075–9078, 1989.
159. GJ Van Berkel, GL Glish, JA McLuckey. Electrospray ionization combined with ion trap mass spectrometry. Anal Chem 62:1284–1295, 1990.
160. BJ Larsen, CN McEwen. An electrospray ion source for magnetic sector mass spectrometers. J Am Soc Mass Spectrom 2:205–211, 1991.
161. RB Cody, J Tamura, BD Musselman. Electrospray ionization/magnetic sector mass spectrometry: Calibration, resolution, and accurate mass measurements. Anal Chem 64:1561–1570, 1992.
162. P Dobberstein, E Shroeder. Accurate mass determination of a high molecular weight protein using electrospray ionization with a magnetic sector instrument. Rapid commun Mass Spectrom 7:861–864, 1993.
163. L Jiang, M Moini, Design and performance of a high resolution electrospray ion source for a magnetic sector mass spectrometer with a heated capillary inlet. J Am Soc Mass Spectrom 6:1256–1261, 1995.

164. (a) CH Sin, ED Lee, ML Lee. Anal Chem 63:2897, 1991. (b) IM Lazar, ED Lee, JCH Sin, AL Rockwood, KD Onuska, ML Lee. Am Lab 32 110–119, 2000. Electrospray ionization time-of-flight mass spectrometric detection for fast liquid phase separations. (c) IM Lazar, AL Rockwood, ED Lee, JC H Sin, ML Lee. High-speed TOFMS detection for capillary electrophoresis. Anal Chem 71:2578–2581, 1999.
165. JG Boyle, CM Whitehouse. An ion-storage time-of-flight mass spectrometer for analysis of electrospray ions. Rapid Commun Mass Spectrom 5:400–405, 1991.
166. JHJ Dawson, M Guilhaus. Rapid Commun Mass Spectrom 3:155–159, 1989.
167. AF Dodonov, IV Chernushevich, VV Laiko. 12 Int Mass Spectrom Conference, Amsterdam, August 1991. Extended Abstracts p. 153.
168. JG Boyle, CM Whitehouse. Time-of-flight mass spectrometry with an electrospray ion beam. Anal Chem 64:2084–2089, 1992.
169. E Clayton, RH Bateman. Rapid Commun Mass Spectrom 6:719–720, 1992.
170. Y Dunayevskiy, P Vouros, T Carrell, EA Wintner, J Rebek Jr. Characterization of the complexity of small-molecule libraries by electrospray ionization mass spectrometry. Anal Chem 67:2906–2915, 1995.
171. R Wiebolt, J Zweigenbaum, JD Henion. Immunoaffinity ultrafiltration with ionspray HPLC/MS for screening small-molecule libraries. Anal Chem 69:1683–1691, 1997.
172. CL Brummel, JC Vickerman, SA Carr, ME Hemling, GD Roberts, W Johnson, J Weinstock, D Gaitanopoulos, SJ Benkovic, N Winograd. Evaluation of mass spectrometric methods applicable to the direct analysis of non-peptide bead-bound combinatorial libraries. Anal Chem 68:237–242, 1996.
173. KF Blom. Strategies and data precision requirements, for the mass spectrometric determination of structures from combinatorial mixtures. Anal Chem 69:4354–4362, 1997.
174. EW Taylor, MG Qian, GD Dollinger. Simultaneous on-line characterization of small organic molecules derived from combinatorial libraries for identity, quantity, and purity by reverse-phase HPLC with chemiluminescent nitrogen, UV, and mass spectrometric detection. Anal Chem 70:3339–3347, 1998.
175. GI Taylor. Proc Roy Soc Lond A 280:283, 1964.
176. MG Ikonomou, AT Blades, P Kebarle. Electrospray mass spectrometry of methanol and water solutions Suppression of electric discharge with SF_6 gas. J Am Soc Mass Spectrom 3:497–505, 1991.
177. FM Wampler III, AT Blades, P Kebarle. Negative ion electrospray mass spectrometry of nucleotides: Ionization from water solution with SF_6 discharge suppression. J Am Soc Mass Spectrom 4:289–295, 1993.
178. DPH Smith. IEEE Trans Ind Appl IA-22:527, 1986.
179. SK Chowdhury, BT Chait. Method for the electrospray ionization of highly conductive aqueous solutions. Anal Chem 63:1660–1664, 1991.
180. H Wang, M Hackett. Ionization within a cylindrical capacitor: Electrospray without an externally applied high voltage. Anal Chem 70:205–212, 1998.

181. A Hirabayashi, M Sakairi, H Koizumi. Sonic spray mass spectrometry. Anal Chem 67:2878–2882, 1995.
182. L Tang, P Kebarle. Effect of the conductivity of the electrosprayed solution on the electrospray current. Factors determining analyte sensitivity in electrospray mass spectrometry. Anal Chem 63:2709–2715, 1991.
183. CG Enke. A predictive model for matrix and analyte effects in electrospray ionization of singly-charged ionic analytes. Anal Chem 69:4885–4893, 1997.
184. AM Striegel, P Piotrowiak, SM Boué, RB Cole. Polarizability and inductive effects contributions to solvent-cation binding observed in electrospray ionization mass spectrometry. J Am Soc Mass Spectrom 10:254–260, 1999.
185. A Chaperouge, L Bigler, A Schäefer, S Bienz. Correlation of stereoselectivity and ion response in electrospray ionization mass-spectrometry. Electrospray ionization-mass spectrometry as a tool to predict chemical behavior? J Am Soc Mass Spectrom 6:207–211, 1995.
186. VQ Nguyen, XG Chen, AL Yergey. Observation of magic numbers in gas-phase hydrates of alkylammonium ions. J Am Soc Mass Spectrom 8:1175–1179, 1997.
187. AT Blades, MG Ikonomou, P Kebarle. Mechanism of electrospray mass spectrometry. Electrospray as an electrolysis cell. Anal Chem 63:2109–2114, 1991.
188. F Charbonnier, C Rolando, F Saru, P Hapiot, J Pinson. Short time-scale observation of an electrospray ion current. Rapid Commun Mass Spectrom 7:707–710, 1993.
189. GJ Van Berkel, SA McLuckey, GL Glish. Electrospray ionization of porphyrins using a quadrupole ion trap for mass analysis. Anal Chem 63:1098–1109, 1991.
190. GJ Van Berkel, SA McLuckey, GL Glish. Preforming ions in solution via charge-transfer complexation for analysis by electrospray ionization mass spectrometry. Anal Chem 63:2064–2068, 1991.
191. GJ Van Berkel, JME Quirke, RA Tigani, AS Dilley, TR Covey. Derivatization for electrospray ionization mass spectrometry. 3. Electrochemically ionizable derivatives. Anal Chem 70:1544–1554, 1998.
192. GJ Van Berkel, SA McLuckey, GL Glish. Electrochemical origin of radical cations observed in electrospray ionization mass spectra. Anal Chem 64:1586–1593, 1992.
193. GJ Van Berkel, KG Asano, SA McLuckey. Observation of gas-phase molecular dications formed from neutral organics in solution via chemical electron-transfer reactions by using electrospray ionization mass spectrometry. J Am Soc Mass Spectrom 5:689–692, 1994.
194. GJ Van Berkel, F Zhou. Chemical electron-transfer reactions in electrospray ionization mass spectrometry: Effective oxidation potentials of electron-transfer reagents in methylene chloride. Anal Chem 66:3408–3415, 1994.
195. GJ Van Berkel, F Zhou. Characterization of an electrospray ion source as a controlled-current electrolytic cell. Anal Chem 67:2916–2923, 1995.
196. GJ Van Berkel, F Zhou. Electrochemistry combined on-line with electrospray mass spectrometry. Anal Chem 67:3643–3649, 1995.

197. GJ Van Berkel, F Zhou. Electrospray as a controlled-current electrolytic cell: Electrochemical ionization of neutral analytes for detection by electrospray mass spectrometry. Anal Chem 67:3958–3964, 1995.
198. GJ Van Berkel, F Zhou. Observation of gas-phase molecular dications formed from neutral organics in solution via the controlled-current electrolytic process inherent to electrospray. J Am Soc Mass Spectrom 7:157–162, 1995.
199. RB van Breemen. Electrospray liquid chromatography-mass spectrometry of carotenoids. Anal Chem 67:2004–2009, 1995.
200. X Xu, W Lu, RB Cole. On-line probe for fast electrochemistry/electrospray mass spectrometry. Investigation of polycyclic aromatic hydrocarbons. Anal Chem 68:4244–4253, 1996.
201. AM Bond, R Colton, AD'Agostino, AJ Downard, JC Traeger. A role for electrospray mass spectrometry in electrochemical studies. Anal Chem 67:1691–1695, 1995.
202. CECA Hop, DA Saulys, DF Gaines. Electrospray mass spectrometry of borane salts: The electrospray needle as an electrochemical cell. J Am Soc Mass Spectrom 6:860–865, 1995.
203. KP Bateman. Electrochemical properties of capillary electrophoresis-nanoelectrospray mass spectrometry. J Am Soc Mass Spectrom 10:309–317, 1999.
204. Lord Rayleigh. Phil Mag 14:404, 1882.
205. A Gomez, K Tang. Phys Fluids 6:404, 1994.
206. K Tang, A Gomez. Phys Fluids 6:2317, 1994.
207. P Kebarle, L Tang. From ions in solution to ions in the gas phase. The mechanism of electrospray mass spectrometry. Anal Chem 65:972A–986A, 1993.
208. P Kebarle, Y Ho. On the mechanism of electrospray mass spectrometry. In: RB Cole, ed. Electrospray Ionization Mass Spectrometry. New York: Wiley, 1997, Chap 7.
209. DB Hager, NJ Dovichi, J Klassen, P Kebarle, Anal Chem 66:3944, 1994. Droplet electrospray mass spectrometry.
210. G Schmelzeisen-Redecker, L Buttering, FW Röllgen. Int J Mass Spectrom Ion Proc 90:139, 1989.
211. JV Iribarne, BA Thomson. On the evaporation of small ions from charge droplets. Chem Phys 64:2287–2294, 1976.
212. BA Thomson, JV Iribarne. Field induced ion evaporation from liquid surfaces at atmospheric pressure. J Chem Phys 71:4451, 1979.
213. JB Fenn, J Roswell, CK Meng. In electrospray ionization, how much pull does an ion need to escape its droplet prison? J Am Soc Mass Spectrom 8:1147–1157, 1997.
214. M Gamero-Castaño, J Fernandez de la Mora. Modulations in the abundance of salt clusters in electrosprays. Anal Chem 72:1426–1429, 2000.
215. M Sakairi, AL Yergey, KWM Siu, JCY Le Blanc, R Guevremont, SS Berman. Electrospray mass spectrometry: Application of ion evaporation theory to amino acids. Anal Chem 63:1488–1490, 1991.

216. AP Bruins. ESI source design and dynamic range considerations. In: RB Cole, ed. Electrospray Ionization Mass Spectrometry. New York: Wiley, 1997, Chap 3.
217. X Feng, GR Agnes. Single isolated droplets with net charge as a source of ions. J Am Soc Mass Spectrom 11:393–399, 2000.
218. SA Shaffer, K Tang, GA Anderson, HR Udseth, RD Smith. A novel ion funnel for focusing ions at elevated pressure using electrospray ionization mass spectrometry. Rapid Commun Mass Spectrom 11:1813–1817, 1997.
219. SA Shaffer, DC Prior, GA Anderson, HR Udseth, RD Smith. An ion funnel interface for improved ion focusing and sensitivity using electrospray ionization mass spectrometry. Anal Chem 70:4111–4119, 1998.
220. (a) SA Shaffer, AV Tolmachev, DC Prior, GA Anderson, HR Udseth, RD Smith. Characterization of an improved electrodynamic ion funnel interface for electrospray ionization mass spectrometry. Anal Chem 71:2957–2964, 1999. (b) T Kim, AV Tolmachev, R Harkewicz, DC Prior, GA Anderson, HR Udseth, RD Smith, TH Bailey, S Rakov, JH Futrell. Design and implementation of a new electrodynamic ion funnel. Anal Chem 72:2247–2255, 2000.
221. ME Belov, MV Gorshkov, HR Udseth, GA Anderson, AV Tolmachev, DC Prior, R Harkewicz, RD Smith. Initial implementation of electrodynamic ion funnel with Fourier transform ion cyclotron resonance mass spectrometry. J Am Soc Mass Spectrom 11:19–23, 2000.
222. BA Thomson. Declustering and fragmentation of protein ions from an electrospray ion source. J Am Soc Mass Spectrom 8:1053–1058, 1997.
223. TY Yen, MJ Charles, RD Voyksner. Processes that affect electrospray ionization-mass spectrometry of nucleobases and nucleosides. J Am Soc Mass Spectrom 7:1106–1108, 1996.
224. RR Ogorzalek Loo, RD Smith. Investigation of the gas-phase structure of electrosprayed proteins using ion-molecule reactions. J Am Soc Mass Spectrom 5:207–220, 1994.
225. D Zhan, J Rosell, JB Fenn. Solvation studies of electrospray ions—Method and early results. J Am Soc Mass Spectrom 9:1241–1247, 1998.
226. SE Rodriguez-Cruz, JS Klassen, ER Williams. Hydration studies of gas-phase Gramicidin S $(M+2H)^{2+}$ ions formed by electrospray: The transition from solution to gas-phase structure. J Am Soc Mass Spectrom 8:565–568, 1997.
227. SE Rodriguez-Cruz, JS Klassen, ER Williams. Hydration of gas-phase ions formed by electrospray ionization. J Am Soc Mass Spectrom 10:958–968, 1999.
228. SW Lee, P Freivogel, T Schindler, JL Beauchamp. J Am Soc Mass Spectrom 10:347–351, 1999.
229. AG Harrison. Energy-resolved mass spectrometry: A comparison of quadrupole cell and cone-voltage collision-induced dissociation. Rapid Commun Mass Spectrom 13:1663, 1999.
230. S Ikegawa, H Okuyama, J Oohashi, N Murao, J Goto. Separation and detection of bile acid 24-glucuronides in human urine by liquid chromatography combined with electrospray ionization mass spectrometry. Anal Sci 15:625–631, 1999.

231. W Weinmann, A Wiedemann, B Eppinger, M Renz, M Svoboda. Screening for drugs in serum by electrospray ionization/collision-induced dissociation and library searching. J Am Soc Mass Spectrom 10:1028–1037, 1999.
232. SA Carr, MJ Huddleston, MF Bean. Selective identification and differentiation of N- and O-linked oligosaccharides in glycoproteins by liquid chromatography-mass spectrometry. Protein Sci 2:183–196, 1993.
233. MJ Huddleston, RS Annan, MF Bean, SJ Carr. Selective detection of phosphopeptides in complex mixtures by electrospray liquid chromatography/mass spectrometry. J Am Soc Mass Spectrom 4:710–717, 1993.
234. PT Jedrzejewski, WD Lehmann. Detection of modified peptides in enzymatic digests by capillary liquid chromatography/electrospray mass spectrometry and a programmable skimmer CID acquisition routine. Anal Chem 69:294–301, 1997.
235. TR Covey, ED Lee, JD Henion. High-speed liquid chromatography/tandem mass spectrometry for the determination of drugs in biological samples. Anal Chem 58:2453, 1986.
236. SK Chowdhury, V Katta, BT Chait. Rapid Commun Mass Spectrom 4:81–87, 1990.
237. AL Rockwood, M Busman, HR Udseth, RD Smith. Thermally induced dissociation of ions from electrospray mass spectrometry. Rapid Commun Mass Spectrom 5:582–585, 1991.
238. MH Allen, ML Vestal. Design and performance of a novel electrospray interface. J Am Soc Mass Spectrom 3:18–26, 1992.
239. M Busman, AL Rockwood, RD Smith. J Phys Chem 96:2397–2400, 1992.
240. MG Ikonomou, P Kebarle. A heated electrospray source for mass spectrometry of analytes from aqueous solutions. J Am Soc Mass Spectrom 5:791–799, 1994.
241. WV Ligon, SB Dorn. Apparatus for interfacing liquid chromatograph with magnetic sector spectrometer. US Patent 5,266,192, Nov 30, 1993.
242. JF Banks, S Shew, CM Whitehouse, JB Fenn. Anal Chem 66:406, 1994.
243. JF Banks, Jr, JP Quinn, CM Whitehouse. LC/ESI-MS determination of proteins using conventional liquid chromatography and ultrasonically assisted electrospray. Anal Chem 66:3688–3695, 1994.
244. MS Kriger, KD Cook, RS Ramsey. Durable gold-coated fused silica capillaries for use in electrospray mass spectrometry. Anal Chem 67:385–389, 1995.
245. (a) YZ Chang, GR Her. Sheathless capillary electrophoresis/electrospray mass spectrometry using a carbon-coated fused-silica capillary. Anal Chem 72:626–630, 2000. (b) EP Maziarz III, SA Lorenz, TD Wood. Polyaniline: A conductive polymer coating for durable nanospray emitters. J Am Soc Mass Spectrom 11:659–663, 2000.
246. MT Davis, DC Stahl, SA Hefta, TD Lee. A microscale electrospray interface for on-line capillary liquid chromatography tandem mass spectrometry of complex peptide mixtures. Anal Chem 67:4549–4556, 1995.
247. JF Kelly, L Ramaley, P Thibault. Capillary zone electrophoresis-electrospray mass spectrometry at submicroliter flow rates: Practical considerations and analytical performance. Anal Chem 69:51–60, 1997.

248. GA Valaskovic, FW McLafferty. Long-lived metallized tips for nanoliter electrospray mass spectrometry. J Am Soc Mass Spectrom 7:1270–1272, 1996.
249. B Feng, RD Smith. A simple nanoelectrospray arrangement with controllable flowrate for mass analysis of submicrolitre protein samples. J Am Soc Mass Spectrom 11:94–99, 2000.
250. CP Kuo, CH Yuan, J Shiea. Generation of electrospray from a solution predeposited on optical fibers. J Am Soc Mass Spectrom 11:464–467, 2000.
251. DP Kirby, JM Thorne, WK Götzinger, BL Karger. A CE/ESI-MS interface for stable low-flow operation. Anal Chem 68:4451–4457, 1996.
252. P Cao, M Moini. A novel sheathless Interface for capillary electrophoresis/electrospray ionization mass spectrometry using an in-capillary electrode. J Am Soc Mass Spectrom 8:561–564, 1997.
253. P Cao, M Moini. Separation and detection of the α- and β-chains of hemoglobin of a single intact red blood cell using capillary electrophoresis/electrospray ionization time-of-flight mass spectrometry. J Am Soc Mass Spectrom 10:184–186, 1999.
254. KWY Fong, TWD Chan. A novel nonmetallized tip for electrospray ionization at nanoliter flow rate. J Am Soc Mass Spectrom 10:72–75, 1999.
255. R Juraschek, T Dülcks, M Karas. Nanoelectrospray—More than just a minimized-flow electrospray ionization source. J Am Soc Mass Spectrom 10:300–308, 1999.
256. RD Smith, CJ Barinaga, HR Udseth. Improved electro-spray ionization interface for capillary zone electrophoresis. Anal Chem 60:1948, 1988.
257. J Wahl, DR Goodlett, HR Udseth, RD Smith. Attomole level capillary electrophoresis-mass spectrometric protein analysis using 5-μm-i.d. capillaries. Anal Chem 64:3194–3196, 1992.
258. K Tsuji, L Baczynskyj, GE Bronson. Capillary electrophoresis-electrospray mass spectrometry for the analysis of recombinant bovine and porcine somatotrophins. Anal Chem 64:1864–1870, 1992.
259. JR Perkins, CE Parker, KB Tomer. Nanoscale separations combined with electrospray ionization mass spectrometry: Sulfonamide determination. J Am Soc Mass Spectrom 3:139–149, 1992.
260. CE Parker, JR Perkins, KB Tomer, Y Shida, K O'Hara, M Kono. Application of nanoscale packed capillary liquid chromatography (75 mm id) and capillary zone electrophoreses/electrospray ionization mass spectrometry to the analysis of macrolide antibiotics. J Am Soc Mass Spectrom 3:563–574, 1992.
261. MA Mosely, JW Jorgenson, J Shabanowitz, DF Hunt, KB Tomer. Optimization of capillary zone electrophoresis/electrospray ionization parameters for the mass spectrometry and tandem mass spectrometry analysis of peptides. J Am Soc Mass Spectrom 3:289–300, 1992.
262. JR Perkins, KB Tomer. Capillary electrophoresis/electrospray ionization mass spectrometry using a high-performance magnetic sector mass spectrometer. Anal Chem 66:2835–2840, 1994.

263. GC Gale, RD Smith. Small volume and low-flow-rate electrospray ionization mass spectrometry of aqueous samples. Rapid Commun Mass Spectrom 7:1017–1021, 1993.
264. PE Andren, MR Emmett, RM Caprioli. Micro-electrospray: Zeptomole/attomole per microliter sensitivity for peptides. J Am Soc Mass Spectrom 5:867–869, 1994.
265. MR Emmett, RM Caprioli. Micro-electrospray mass spectrometry: Ultra-high-sensitivity analysis of peptides and proteins. J Am Soc Mass Spectrom 5:605–613, 1994.
266. M Wilm, M Mann. Int J Mass Spectrom Ion Process 136:167–180, 1994.
267. Y Takada, M Sakairi, H Koizumi. Atmospheric pressure chemical ionization interface for capillary electrophoresis. mass spectrometry. Anal Chem 67:1474–1476, 1995.
268. L Licklider, WE Kuhr, MP Lacey, T Keough, MP Purdon, R Takigiku. On-line microreactor/capillary electrophoresis/mass spectrometry for the analysis of proteins and peptides. Anal Chem 67:4170–4177, 1995.
269. MS Wilm, M Mann. Analytical properties of the nanoelectrospray ion source. Anal Chem 68:1–8, 1996.
270. TA Fligge, J Kast, K Bruns, M Przybylski. Direct monitoring of protein-chemical reactions utilizing nanoelectrospray mass spectrometry. J Am Soc Mass Spectrom 10:112–118, 1999.
271. J Abian, AJ Oostercamp, E Gelpi. Comparison of conventional, narrow-bore, and capillary liquid chromatography/mass spectrometry for electrospray ionization mass spectrometry: Practical considerations. J Mass Spectrom 34:244–254, 1999.
272. Q Tang, AK Harrata, CS Lee. Capillary isoelectric focusing-electrospray mass spectrometry for protein analysis. Anal Chem 67:3515–3519, 1995.
273. W Lu, GK Poon, PL Carmichael, RB Cole. Analysis of tamoxifen and its metabolites by on-line capillary electrophoresis-electrospray ionization mass spectrometry employing nonaqueous media containing surfactants. Anal Chem 68:668–674, 1996.
274. J Ding, P Vouros. Capillary electrochromatography and capillary electrochromatography-mass spectrometry for the analysis of DNA adduct mixtures. Anal Chem 69:379–384, 1997.
275. L Fang, R Zhang, ER Williams, RN Zare. On-line time-of-flight mass spectrometric analysis of peptides separated by capillary electrophoresis. Anal Chem 66:3696–3701, 1994.
276. JT Wu, P Huang, MX Li, MG Qian, DM Lubman. Open-tubular capillary electrochromatography with an on-line ion trap storage/reflectron time-of-flight mass detector for ultrafast peptide mixture analysis. Anal Chem 69:320–326, 1997.
277. MB Barroso, AP de Jong. Sheathless preconcentration capillary zone electrophoresis applied to peptide analysis. J Am Soc Mass Spectrom 10:1271–1278, 1999.

278. B Feng, RD Smith. A simple nanoelectrospray arrangement with controllable flowrate for mass analysis of submicroliter protein samples. J Am Soc Mass Spectrom 11:94–99, 2000.
279. M Serwe, GA Ross. A comparison of CE-MS and LC-MS for peptide samples. LCGC 18:46–55, 2000.
280. JC Le, J Hui, M Haniu, V Katta, MF Rohde. Development and evaluation of on-line nanoliter flow analysis of protein digests by pneumatic-splitter electrospray liquid chromatography mass spectrometry. J Am Soc Mass Spectrom 8:703–712, 1997.
281. CX Zhang, F Xiang, L Paša Tolić, GA Anderson, TD Veenstra, RD Smith. Stepwise mobilization of focused proteins in capillary isoelectric focusing mass spectrometry. Anal Chem 72:1462–1468, 2000.
282. IM Lazar, ML Lee. Effect of electrospray needle voltage on electroosmotic flow in capillary electrophoresis/mass spectrometry. J Am Soc Mass Spectrom 10:261–264, 1999.
283. Q Xue, F Foret, YM Dunayevskiy, PM Zavracky, NE McGruer, BL Karger. Multichannel microchip electrospray mass spectrometry. Anal Chem 69:426–430, 1997.
284. L Licklider, XQ Wang, A Desai, YC Tai, TD Lee. A micromachined chip-based electrospray source for mass spectrometry. Anal Chem 72:367–375, 2000.
285. J Li, JF Kelly, I Chernushevich, DJ Harrison, P Thibault. Separation and identification of peptides from gel-isolated membrane proteins using a microfabricated device for combined capillary electrophoresis/nanoelectrospray mass spectrometry. Anal Chem 72:599–609, 2000.
286. (a) DB Kassel, T Wang, Z Lu. Parallel fluid electrospray mass spectrometer. Intl Patent Appl PCT/US99/12905, June 8, 1998. (b) RH Bateman, JAD Hickson. Multi-inlet mass spectrometer. Eur Patent Appl 99304746.3, June 17, 1999. (c) RB Cody, patent pending.
287. L Romanyshyn, PR Tiller, CECA Hop. Bioanalytical applications of "fast chromatography" to high-throughput liquid chromatography/tandem mass spectrometric quantitation. Rapid Commun Mass Spectrom 14:1662–1668, 2000.
288. C Masselon, GA Anderson, R Harkewicz, JE Bruce, L Paša Tolić, RD Smith. Accurate mass multiplexed tandem mass spectrometry for high-throughput polypeptide identification from mixtures. Anal Chem 72:1918–1924, 2000.
289. SK Chowdhury, V Katta, RC Beavis, BT Chait. Removal of adducts (molecular mass = 98 u) attached to peptide and protein ions in electrospray ionization mass spectra. J Am Soc Mass Spectrom 1:382. 1990.
290. JT Stults, JT Marsters. Improved electrospray ionization of synthetic oligodeoxynucleotides. Rapid Commun Mass Spectrom 5:359–363, 1991.
291. M Greig RH Griffey. Utility of organic bases for improved electrospray mass spectrometry of oligonucleotides. Rapid Commun Mass Spectrom 9:97–102, 1995.
292. N Torto, A Hofte, R van der Hoeven, U Tjaden, L Gorton, G Marko-Varga, C Bruggink, J van der Greef. Microdialysis introduction high-performance anion-

exchange chromatography/ion spray mass spectrometry for monitoring on-line desalted carbohydrate hydrolysates. J Mass Spectrom 33:334–341, 1998.
293. RAM van der Hoeven, UR Tjaden, J van def Greef, WHM van Casteren, HA Schols, AGJ Voragen. Recent progress in high-performance anion-exchange chromatography/ionspray mass spectrometry for molecular mass determination and characterization of carbohydrates using static and scanning array detection. J Mass Spectrom 33:377–386, 1998.
294. Commercially available from the Millipore Corporation as Zip Tips.
295. MG Ikonomou, AT Blades, P Kebarle. Anal Chem 62:957–967, 1990.
296. DL Buhrman, PL Price, PJ Rudewicz. Quantitation of SR27417 in human plasma using electrospray liquid chromatography-tandem mass spectrometry: A study of ion suppression. J Am Soc Mass Spectrom 7:1099–1105, 1996.
297. TL Constantopoulos, GS Jackson, CG Enke. Effects of salt concentration on analyte response using electrospray ionization mass spectrometry. J Am Soc Mass Spectrom 10:625–634, 1999.
298. KL Rundlett, DW Armstrong. Mechanism of signal suppression by anionic surfactants in capillary electrophoresis-electrospray ionization mass spectrometry. Anal Chem 68:3493–3497, 1996.
299. FE Kuhlmann, A Apffel, SM Fischer, G Goldberg, PC Goodley. Signal enhancement for gradient reverse-phase high-performance liquid chromatography-electrospray ionization mass spectrometry with trifluoroacetic and other strong acid modifiers by postcolumn addition of propionic acid and isopropanol. J Am Soc Mass Spectrom 6:1221–1225, 1995.
300. M Stefansson, PJR Sjöberg, KE Markides. Regulation of multimer formation in electrospray mass spectrometry. Anal Chem 68:1792–1797, 1996.
301. L Drahos, RMA Heeren, C Collette, E De Pauw, K Vékey. Thermal energy distribution observed in electrospray ionization. J Mass Spectrom 34:1373–1379, 1999.
302. G Wang, RB Cole. Solution, gas-phase, and instrumental parameter influences on charge-state distributions in electrospray ionization mass spectrometry. In: RB Cole, ed. Electrospray ionization Mass Spectrometry. New York: Wiley, 1997, Chap 4.
303. JA Loo, CG Edmonds, HR Udseth, RD Smith. Effect of reducing disulfide-containing proteins on electrospray ionization mass spectra. Anal Chem 62:693–698, 1990.
304. IA Kaltashov, CC Fenselau. A direct comparison of first and second gas-phase basicities of the octapeptide RPPGFSPF. J Am Chem Soc 117:9906–9910, 1995.
305. IA Kaltashov, CC Fenselau. Thermochemistry of multiply charged mellitin in the gas phase determined by modified kinetic method. Rapid Commun Mass Spectrom 10:857–861, 1996.
306. ER Williams. Proton transfer reactivity of large multiply charged ions. J Mass Spectrom 31:831, 1996.
307. (a) K Vekey. Adv Mass Spectrom 13:537, 1995. (b) K Vekey. Multiply charged ions. Mass Spectrom Rev 14:195–225, 1995.

308. S Gronert. Determining gas-phase properties and reactivities of multiply charged ions. J Mass Spectrom 34:787–796, 1999.
309. G Wang, RB Cole. Effect of solution ionic strength on analyte charge state distributions in positive and negative ion electrospray ionization mass spectrometry. Anal Chem 66:3702–3708, 1994.
310. UA Mirza, BT Chait. Effect of anions on the positive ion electrospray ionization mass spectra of peptides and proteins. Anal Chem 66:2898–2904, 1994.
311. G Wang, RB Cole. Mechanistic interpretation of the dependence of charge state distributions on analyte concentrations in electrospray ionization mass spectrometry. Anal Chem 67:2892–2900, 1995.
312. G Wang, RB Cole. Effects of solvent and counterion on ion pairing and observed charge states of diquaternary ammonium ions in electrospray ionization mass spectrometry. Anal Chem 7:1050–1058, 1996.
313. RH Griffey, H Sasmor, MJ Greig. Oligonucleotide charge states in negative ionization electrospray mass spectrometry are a function of solution ammonium ion concentration. J Am Soc Mass Spectrom 8:155–160, 1997.
314. PD Schnier, DS Gross, ER Williams. On the maximum charge state and proton transfer reactivity of peptide and protein ions formed by electrospray ionization. J Am Soc Mass Spectrom 6:1086–1097, 1995.
315. B Ganem, YT Li, JT Henion. Detection of noncovalent receptor-ligand complexes by mass spectrometry. J Am Chem Soc 113:6294–6296, 1991.
316. B Ganem, YT Li, JT Henion. Observation of noncovalent enzyme-substrate and enzyme-product complexes by ion-spray mass spectrometry. J Am Chem Soc 113:7818–7819, 1991.
317. AK Ganguly, BN Pramanik, A Tsarbopoulos, TR Covey, E Huang, SA Fuhrman. J Am Chem Soc 114:3992, 1992. (b) BN Pramanik, PL Bartner, UA Mirza, YH Liu, AK Ganguly. Electrospray ionization mass spectrometry for the study of non-covalent complexes: An emerging technology. J Mass Spectrom 33(10):911–920, 1998. (c) AK Ganguly, BN Pramanik, EC Huang, A Tsarbopoulos, VM Girijavallabhan, S Liberles. Studies of the ras-GDP and ras-GTP noncovalent complexes by electrospray mass spectrometry. Tetrahedron 49:7985–7996, 1993. (d) AK Ganguly, BN Pramanik, A Tsarbopoulos, TR Covey, E Huang, SA Fuhrman. Mass spectrometric detection of the noncovalent GDP-bound conformational state of the human H-ras protein. J Am Chem Soc 114:6559–6560, 1992.
318. E Huang, BN Pramanik, A Tsarbopoulos, P Reichert, AK Ganguly, PP Trotta, TL Nagabhushan, TR Covey. Application of electrospray mass spectrometry in probing protein–protein and protein–ligand noncovalent interactions. J Am Chem Soc 4:624–630, 1993.
319. (a) YT Li, YL Hsieh, JD Henion, MW Senko, FW McLafferty, B Ganem. Mass spectrometric studies on noncovalent dimers of leucine zipper peptides. J Am Chem Soc 115:8409–8413, 1993. (b) B Ganem, YT Li, JD Henion. Detection of oligonucleotide duplex forms by ion-spray mass spectrometry. Tetrahedron Lett 34:1445–1448, 1993. (c) YT Li, YL Hsieh, JD Henion, B Ganem. Studies on

haem binding in myoglobin, haemoglobin and cytochrome c by ion spray mass spectrometry. J Am Soc Mass Spectrom 4:631–637, 1993.
320. B Ganem, YT Li, JD Henion. Chemtracts 6:1, 1993.
321. JA Loo, AB Giordani, H Muenster. Observation of intact heme-bound myoglobin by electrospray ionization on a double-focusing mass spectrometer. Rapid Commun Mass Spectrom 7:186–189, 1993.
322. BL Schwartz, KJ Light-Wahl, RD Smith. Observation of noncovalent complexes to the avidin tetramer by electrospray ionization mass spectrometry. J Am Soc Mass Spectrom 5:201–204, 1994.
323. YL Hsieh, J Cai, YT Li, JD Henion, B Ganem. Detection of noncovalent FKBP-FK506 and FKBP-rapamycin complexes by capillary electrophoresis-mass spectrometry and capillary electrophoresis-tandem mass spectrometry. J Am Soc Mass Spectrom 6:95, 1995.
324. R Feng, AL Castelhano, R Billedeau, Z Yuan. Study of noncovalent enzyme-inhibitor complexes and metal binding stoichiometry of matrilysin by electrospray ionization mass spectrometry. J Am Soc Mass Spectrom 6:1105–1111, 1995.
325. R Ramanathan, L Prokai. Electrospray ionization mass spectrometric study of encapsulation of amino acids by cyclodextrins. J Am Soc Mass Spectrom 6:866–871, 1995.
326. SG Penn, F He, MK Green, CB Lebrilla. The use of heated capillary dissociation and collision-induced dissociation to determine the strengths of noncovalent bonding interactions in gas-phase peptide-cyclodextrin complexes. J Am Soc Mass Spectrom 8:244–252, 1997.
327. JB Cunniff, P Vouros. False positives and the detection of cyclodextrin inclusion complexes by electrospray mass spectrometry. J Am Soc Mass Spectrom 6:437–447, 1995.
328. G Hopfgartener, C Piguet, JD Henion. Ion spray-tandem mass spectrometry of supramolecular coordination complexes. J Am Soc Mass Spectrom 5:748–756, 1994.
329. T Hayashi, Y Hitomi, H Ogoshi. Artificial protein-protein complexation between a reconstituted myoglobin and cytochrome c. J Am Chem Soc 120:4910–4915, 1998.
330. D Fabris, C Fenselau. Anal Chem 71:384–387, 1999.
331. KX Wan, ML Gross, T Shibue. Gas-phase stability of double-stranded oligodeoxynucleotides and their noncovalent complexes with DNA-binding drugs as revealed by collisional activation in an ion trap. J Am Soc Mass Spectrom 11:450–457, 2000.
332. R Guevremont, KWM Siu, JCY Le Blanc, SS Berman. Are the electrospray mass spectra of proteins related to their aqueous solution chemistry? J Am Soc Mass Spectrom 3:216–224, 1992.
333. JA Loo, HR Udseth, RD Smith. Rapid Commun Mass Spectrom 2:207–210, 1988.
334. MA Kelly, MM Vestling, CC Fenselau. Org Mass Spectrom 27:1143–1147, 1992.

335. G Wang, RB Cole. Disparity between solution-phase equilibria and charge state distributions in positive-ion electrospray mass spectrometry. Org Mass Spectrom 29:419–427, 1994.
336. AT Blades, P Kebarle. J Am Chem Soc 116:10761, 1994.
337. RB Cody. Observation of remote-site fragmentation at low collision energy in Fourier transform mass spectrometry. Rapid Commun Mass Spectrom 2:260–261, 1988.
338. RB Cody, A Kusai, Y Ueda, T Morita. GC/MS and LC/MS analysis of endocrine disruptors. Proc 47th ASMS Conf on Mass Spectrometry and Allied Topics, Dallas, TX, June 13–17, 1999.
339. CN McEwen, BS Larsen. Electrospray ionization on quadrupole and magnetic-sector mass spectrometers. In: RB Cole, ed. Electrospray Ionization Mass Spectrometry. New York: Wiley, 1997, Chap 5.
340. AN Tyler, E Clayton, BN Green. Exact mass measurement of polar organic molecules at low resolution using electrospray ionization and a quadrupole mass spectrometer. Anal Chem 68:3561–3569, 1996.
341. V Katta, SK Chowdhury, BT Chait. Use of a single quadrupole mass spectrometer for collision-induced dissociation studies of multiply charged peptide ions produced by electrospray ionization. Anal Chem 63:174–178, 1991.
342. R Feng, Y Konishi. Collisionally-activated dissociation of multiply charged 150-kDa antibody ions. Anal Chem 65:645–649, 1993.
343. N Sproch, T Kruger. Construction and characterization of a small-bore electrospray ionization source. J Am Soc Mass Spectrom 4:964–967, 1993.
344. LQ Huang, A Paiva, R Bhat, M Wong. Characterization of large, heterogeneous proteins by electrospray ionization-mass spectrometry. J Am Soc Mass Spectrom 7:1219–1226, 1996.
345. RR Ogorzalek Loo, BE Winger, RD Smith. Proton transfer reaction studies of multiply charged proteins in a high mass-to-charge ratio quadrupole mass spectrometer. J Am Soc Mass Spectrom 5:1064–1071, 1994.
346. MH Amad, RS Houk. Mass resolution of 11,000 to 22,000 with a multiple pass quadrupole mass analyzer. J Am Soc Mass Spectrom 11:407–415, 2000.
347. AM Starrett, GC DiDonato. High resolution accurate mass measurement of product ions formed in an electrospray source on a sector instrument. Rapid Commun Mass Spectrom 7:12–15, 1993.
348. RB Cody, J Tamura, JW Finch, BD Musselman. Improved detection limits for electrospray ionization on a magnetic sector mass spectrometer by using an array detector. J Am Soc Mass Spectrom 5:194–200, 1994.
349. JA Loo, RR Ogorzalek Loo. Applying charge discrimination with electrospray ionization-mass spectrometry to protein analysis. J Am Soc Mass Spectrom 6:1098–1104, 1995.
350. (a) J Finch, BD Musselman, GR Sims, F Kunihiro. Proc 43rd Annual Conf on Mass Spectrometry and Allied Topics, Atlanta, GA, May 21–26, 1995. (b) C Martin, EG Owen, RB Cody, unpublished data.

351. VSK Kolli, R Orlando. Complete sequence determination of large peptides by high energy collisional activation of multiply protonated ions. J Am Soc Mass Spectrom 6:234–241, 1995.
352. MC Sullards, JA Reiter. On the use of scans at constant ratio of B/E for studying decompositions of peptide metal(II)-ion complexes formed by electrospray ionization. J Am Soc Mass Spectrom 6:608–610, 1995.
353. J Adams, FH Strobel, A Reiter, MC Sullards. The importance of charge-separation reactions in tandem mass spectrometry of doubly protonated angiotensin II formed by electrospray ionization. Experimental considerations and structural implications. J Am Soc Mass Spectrom 7:30–41, 1996.
354. Y Takahashi, S Fujimaki, T Kobayashi, T Morita, T Higuchi. Accurate mass determination by multiple sprayers nano-electrospray mass spectrometry on a magnetic sector instrument. Rapid Commun Mass Spectrom 14:947–949, 2000.
355. DA Laude, E Stevenson, JM Robinson. Electrospray ionization/Fourier transform ion cyclotron resonance mass spectrometry. In: RB Cole, ed. Electrospray Ionization Mass Spectrometry. New York: Wilen, 1997, Chap 8.
356. NL Kelleher, MW Senko, MM Siegel, FW McLafferty. Unit resolution mass spectra of 112 kDa molecules with 3 Da accuracy. J Am Soc Mass Spectrom 8:380–383, 1997.
357. RB Cody, BS Freiser. Collision-induced dissociation in a Fourier transform mass spectrometer. Int J Mass Spectrom Ion Process 41:199–204, 1982.
358. RB Cody, RC Burnier, BS Freiser. Collision-induced dissociation with Fourier transform mass spectrometry. Anal Chem 52:54, 96–101, 1982.
359. JW Gauthier, TR Trantman, DB Jacobson. Anal Chim Acta 246:211–225, 1991.
360. MW Senko, JP Speir, FW McLafferty. Collisional activation of large multiply charged ions using Fourier transform mass spectrometry. Anal Chem 66:2801–2808, 1994.
361. SA Hofstadler, JH Wahl, R Bakhtiar, GA Anderson, JE Bruce, RD Smith. Capillary electrophoresis Fourier transform ion cyclotron resonance mass spectrometry with sustained off-resonance irradiation for the characterization of protein and peptide mixtures. J Am Soc Mass Spectrom 5:894–899, 1994.
362. Q Wu, S Van Orden, X Cheng, R Bakhtiar, RD Smith. Characterization of cytochrome c-variants with high-resolution FTICR mass spectrometry: Correlation of fragmentation and structure. Anal Chem 67:2498–2509, 1995.
363. AR Heck, PJ Derrick. Ultrahigh mass accuracy in isotope-selective collision-induced dissociation using correlated sweep excitation and sustained off-resonance irradiation: A Fourier transform ion cyclotron resonance mass spectrometry case study on the $[M+2H]^{2+}$ ion of bradykinin. Anal Chem 69:3603–3607, 1997.
364. L Paša Tolić, JE Bruce, QP Lei, GA Anderson, RD Smith. In-trap cleanup of proteins from electrospray ionization using soft sustained off-resonance irradiation with Fourier transform ion cyclotron resonance mass spectrometry. Anal Chem 70:405–408, 1998.

365. JP Speir, MW Senko, DP Little, JA Loo, FW McLafferty. High-resolution tandem mass spectra of 37–67 kDa proteins. J Mass Spectrom 30:39–42, 1995.
366. RB Cody, RC Burnier, CJ Cassady, BS Freiser. Consecutive collision induced dissociations in Fourier transform mass spectrometry. Anal Chem 54:2225–2228, 1982.
367. ER Williams, KD Henry, FW McLafferty. J Am Chem Soc 112:6157–6162, 1990.
368. Z Guan, SA Hofstadler, DA Laude. Jr. Remeasurement of electrosprayed proteins in the trapped ion cell of a Fourier transform ion cyclotron resonance mass spectrometer. Anal Chem 65:1588–1593, 1993.
369. VL Campbell, Z Guan, DA Laude. Remeasurement at high resolving power in Fourier transform ion cyclotron resonance mass spectrometry. J Am Soc Mass Spectrom 6:564–570, 1995.
370. KD Henry, JP Quinn, FW McLafferty. High-resolution electrospray mass spectra of large molecules. J Am Chem Soc 113:5447–5449, 1991.
371. MV Buchanan, RL Hettich. Fourier transform mass spectrometry of high-mass biomolecules. Anal Chem 65:245A–258A, 1993.
372. SC Beu, MW Senko, P Quinn, FW McLafferty. Improved Fourier transform ioncyclotron resonance mass spectrometry of large biomolecules. J Am Soc Mass Spectrom 4:190–192, 1993.
373. SC Beu, MW Senko, P Quinn, FM Wampler, FW McLafferty. Fourier-transform electrospray instrumentation for tandem high-resolution mass spectrometry of large molecules. J Am Soc Mass Spectrom 4:557–565, 1993.
374. BE Winger, SA Hofstadler, JE Bruce, HR Udseth, RD Smith. High-resolution accurate mass measurements of biomolecules using a new electrospray ionization ion cyclotron resonance mass spectrometer. J Am Soc Mass Spectrom 4:566–577, 1993.
375. DP Little, JP Speir, MW Senko, PB O'Connor, FW McLafferty. Infrared multiphoton dissociation of large multiply charged ions for biomolecule sequencing. Anal Chem 66:2809–2815, 1994.
376. X Cheng, R Bakhtiar, S Van Orden, RD Smith. Charge-state shifting of individual multiply charged ions of bovine serum albumin dimer and molecular weight determination using an individual-ion approach. Anal Chem 66:2084–2087, 1994.
377. EF Gordon, BA Mansoori, CF Carroll, DC Muddiman. Hydropathic influences on the quantification of equine heart cytochrome c using relative ion abundance measurements by electrospray ionization Fourier transform ion cyclotron resonance mass spectrometry. J Mass Spectrom 34:1055–1062, 1999.
378. VL Campbell, Z Guan, DA Laude Jr. Selective generation of charge-dependent/independent ion energy distributions from a heated capillary electrospray source. J Am Soc Mass Spectrom 5:221–229, 1994.
379. JA Hofstadler, FD Swanek, DC Gale, AG Ewing, RD Smith. Capillary electrophoresis-electrospray ionization Fourier transform ion cyclotron resonance mass spectrometry for direct analysis of cellular proteins. Anal Chem 67:1477–1480, 1995.

380. JH Wahl, SA Hofstadler, RD Smith. Direct electrospray ion current monitoring detection and its use with on-line capillary electrophoresis mass spectrometry. Anal Chem 67:462–465, 1995.
381. GA Valaskovic, NL Kelleher, DP Little, DJ Aaserud, FW McLafferty. Attomole-sensitivity electrospray source for large-molecule mass spectrometry. Anal Chem 67:3802–3805, 1995.
382. PB O'Connor, FW McLafferty. High-resolution ion isolation with the ion cyclotron resonance capacitively coupled open cell. J Am Soc Mass Spectrom 6:533–535, 1995.
383. DP Little, FW McLafferty. Infrared photodissociation of non-covalent adducts of electrosprayed nucleotide ions. J Am Soc Mass Spectrom 7:209–210, 1996.
384. DC Muddiman, DS Wunschel, C Liu, L Paša Tolić, KF Fox, A Fox, GA Anderson, RD Smith. Characterization of PCR products from bacilli using electrospray ionization FTICR mass spectrometry. Anal Chem 68:3705–3712, 1996.
385. JA Carroll, SG Penn, ST Fannin, J Wu, MT Cancilla, MK Green, CB Lebrilla. A dual vacuum chamber Fourier transform mass spectrometer with rapidly interchangeable LSIMS, MALDI, and ESI sources: Initial results with LSIMS and MALDI. Anal Chem 68:1798–1804, 1996.
386. Q Wu, GA Anderson, HR Udseth, MG Sherman, S Van Orden, R Chen, SA Hofstadler, MV Gorshkov, DW Mitchell, AL Rockwood, RD Smith. A high-performance low magnetic field internal electrospray ionization-Fourier transform ion cyclotron resonance mass spectrometer. J Am Soc Mass Spectrom 7:915–922, 1996.
387. DJ Aaserud, NL Kelleher, DP Little, FW McLafferty. Accurate base composition of double-stranded DNA by mass spectrometry. J Am Soc Mass Spectrom 7:1266–1269, 1996.
388. SA Hofstadler, DA Laude Jr. Electrospray ionization in the strong magnetic field of a Fourier transform ion cyclotron resonance mass spectrometer. Anal Chem 64:572–575, 1992.
389. MV Gorshkov, L Paša Tolić, JE Bruce, GA Anderson, RD Smith. A dual-trap design and its applications in electrospray ionization FTICR mass spectrometry. Anal Chem 69:1307–1314, 1997.
390. DS Wunschel, DC Muddiman, KF Fox, A Fox, RD Smith. Heterogeneity in *Bacillus cereus* PCR products detected by ESI-FTICR mass spectrometry. Anal Chem 70:1203–1207, 1998.
391. RD Burton, KP Matuszak, CH Watson, JR Eyler. Exact mass measurements using a 7 tesla Fourier transform ion cyclotron resonance mass spectrometer in a good laboratory practices-regulated environment. J Am Soc Mass Spectrom 10:1291–1297, 1999.
392. EP Maziarz III, GA Baker, SA Lorenz, TD Wood. External ion accumulation of low molecular weight poly(ethylene glycol) by electrospray ionization Fourier transform mass spectrometry. J Am Soc Mass Spectrom 10:1298–1304, 1999.

393. JL Sterner, MV Johnston, GR Nicol, DP Ridge. Signal suppression in electrospray ionization Fourier transform mass spectrometry of multi-component samples. J Mass Spectrom 35:385–391, 2000.
394. JA Loo, CG Edmonds, RD Smith. Tandem mass spectrometry of very large molecules: Serum albumin sequence information from multiply charged ions formed by electrospray ionization. Anal Chem 63:2488–2499, 1991.
395. SA Hofstadler, DA Laude. Trapping and detection of ions generated in a high magnetic field electrospray ionization Fourier transform ion cyclotron resonance mass spectrometer. J Am Soc Mass Spectrom 3:615–623, 1992.
396. JE Bruce, GA Anderson, SA Hofstadler, BE Winger, RD Smith. Time-base modulation for the correction of cyclotron frequency shifts observed in long-lived transients from Fourier-transform ion cyclotron resonance mass spectrometry of electrosprayed biopolymers. Rapid Commun Mass Spectrom 7:700–703, 1993.
397. JE Bruce, GA Anderson, SA Hofstadler, SA Van Orden, MS Sherman, AL Rockwood, RD Smith. Selected-ion accumulation from an external electrospray ionization source with a Fourier-transform ion cyclotron resonance mass spectrometer. Rapid Commun Mass Spectrom 7:914–919, 1993.
398. MW Senko, CL Hendrickson, MR Emmett, SDH Shi, AG Marshall. External accumulation of ions for enhanced electrospray ionization Fourier transform ion cyclotron resonance mass spectrometry. J Am Soc Mass Spectrom 8:970–976, 1997.
399. KA Sannes-Lowery, SA Hofstadler. Characterization of multipole storage assisted dissociation: Implications for electrospray ionization mass spectrometry characterization of biomolecules. J Am Soc Mass Spectrom 11:1–9, 2000.
400. K Håkansson, J Axelsson, M Palmblad, P Håkansson. Mechanistic studies of multipole storage assisted dissociation. J Am Soc Mass Spectrom 11:210–217, 2000.
401. R Chen, Q Wu, DW Mitchell, SA Hofstadler, AL Rockwood, RD Smith. Direct charge number and molecular weight determination of large individual ions by electrospray ionization Fourier transform ion cyclotron resonance mass spectrometry. Anal Chem 66:3964–3969, 1994.
402. R Chen, X Cheng, DW Mitchell, SA Hofstadler, Q Wu, AL Rockwood, MG Sherman, RD Smith. Trapping, detection, and mass determination of coliphage T4 DNA ions of 10^8 Da by electrospray ionization Fourier transform ion cyclotron resonance mass spectrometry. Anal Chem 67:1159–1163, 1995.
403. NL Kelleher, MW Senko, DP Little, PB O'Connor, FW McLafferty. Complete large-molecule high-resolution mass spectra from 50-femtomole microvolume injection. J Am Soc Mass Spectrom 6:220–221, 1995.
404. ER Williams, Tandem FTMS of large biomolecules. Anal Chem 70:179A–185A, 1998.
405. MV Gorshkov, L Paša Tolić, HR Udseth, GA Anderson, BM Huang, JE Bruce, DC Prior, JA Hofstadler, L Tang, LZ Chen, JA Willett, AL Rockwood, MS Sherman, RD Smith. Electrospray ionization Fourier transform ion cyclotron

resonance mass spectrometry at 11.5 tesla: Instrument design and initial results. J Am Soc Mass Spectrom 9:692–700, 1998.

406. (a) JE Bruce, GA Anderson, MD Brands, L Paša Tolić, RD Smith. Obtaining more accurate Fourier transform ion cyclotron resonance mass measurements without internal standards using multiply charged ions. J Am Soc Mass Spectrom 11:416–421, 2000. (b) RB Cody. Method for external calibration of ion cyclotron resonance mass spectrometers. US Patent 4,933,547, June 12, 1990.

407. KA Johnson, MFJM Verhage, PS Brereton, MWW Adams, IJ Amster. Probing the stoichiometry and oxidation states of metal centers in iron-sulfur proteins using electrospray FTICR mass spectrometry. Anal Chem 72:1410–1418, 2000.

408. RA Zubarev, DM Horn, EK Fridriksson, NL Kelleher, NA Kruger, MA Lewis, BK Carpenter, FW McLafferty. Electron capture dissociation for structural characterization of multiply charged protein cations. Anal Chem 72:563–573, 2000.

409. N Huang, MM Siegel, GH Kruppa, FH Laukien. Automation of a Fourier transform ion cyclotron resonance mass spectrometer for acquisition, analysis and e-mailing of high-resolution exact-mass electrospray ionization mass spectral data. J Am Soc Mass Spectrom 10:1166–1173, 1999.

410. SJ Eyles, JP Speir, GH Kruppa, LM Gierasch, IA Kaltashov. Protein conformational stability probed by Fourier transform ion cyclotron resonance mass spectrometry. J Am Chem Soc 122:495–500, 2000.

411. ME Bier, JC Schwartz. Electrospray ionization quadrupole ion-trap mass spectrometry. In: RB Cole, ed. Electrospray Ionization Mass Spectrometry. New York: Wiley, 1997, Chap 7.

412. FA Londry, GJ Wells, RE March. Enhanced mass resolution in a quadrupole ion trap. Rapid Commun Mass Spectrom 7:43–45, 1993.

413. JD Williams, RG Cooks. Improved accuracy of mass measurement with a quadrupole ion-trap mass spectrometer. Rapid Commun Mass Spectrom 6:524–527, 1992.

414. KJ Hart, SA McLuckey, GL Glish. Evidence of isomerization during isolation in the quadrupole ion trap. J Am Soc Mass Spectrom 3:680–682, 1992.

415. N Sadagopan, JT Watson. Investigation of the tris(trimethoxyphenyl)phosphonium acetyl charged derivatives of peptides by electrospray ionization mass spectrometry and tandem mass spectrometry. J Am Soc Mass Spectrom 11:107–119, 2000. ·

416. MJ Doktycz, S Habibi-Goudarzi, SA McLuckey. Accumulation and storage of ionized duplex DNA molecules in a quadrupole ion trap. Anal Chem 66:3416–3422, 1994.

417. SA McLuckey, RS Ramsey. Gaseous myoglobin ions stored at greater than 300 KJ Am Soc Mass Spectrom 5:324–327, 1994.

418. TG Schaaff, JL Stephenson Jr, SA McLuckey. Gas phase H/D exchange kinetics: DI vs D_2O. J Am Soc Mass Spectrom 11:167–171, 2000.

419. (a) WJ Herron, DE Goeringer, SA McLuckey. Ion-ion reactions in the gas phase: Proton transfer reactions of protonated pyridine with multiply charged oligonucleotide anions. J Am Soc Mass Spectrom 6:529–532, 1995. (b) SA McLuckey,

JL Stephenson Jr. Ion/ion chemistry of high-mass multiply charged ions. Mass Spectrom Rev 17:369–407, 1998.
420. WH Benner. A gated electrostatic ion trap to repetitiously measure the charge and m/z of large electrosprayed ions. Anal Chem 69:4162–4168, 1997.
421. SA McLuckey, GJ Van Berkel, GL Glish, EC Huang, JD Henion. Ion spray liquid chromatography/ion trap mass spectrometry determination of biomolecules. Anal Chem 63:375–385, 1991.
422. JD Henion, AV Mordehai, J Cai. Quantitative capillary electrophoresis-ion spray mass spectrometry on a benchtop ion trap for the determination of isoquinoline alkaloids. Anal Chem 66:2103–2109, 1994.
423. SA McLuckey, GL Van Berkel, DE Goeringer, GL Glish, Ion trap mass spectrometry using high-pressure ionization. Anal Chem 66:737A–743A, 1994.
424. SA McLuckey, S Habibi-Goudarzi. Ion trap tandem mass spectrometry applied to small multiply charged oligonucleotides with a modified base. J Am Soc Mass Spectrom 5:740–747, 1994.
425. SA McLuckey, GJ Van Berkel, GL Glish. J Am Chem Soc 112:5668–5670, 1990.
426. SA McLuckey, DE Goeringer. Ion/molecule reactions for improved effective mass resolution in electrospray mass spectrometry. Anal Chem 67:2493–2497, 1995.
427. JL Stephenson Jr, SA McLuckey. Reactions of poly(ethylene glycol) cations with iodide and perfluorocarbon anions. J Am Soc Mass Spectrom 9:957–965, 1998.
428. R Körner, M Wilm, K Morand, M Schubert, M Mann. Nano electrospray combined with a quadrupole ion trap for the analysis of peptides and protein digests. J Am Soc Mass Spectrom 7:150–156, 1996.
429. JL Stephenson Jr, SA McLuckey, Anion effects on storage and resonance ejection of high mass-to-charge cations in quadrupole ion trap mass spectrometry. Anal Chem 69:3760–3766, 1997.
430. AV Mordehai, JD Henion. A novel differentially pumped design for atmospheric pressure ionization-ion trap mass spectrometry. Rapid Commun Mass Spectrom 7:205–209, 1993.
431. LCM Ngoka, ML Gross. Multistep tandem mass spectrometry for sequencing cyclic peptides in an ion-trap mass spectrometer. J Am Soc Mass Spectrom 10:732–746, 1999.
432. CS Hoagland, SJ Valentine, DE Clemmer. An ion trap interface for ESI-ion mobility experiments. Anal Chem 69:4156–4161, 1997.
433. IV Chernushevich, W Ens, KG Standing, Electrospray ionization time-of-flight mass spectrometry. In: RB Cole, ed. Electrospray Ionization Mass Spectrometry. New York: Wiley, 1997, Chap 6.
434. AN Verentchikov, W Ens, KG Standing. Anal Chem 66:126, 1994.
435. JF Banks, T Dresch. Detection of fast capillary electrophoresis peptide and protein separations using electrospray ionization with a time-of-flight mass spectrometer. Anal Chem 68:1480–1485, 1996.

436. AN Krutchinsky, IV Chernushevich, VL Spicer, W Ens, KG Standing. Collisional damping interface for an electrospray ionization time-of-flight mass spectrometer. J Am Soc Mass Spectrom 9:569–579, 1998.
437. IV Chernushevich, AN Verentchikov, W Ens, KG Standing. Effect of ion-molecule collisions in the vacuum chamber of an electrospray ionization-time-of-flight mass spectrometer on mass spectra of proteins. J Am Soc Mass Spectrom 7:342–349, 1996.
438. DC Barbacci, DH Russell, JA Schultz, J Holocek, S Ulrich, W Burton, M Van Stipdonck. Multi-anode detection in electrospray ionization time-of-flight mass spectrometry. J Am Soc Mass Spectrom 9:1328–1333, 1998.
439. AL Rockwood. Presented at the 34th Annual Conf on Mass Spectrometry and Allied Topics, Cincinnati, OH, June 8–13, 1986.
440. MG Quian, DM Lubman. A marriage made in MS. Anal Chem 67:235A–242A, 1995.
441. (a)A Brock, N Rodriguez, RN Zare. Hadamard transform time-of-flight mass spectrometry. Anal Chem 70:3735–3741, 1998. (b) A Brock, N Rodriguez, RA Zare. Rev Sci Instrum 71:1306–1318, 2000.
442. RW Purves, L Li, Development and characterization of an electrospray ionization ion trap/linear time-of-flight mass spectrometer. J Am Soc Mass Spectrom 8:1085–1093, 1997.
443. SM Michael, BM Chien, DM Lubman. An ion trap storage/time-of-flight mass spectrometer. Rev Sci Instrum 63:4277–4284, 1992.
444. SM Michael, BM Chien, DM Lubman. Detection of electrospray ionization using a quadrupole ion trap storage/reflectron time-of-flight mass spectrometer. Anal Chem 65:2614–2620, 1993.
445. JT Wu, MG Qian, MX Xi, L Liu, DM Lubman. Use of an ion trap storage/reflectron time-of-flight mass spectrometer as a rapid and sensitive detector for capillary electrophoresis in protein digest analysis. Anal Chem 68:3388–3396, 1996.
446. MX Li, JT Wu, L Liu, DM Lubman. The use of on-line capillary electrophoresis/electrospray ionization with detection via an ion trap storage/reflectron time-of-flight mass spectrometer for rapid mutation-site analysis of hemoglobin variants. Rapid Commun Mass Spectrom 11:99–108, 1997.
447. HR Morris, T Paxton, A Dell, J Langhorne, M Berg, RS Bordoli, J Hoyes, RH Bateman. High sensitivity collisionally-activated decomposition tandem mass spectrometry on a novel quadrupole/orthogonal-acceleration time-of-flight mass spectrometer. Rapid Commun Mass Spectrom 10:889–896, 1996.
448. A Shevchenko, I Chernushevich, W Ens, KG Standing, B Thompson, M Wilm, M Mann. Rapid de novo peptide sequencing by a combination of nanoelectrospray, isotopic labelling and a quadrupole/time-of-flight mass spectrometer. Rapid Commun Mass Spectrom 11:1015–1024, 1997.
449. G Hopfgartner, IV Chernusevich, T Covey, JB Plomley, R Bonner. Exact mass measurement of product ions for the structural elucidation of drug metabolites with a tandem quadrupole orthogonal-acceleration time-of-flight mass spectrometer. J Am Soc Mass Spectrom 10:1305–1314, 1999.

450. FG Hanisch, BN Green, R Bateman, J Peter-Katalinic. Localization of O-glycosylation sites of MUCI tandem repeats by QTOF ESI mass spectrometry. J Mass Spectrom 33:358–362, 1998.
451. H Yoshitsugu, T Fukuhara, M Ishibashi, T Nanbo, N Kagi. Key fragments for identification of positional isomer pair in glucuronides from the hydroxylated metabolites of RT-3003 (vintoperol) by liquid chromatography/electrospray ionization mass spectrometry. J Mass Spectrom 34:1063–1068, 1999.
452. E Nikolaev, A Somogyi, V Wysocki, C Martin, G Samuelson. Implementation of SID and CID on a sector/time-of-flight hybrid mass spectrometer. Proc 47th Annual Conf on Mass Spectometry and Allied Topics, Dallas, TX, 1999.
453. R Feng, Y Konishi, AW Bell. High accuracy molecular weight determination and variation characterization of proteins up to 80 ku by ionspray mass spectrometry. J Am Soc Mass Spectrom 2:387–401, 1991.
454. J Zaia, RS Annan, K Biemann. The correct molecular weight of myoglobin, a common calibrant for mass spectrometry. Rapid Commun Mass Spectrom 6:32–36, 1992.
455. (a) RB Cody, JW Finch. Electrospray ionization mass spectra of jeffamine polymers and the use of these polymers as reference standards for accurate mass measurements. Proc 42nd ASMS Conf on Mass Spectrometry and Allied Topics, Chicago, IL, May 29–June 3, 1994. (b) RB Cody, unpublished data.
456. Jeffamine™ EO/PO copolymers are available from the Huntsman Corporation Salt Lake City, UT, and aminated PEGs are available from Molecular Probes, Inc. (Eugene, OR).
457. KL Olson, KL Rinehart Jr, JC Cook Jr. Biomed Mass Spectrom 5:284–290, 1977.
458. Lancaster Synthesis, Salem, NH.
459. (a) L Jiang, M Moini. Ultramark 1621 as a reference compound for positive and negative ion fast-atom bombardment high-resolution mass spectrometry. J Am Soc Mass Spectrom 3:842–846, 1992. (b) M Moini. Ultramark 1621 as a calibration/reference compound for mass spectrometry. II. Positive- and negative-ion electrospray ionization. Rapid Commun Mass Spectrom 8:711–714, 1994.
460. M Moini, BL Jones, RM Rogers, L Jiang. Sodium trifluoracetate as a tune/calibration compound for positive- and negative-ion electrospray ionization mass spectrometry in the mass range of 100–4000 Da. J Am Soc Mass Spectrom 9:977–980, 1998.
461. S König, HM Fales. Calibration of mass ranges up to m/z 10,000 in electrospray mass spectrometers. J Am Soc Mass Spectrom 10:273–276, 1999.
462. (a) DW Ledman, RO Fox. Water cluster calibration reduces mass error in electrospray ionization mass spectrometry of proteins. J Am Soc Mass Spectrom 8:1158–1164, 1997. (b) Y Xu, S Crutchfield. 45th ASMS Conf on Mass Spectrometry and Allied Topics, Poster TP1271, Palm Springs, 1997. A practical calibration method for liquid chromatograph mass spectrometry systems.
463. K Yamaguchi, S Sakamoto, T Imamoto, T Ishikawa. Internal calibrant for an exact mass measurement with electrospray ionization mass spectrometry. Anal Sci 15:1037–1038, 1999.

464. (a) CECA Hop. Generation of high molecular weight cluster ions by electrospray ionization: Implications for mass calibration. J Mass Spectrom 31:1314–1316, 1996. (b) JF Anacleto, S Pleasance, RK Boyd. Calibration of ion spray mass spectra using cluster ions. Org Mass Spectrom 27:660–666, 1992. (c) S Pleasance, P Thibault, PG Sim, RK Boyd. Caesium iodide clusters as mass calibrants in ionspray mass spectrometry. Rapid Commun Mass Spectrom 5:307–308, 1991.
465. H Fujiwara, RC Chott, RG Nadeau. Accurate mass determination: Sensitive and volatile references for positive-ion chemical ionization and negative-ion electrospray mass spectrometry. Rapid Commun Mass Spectrom 11:1547–1553, 1997.
466. RA Zubarev, PA Demirev, P Håkansson, B Sundqvist. Approaches and limits for accurate mass characterization of large biomolecules. Anal Chem 67:3793–3798, 1995.
467. MW Senko, SC Beu, FW McLafferty. Automated assignment of charge states from resolved isotopic peaks for multiply charged ions. J Am Soc Mass Spectrom 6:52–56, 1995.
468. MW Senko, SC Beu, FW McLafferty. Determinations of monoisotopic masses and ion populations for large biomolecules from resolved isotopic distributions. J Am Soc Mass Spectrom 6:229–233, 1995.
469. MW Senko, SC Beu, FW McLafferty. Mass and charge assignment for electrospray ions by cation adduction. Anal Chem 4:828–830, 1993.
470. JB Cunniff, P Vouros. Mass and charge state assignment for proteins and peptide mixtures via noncovalent adduction in electrospray mass spectrometry. J Am Soc Mass Spectrom 6:1175–1182, 1995.
471. DM Horn, RA Zubarev, FW McLafferty. Automated reduction and interpretation of high resolution electrospray mass spectra of large molecules. J Am Soc Mass Spectrom 11:320–332, 2000.
472. BB Reinhold, VN Reinhold. Electrospray ionization mass spectrometry: Deconvolution by an entropy-based algorithm. J Am Soc Mass Spectrom 3:207–215, 1992.
473. M Labowski, CM Whitehouse, JB Fenn. Three-dimensional deconvolution of multiply charged spectra. Rapid Commun Mass Spectrom 7:71–84, 1993.
474. (a) JJ Hagen, CA Monnig. Method for estimating molecular mass from electrospray spectra. Anal Chem 66:1877–1883, 1994. (b) AG Brenton, CM Lock. A new method for the rapid deconvolution of partially resolved spectra. Rapid Commun Mass Spectrom 9:143–149, 1995.
475. H Kato, M Ishihara. Presented at the 14th Int Mass Spectrometry Conf, Tampere, Finland, 1998. New method for ESI deconvolution, partial correlation method (PCM).
476. Z Zhang, AG Marshall. A universal algorithm for fast and automated charge state deconvolution of electrospray mass-to-charge ratio spectra. J Am Soc Mass Spectrom 9:225–233, 1998.
477. GJ Daniel, SF Gull. Maximum entropy algorithm applied to image enhancement. IEEE Proc 127E:170, 1980.

478. SF Gull, J Skilling. Maximum entropy method in image processing. IEEE Proc 131F:646–659, 1984.
479. AG Ferrige, MJ Seddon, BN Green, SA Jarvis, J Skilling. Disentangling electrospray spectra with maximum entropy. Rapid Commun Mass Spectrom 6:707–711, 1992.
480. AG Ferrige, MJ Seddon, SA Jarvis. Maximum entropy deconvolution in electrospray mass spectrometry. Rapid Commun Mass Spectrom 5:374–379, 1991.
481. AG Ferrige, MJ Seddon, J Skilling, N Ordsmith. The application of MaxEnt to high resolution mass spectrometry. Rapid Commun Mass Spectrom 6:765, 1992.
482. Z Zhang, S Guan, AG Marshall. Enhancement of the effective resolution of mass spectra of high-mass biomolecules by maximum entropy-based deconvolution to eliminate the isotopic natural abundance distribution. J Am Soc Mass Spectrom 8:659–670, 1997.
483. RP Evershed, DHL Robertson, RJ Beynon, BN Green. Application of electrospray ionization mass spectrometry with maximum entropy analysis to allelic "fingerprinting" of major urinary proteins. Rapid Commun Mass Spectrom 7:882–886, 1993.
484. W Windig, JM Phalp, AW Payne. A noise and background reduction method for component detection in liquid chromatography/mass spectrometry. Anal Chem 68:3602, 1996.
485. DC Muddiman, AL Rockwood, Q Gao, JC Severs, HR Udseth, RD Smith. Application of sequential paired covariance to capillary electrophoresis electrospray ionization time-of-flight mass spectrometry: Unraveling the signal from the noise in the electropherogram. Anal Chem 67:4371–4375, 1995.
486. (a) A Carrier, D Zidarov, M Bertrand. MSTBA: A new background treatment algorithm for LC/MS experiments. Proc 42nd ASMS Conf on Mass Spectrometry and Allied Topics, Chicago, IL, May 29–June3, 1994. (b) G Sanchez D Zidarov, MJ Bertrand. Software for information enhancement and compound identification in qualitative GC-LC/MS analysis. Proc 47th ASMS Conf on Mass Spectrometry and Allied Topics, Dallas, TX, June 13–17, 1999.

APPENDIX: MASS-TO-CHARGE RATIOS FOR SEVERAL COMMON REFERENCE STANDARDS USED TO CALIBRATE THE MASS SPECTROMETER *M/Z* SCALE FOR ESI

A1 Singly Charged Positive Ion Cesium Iodide Clusters
A2 Singly Charged Negative Ion Cesium Iodide Clusters
A3 Singly Charged Ions: Poly(ethylene oxide)
A4 Singly Charged Ions: Poly(propylene oxide)
A5 Mass-to-Charge Ratios for a D-Type Jeffamine

ESI-MS: History, Theory, Instrumentation

A6 Negative Ions: Mass-to-Charge Ratios for Alkyl Ethoxy Sulfate
A7 Negative Ions: Sodium Trifluoroacetate Clusters
A8 Positive Ions: Sodium Trifluoroacetate Clusters
A9 Positive Ions: Perfluoroalkylphosphazine
A10 Negative Ions: Perfluoroalkylphosphazine
A11 Water Clusters

A1 Singly Charged Positive Ion Cesium Iodide Clusters

n	m/z
0	132.90542
1	392.71533
2	652.52524
3	912.33514
4	1172.14505
5	1431.95496
6	1691.76486
7	1951.57477
8	2211.38467
9	2471.19458
10	2731.00449
11	2990.81439
12	3250.62430
13	3510.43420

A2 Singly Charged Negative Ion Cesium Iodide Clusters

n	m/z
0	126.90447
1	386.71439
2	646.52429
3	906.33420
4	1166.14410
5	1425.95401
6	1685.76392
7	1945.57382
8	2205.38373
9	2465.19363
10	2725.00354
11	2984.81345
12	3244.62335
13	3504.43326

A3 Singly Charged Ions: Poly(ethylene oxide)[a]

n	Fragment[b]	$[M+H]^+$	$[M+NH_4]^+$	$[M+Na]^+$	$[M+K]^+$	$[M-H]^-$
1	45.03404					
2	89.06025					
3	133.08647					
4	177.11268	195.12325	212.14980	217.10519	233.07913	193.10760
5		239.14946	256.17601	261.13141	277.10535	237.13381
6		283.17568	300.20223	305.15762	321.13156	281.16003
7		327.20189	344.22844	349.18384	365.15778	325.18624
8		371.22811	388.25466	393.21005	409.18399	369.21246
9		415.25432	432.28087	437.23627	453.21021	413.23867
10		459.28054	476.30709	481.26248	497.23642	457.26489
11		503.30675	520.33330	525.28870	541.26264	501.29110
12		547.33297	564.35952	569.31491	585.28885	545.31732
13		591.35918	608.38573	613.34113	629.31506	589.34353
14		635.38540	652.41195	657.36734	673.34128	633.36975
15		679.41161	696.43816	701.39356	717.36749	677.39596
16		723.43783	740.46438	745.41977	761.39371	721.42218

(*Continues*)

A3 *Continued.*

n	Fragment[b]	[M+H]$^+$	[M+NH$_4$]$^+$	[M+Na]$^+$	[M+K]$^+$	[M−H]$^-$
17		767.46404	784.49059	789.44599	805.41992	765.44839
18		811.49026	828.51681	833.47220	849.44614	809.47461
19		855.51647	872.54302	877.49841	893.47235	853.50082
20		899.54269	916.56923	921.52463	937.49857	897.52704
21		943.56890	960.59545	965.55084	981.52478	941.55325
22		987.59512	1004.62166	1009.57706	1025.55100	985.57947
23		1031.62133	1048.64788	1053.60327	1069.57721	1029.60568
24		1075.64754	1092.67409	1097.62949	1113.60343	1073.63189
25		1119.67376	1136.70031	1141.65570	1157.62964	1117.65811
26		1163.69997	1180.72652	1185.68192	1201.65586	1161.68432
27		1207.72619	1224.75274	1229.70813	1245.68207	1205.71054
28		1251.75240	1268.77895	1273.73435	1289.70829	1249.73675
29		1295.77862	1312.80517	1317.76056	1333.73450	1293.76297
30		1339.80483	1356.83138	1361.78678	1377.76072	1337.78918
31		1383.83105	1400.85760	1405.81299	1421.78693	1381.81540
32		1427.85726	1444.88381	1449.83921	1465.81315	1425.84161
33		1471.88348	1488.91003	1493.86542	1509.83936	1469.86783
34		1515.90969	1532.93624	1537.89164	1553.86558	1513.89404
35		1559.93591	1576.96246	1581.91785	1597.89179	1557.92026
36		1603.96212	1620.98867	1625.94407	1641.91801	1601.94647
37		1647.98834	1665.01489	1669.97028	1685.94422	1645.97269
38		1692.01455	1709.04110	1713.99650	1729.97043	1689.99890
39		1736.04077	1753.06732	1758.02271	1773.99665	1734.02512
40		1780.06698	1797.09353	1802.04893	1818.02286	1778.05133
41		1824.09320	1841.11975	1846.07514	1862.04908	1822.07755
42		1868.11941	1885.14596	1890.10135	1906.07529	1866.10376
43		1912.14563	1929.17217	1934.12757	1950.10151	1910.12998
44		1956.17184	1973.19839	1978.15378	1994.12772	1954.15619
45		2000.19806	2017.22460	2022.18000	2038.15394	1998.18241
46		2044.22427	2061.25082	2066.20621	2082.18015	2042.20862
47		2088.25048	2105.27703	2110.23243	2126.20637	2086.23483
48		2132.27670	2149.30325	2154.25864	2170.23258	2130.26105
49		2176.30291	2193.32946	2198.28486	2214.25880	2174.28726
50		2220.32913	2237.35568	2242.31107	2258.28501	2218.31348

[a] Also known as polyethylene glycol or PEG. [HO(C$_2$H$_4$O)$_n$H + adduct]$^+$ or [HO(C$_2$H$_4$O)$_n$]$^-$.
[b] Positive ions: [(C$_2$H$_4$O)$_n$H]$^+$.

A4 Singly Charged Ions: Poly(propylene oxide)[a]

n	Fragment[b]	[M+H]$^+$	[M+NH$_4$]$^+$	[M+Na]$^+$	[M+K]$^+$	[M−H]$^−$
1	59.04969					
2	117.09156					
3	175.13342					
4	233.17529	251.18586	268.21240	273.16780	289.14174	249.17020
5		309.22772	326.25427	331.20967	347.18360	307.2121
6		367.26959	384.29614	389.25153	405.22547	365.2539
7		425.31145	442.33800	447.29340	463.26734	423.2958
8		483.35332	500.37987	505.33526	521.30920	481.3377
9		541.39519	558.42174	563.37713	579.35107	539.3795
10		599.43705	616.46360	621.41900	637.39293	597.4214
11		657.47892	674.50547	679.46086	695.43480	655.4633
12		715.52079	732.54733	737.50273	753.47667	713.5051
13		773.56265	790.58920	795.54460	811.51853	771.547
14		831.60452	848.63107	853.58646	869.56040	829.5889
15		889.64638	906.67293	911.62833	927.60227	887.6307
16		947.68825	964.71480	969.67019	985.64413	945.6726
17		1005.73012	1022.75667	1027.71206	1043.68600	1003.714
18		1063.77198	1080.79853	1085.75393	1101.72787	1061.756
19		1121.81385	1138.84040	1143.79579	1159.76973	1119.798
20		1179.85572	1196.88226	1201.83766	1217.81160	1177.84
21		1237.89758	1254.92413	1259.87953	1275.85346	1235.882
22		1295.93945	1312.96600	1317.92139	1333.89533	1293.924
23		1353.98131	1371.00786	1375.96326	1391.93720	1351.966
24		1412.02318	1429.04973	1434.00512	1449.97906	1410.008
25		1470.06505	1487.09160	1492.04699	1508.02093	1468.049

(*Continues*)

A3 *Continued.*

A4 *Continued.*

n	Fragment[b]	[M+H]$^+$	[M+NH$_4$]$^+$	[M+Na]$^+$	[M+K]$^+$	[M−H]$^−$
26		1528.10691	1545.13346	1550.08886	1566.06280	1526.091
27		1586.14878	1603.17533	1608.13072	1624.10466	1584.133
28		1644.19065	1661.21720	1666.17259	1682.14653	1642.175
29		1702.23251	1719.25906	1724.21446	1740.18839	1700.217
30		1760.27438	1777.30093	1782.25632	1798.23026	1758.259
31		1818.31625	1835.34279	1840.29819	1856.27213	1816.301
32		1876.35811	1893.38466	1898.34006	1914.31399	1874.342
33		1934.39998	1951.42653	1956.38192	1972.35586	1932.384
34		1992.44184	2009.46839	2014.42379	2030.39773	1990.426
35		2050.48371	2067.51026	2072.46565	2088.43959	2048.468
36		2108.52558	2125.55213	2130.50752	2146.48146	2106.51
37		2166.56744	2183.59399	2188.54939	2204.52332	2164.552
38		2224.60931	2241.63586	2246.59125	2262.56519	2222.594
39		2282.65118	2299.67772	2304.63312	2320.60706	2280.636
40		2340.69304	2357.71959	2362.67499	2378.64892	2338.677
41		2398.73491	2415.76146	2420.71685	2436.69079	2396.719
42		2456.77677	2473.80332	2478.75872	2494.73266	2454.761
43		2514.81864	2531.84519	2536.80058	2552.77452	2512.803
44		2572.86051	2589.88706	2594.84245	2610.81639	2570.845
45		2630.90237	2647.92892	2652.88432	2668.85826	2628.887
46		2688.94424	2705.97079	2710.92618	2726.90012	2686.929
47		2746.98611	2764.01266	2768.96805	2784.94199	2744.97
48		2805.02797	2822.05452	2827.00992	2842.98385	2803.012
49		2863.06984	2880.09639	2885.05178	2901.02572	2861.054
50		2921.11170	2938.13825	2943.09365	2959.06759	2919.096

[a] Also known as polypropylene glycol or PPG. [HO(C$_3$H$_6$O)$_n$ H + adduct]$^+$ or [HO(C$_3$H$_6$O)$_n$]$^-$.
[b] Positive ions: [(C$_3$H$_6$O)$_n$H]$^+$.

A5 Mass-to-Charge Ratios for a D-Type Jeffamine
$H_2NCH(CH_3)CH_2-[OCH_2CH(CH_3)]_n-NH_2$

n	$[M+H]^+$	$[M+2H]^{2+}$
1	—	67.07096
2	191.17595	96.09189
3	249.21782	125.11282
4	307.25969	154.13376
5	365.30155	183.15469
6	423.34342	212.17562
7	481.38529	241.19656
8	539.42715	270.21749
9	597.46902	299.23842
10	655.51088	328.25935
11	713.55275	357.28029
12	771.59462	386.30122
13	829.63648	415.32215
14	887.67835	444.34309
15	945.72022	473.36402
16	1003.76208	502.38495
17	1061.80395	531.40589
18	1119.84581	560.42682
19	1177.88768	589.44775
20	1235.92955	618.46869
21	1293.97141	647.48962
22	1352.01328	676.51055
23	1410.05515	705.53149
24	1468.09701	734.55242
25	1526.13888	763.57335
26	1584.18075	792.59429
27	1642.22261	821.61522
28	1700.26448	850.63615
29	1758.30634	879.65708
30	1816.34821	908.67802
31	1874.39008	937.69895
32	1932.43194	966.71988
33	1990.47381	995.74082
34	2048.51568	1024.76175
35	2106.55754	1053.78268
36	2164.59941	1082.80362
37	2222.64127	1111.82455
38	2280.68314	1140.84548
39	2338.72501	1169.86642
40	2396.76687	1198.88735

Source: Ref. 455

A6 Negative Ions: Mass-to-Charge Ratios for Akyl Ethoxy Sulfate[a]

m/z	Composition
265.147339	$C_{12}H_{25}SO_4^-$
279.163025	$C_{13}H_{27}SO_4^-$
309.173554	$[C_2H_4O+C_{12}H_{25}SO_4]^-$
323.189240	$[C_2H_4O+C_{13}H_{27}SO_4]^-$
353.199768	$[(C_2H_4O)_2+C_{12}H_{25}SO_4]^-$
367.215454	$[(C_2H_4O)_2+C_{13}H_{27}SO_4]^-$
397.225983	$[(C_2H_4O)_3+C_{12}H_{25}SO_4]^-$
411.241669	$[(C_2H_4O)_3+C_{13}H_{27}SO_4]^-$
441.252198	$[(C_2H_4O)_4+C_{12}H_{25}SO_4]^-$
455.267884	$[(C_2H_4O)_4+C_{13}H_{27}SO_4]^-$
485.278413	$[(C_2H_4O)_5+C_{12}H_{25}SO_4]^-$
499.294099	$[(C_2H_4O)_5+C_{13}H_{27}SO_4]^-$
529.304628	$[(C_2H_4O)_6+C_{12}H_{25}SO_4]^-$
543.320314	$[(C_2H_4O)_6+C_{13}H_{27}SO_4]^-$
573.330842	$[(C_2H_4O)_7+C_{12}H_{25}SO_4]^-$
587.346528	$[(C_2H_4O)_7+C_{13}H_{27}SO_4]^-$
617.357057	$[(C_2H_4O)_8+C_{12}H_{25}SO_4]^-$
631.372743	$[(C_2H_4O)_8+C_{13}H_{27}SO_4]^-$
661.383272	$[(C_2H_4O)_9+C_{12}H_{25}SO_4]^-$
675.398958	$[(C_2H_4O)_9+C_{13}H_{27}SO_4]^-$
705.409487	$[(C_2H_4O)_{10}+C_{12}H_{25}SO_4]^-$
719.425173	$[(C_2H_4O)_{10}+C_{13}H_{27}SO_4]^-$
749.435702	$[(C_2H_4O)_{11}+C_{12}H_{25}SO_4]^-$
763.451388	$[(C_2H_4O)_{11}+C_{13}H_{27}SO_4]^-$
793.461916	$[(C_2H_4O)_{12}+C_{12}H_{25}SO_4]^-$
807.477602	$[(C_2H_4O)_{12}+C_{13}H_{27}SO_4]^-$
837.488131	$[(C_2H_4O)_{13}+C_{12}H_{25}SO_4]^-$
851.503817	$[(C_2H_4O)_{13}+C_{13}H_{27}SO_4]^-$
881.514346	$[(C_2H_4O)_{14}+C_{12}H_{25}SO_4]^-$
895.530032	$[(C_2H_4O)_{14}+C_{13}H_{27}SO_4]^-$
925.540561	$[(C_2H_4O)_{15}+C_{12}H_{25}SO_4]^-$
939.556247	$[(C_2H_4O)_{15}+C_{13}H_{27}SO_4]^-$
983.582462	$[(C_2H_4O)_{16}+C_{13}H_{27}SO_4]^-$
969.566776	$[(C_2H_4O)_{16}+C_{12}H_{25}SO_4]^-$
1013.592990	$[(C_2H_4O)_{17}+C_{12}H_{25}SO_4]^-$
1027.608676	$[(C_2H_4O)_{17}+C_{13}H_{27}SO_4]^-$

[a] Ivory Liquid dishwashing soap. (Ivory Liquid is a registered trademark of the Procter and Gamble Company.)

A7 Negative Ions: Sodium Trifluoracetate Clusters
$[(CF_3CO_2Na)_nCF_3CO_2]^-$

n	m/z
	59.01330[a]
0	112.98504
1	248.95985
2	384.93465
3	520.90946
4	656.88427
5	792.85908
6	928.83389
7	1064.80869
8	1200.78350
9	1336.75831
10	1472.73312
11	1608.70792
12	1744.68273
13	1880.65754
14	2016.63235
15	2152.60715
16	2288.58196
17	2424.55677
18	2560.53158
19	2696.50639
20	2832.48119
21	2968.45600
22	3104.43081
23	3240.40562
24	3376.38042
25	3512.35523
26	3648.33004
27	3784.30485
28	3920.27966
29	4056.25446
30	4192.22927

[a] Acetate (from impurity or added acetic acid)
Source: Ref. 460.

A8 Positive Ions: Sodium Trifluoracetate Clusters $[(CF_3CO_2Na)_nNa]^+$

n	m/z
0	22.98977
1	158.96458
2	294.93938
3	430.91419
4	566.88900
5	702.86381
6	838.83862
7	974.81342
8	1110.78823
9	1246.76304
10	1382.73785
11	1518.71265
12	1654.68746
13	1790.66227
14	1926.63708
15	2062.61189
16	2198.58669
17	2334.56150
18	2470.53631
19	2606.51112
20	2742.48592
21	2878.46073
22	3014.43554
23	3150.41035
24	3286.38515
25	3422.35996
26	3558.33477
27	3694.30958
28	3830.28439
29	3966.25919
30	4102.23400

Source: Ref. 460.

A9 Positive Ions: Perfluoroalkylphosphazine[a]

Exact Mass	Composition
869.9979	$C_{17}H_{17}O_6N_3P_3F_{22}$
922.01035	$C_{18}H_{19}O_6N_3P_3F_{24}$
969.99151	$C_{19}H_{17}O_6N_3P_3F_{26}$
1022.00397	$C_{20}H_{19}O_6N_3P_3F_{28}$
1069.98512	$C_{21}H_{17}O_6N_3P_3F_{30}$
1121.99758	$C_{22}H_{19}O_6N_3P_3F_{32}$
1169.97874	$C_{23}H_{17}O_6N_3P_3F_{34}$
1221.99119	$C_{24}H_{19}O_6N_3P_3F_{36}$
1269.97235	$C_{25}H_{17}O_6N_3P_3F_{38}$
1321.98481	$C_{26}H_{19}O_6N_3P_3F_{40}$
1369.96596	$C_{27}H_{17}O_6N_3P_3F_{42}$
1421.97842	$C_{28}H_{19}O_6N_3P_3F_{44}$
1469.95958	$C_{29}H_{17}O_6N_3P_3F_{46}$
1521.97203	$C_{30}H_{19}O_6N_3P_3F_{48}$
1569.95319	$C_{31}H_{17}O_6N_3P_3F_{50}$
1621.96564	$C_{32}H_{19}O_6N_3P_3F_{52}$
1669.9468	$C_{33}H_{17}O_6N_3P_3F_{54}$
1721.95926	$C_{34}H_{19}O_6N_3P_3F_{56}$
1769.94041	$C_{35}H_{17}O_6N_3P_3F_{58}$
1821.95287	$C_{36}H_{19}O_6N_3P_3F_{60}$
1869.93403	$C_{37}H_{17}O_6N_3P_3F_{62}$
1921.94648	$C_{38}H_{19}O_6N_3P_3F_{64}$

[a] Ultramark 1621, Lancaster Synthesis.
Source: Ref. 459a.

A10 Negative Ions: Perfluoroalkylphosphazine[a]

m/z	Composition
691.97618	$C_{12}H_{13}O_6N_3P_3F_{16}$
791.96979	$C_{14}H_{13}O_6N_3P_3F_{20}$
805.98544	$C_{15}H_{15}O_6N_3P_3F_{20}$
891.9634	$C_{16}H_{13}O_6N_3P_3F_{24}$
905.97905	$C_{17}H_{15}O_6N_3P_3F_{24}$
991.95702	$C_{18}H_{13}O_6N_3P_3F_{28}$
1005.97267	$C_{19}H_{15}O_6N_3P_3F_{28}$
1091.95063	$C_{20}H_{13}O_6N_3P_3F_{32}$
1105.96628	$C_{21}H_{15}O_6N_3P_3F_{32}$
1191.94424	$C_{22}H_{13}O_6N_3P_3F_{36}$
1205.95989	$C_{23}H_{15}O_6N_3P_3F_{36}$
1291.93786	$C_{24}H_{13}O_6N_3P_3F_{40}$
1305.95351	$C_{25}H_{15}O_6N_3P_3F_{40}$
1391.93147	$C_{26}H_{13}O_6N_3P_3F_{44}$
1405.94712	$C_{27}H_{15}O_6N_3P_3F_{44}$
1505.94073	$C_{29}H_{15}O_6N_3P_3F_{48}$
1605.93434	$C_{31}H_{15}O_6N_3P_3F_{52}$
1705.92796	$C_{33}H_{15}O_6N_3P_3F_{56}$
1805.92157	$C_{35}H_{15}O_6N_3P_3F_{60}$

[a] Ultramark 1621, Lancaster Synthesis.
Source: Ref. 459a.

A11 Water Clusters $[H_2O]_n$ + adduct]$^+$

n	Adduct: H^+	Adduct: NH_4^+
3	55.03952	54.05550
4	73.05008	72.06607
5	91.06065	90.07663
6	109.07121	108.08720
7	127.08178	126.09776
8	145.09234	144.10833
9	163.10291	162.11889
10	181.11347	180.12946
11	199.12404	198.14002
12	217.13460	216.15059
13	235.14517	234.16115
14	253.15573	252.17172
15	271.16630	270.18228
16	289.17686	288.19285
17	307.18743	306.20341
18	325.19799	324.21398
19	343.20856	342.22454
20	361.21912	360.23511
21	379.22969	378.24567
22	397.24025	396.25624
23	415.25082	414.26680
24	433.26138	432.27736
25	451.27195	450.28793
26	469.28251	468.29849
27	487.29307	486.30906
28	505.30364	504.31962
29	523.31420	522.33019
30	541.32477	540.34075
31	559.33533	558.35132
32	577.34590	576.36188
33	595.35646	594.37245
34	613.36703	612.38301
35	631.37759	630.39358
36	649.38816	648.40414
37	667.39872	666.41471
38	685.40929	684.42527
39	703.41985	702.43584
40	721.43042	720.44640

(*Continues*)

A11 *Continued.*

n	Adduct: H^+	Adduct: NH_4^+
41	739.44098	738.45697
42	757.45155	756.46753
43	775.46211	774.47810
44	793.47268	792.48866
45	811.48324	810.49923
46	829.49381	828.50979
47	847.50437	846.52036
48	865.51494	864.53092
49	883.52550	882.54148
50	901.53607	900.55205
51	919.54663	918.56261
52	937.55719	936.57318
53	955.56776	954.58374
54	973.57832	972.59431
55	991.58889	990.60487
56	1009.59945	1008.61544
57	1027.61002	1026.62600
58	1045.62058	1044.63657
59	1063.63115	1062.64713
60	1081.64171	1080.65770
61	1099.65228	1098.66826
62	1117.66284	1116.67883
63	1135.67341	1134.68939
64	1153.68397	1152.69996
65	1171.69454	1170.71052
66	1189.70510	1188.72109
67	1207.71567	1206.73165
68	1225.72623	1224.74222
69	1243.73680	1242.75278
70	1261.74736	1260.76335
71	1279.75793	1278.77391
72	1297.76849	1296.78448
73	1315.77906	1314.79504
74	1333.78962	1332.80560
75	1351.80019	1350.81617
76	1369.81075	1368.82673
77	1387.82131	1386.83730
78	1405.83188	1404.84786

(*Continues*)

A11 *Continued.*

n	Adduct: H^+	Adduct: NH_4^+
79	1423.84244	1422.85843
80	1441.85301	1440.86899
81	1459.86357	1458.87956
82	1477.87414	1476.89012
83	1495.88470	1494.90069
84	1513.89527	1512.91125
85	1531.90583	1530.92182
86	1549.91640	1548.93238
87	1567.92696	1566.94295
88	1585.93753	1584.95351
89	1603.94809	1602.96408
90	1621.95866	1620.97464
91	1639.96922	1638.98521
92	1657.97979	1656.99577
93	1675.99035	1675.00634
94	1694.00092	1693.01690
95	1712.01148	1711.02747
96	1730.02205	1729.03803
97	1748.03261	1747.04860
98	1766.04318	1765.05916
99	1784.05374	1783.06973
100	1802.06431	1801.08029
101	1820.07487	1819.09085
102	1838.08544	1837.10142
103	1856.09600	1855.11198
104	1874.10656	1873.12255
105	1892.11713	1891.13311
106	1910.12769	1909.14368
107	1928.13826	1927.15424
108	1946.14882	1945.16481
109	1964.15939	1963.17537
110	1982.16995	1981.18594
111	2000.18052	1999.19650
112	2018.19108	2017.20707
113	2036.20165	2035.21763
114	2054.21221	2053.22820
115	2072.22278	2071.23876

2
Nanospray Electrospray Ionization Development

LC/MS, CE/MS Application

Tom R. Covey
MDS Sciex, Concord, Ontario, Canada

Devanand Pinto
National Research Council of Canada, Institute of Marine Biosciences, Halifax, Nova Scotia, Canada

I. INTRODUCTION

Since the mid-1980s, when the potential of electrospray ionization (ESI) was beginning to be realized, the development of ion sources based on this desorption process has taken two distinct routes. One involves the optimization of the instrumentation to accept, with as high efficiency as possible, liquid flow rates in the microliter to milliliter per minute range. The other involves quite the opposite—instrumentation for the spraying and ionization of samples in the nanoliter per minute regime, which is orders of magnitude lower (1).

High performance liquid chromatography (HPLC) instrumentation and methodologies have been evolving for over three decades and, for good reasons, have focused on the development of operational parameters in the high flow (mL/min) regime. The production of reliable and precise pumping systems for this flow regime can be achieved within the scope of present-day engineering and manufacturing capabilities. Typical sample-handling volumes encountered with methods operating at these flows are readily amenable to high degrees of automation with precision and reliability. Column designs, stationary-phase chemistries, and packing procedures have been finely tuned to provide optimum separation efficiencies and analysis speeds at these flow rates. At the dawn of the electrospray age during the late 1980s, it was immediately obvious that a useful and routine coupling of liquid chromatography with mass spectrometry (LC/MS) required a seamless dovetailing of mass spectrometry with the finely tuned chromatographic technology already in existence. The primary focus of the mass spectrometry instrumentation manufacturers was to make this happen, and the bulk of the applications centered on the quantitative and qualitative analysis of small molecules (e.g., drug candidates) at high chromatographic flow rates. Atmospheric pressure ionization appeared to be the best approach to accommodate this flow regime, and once this was recognized the sun began to set on earlier ionization techniques developed to couple liquid chromatography and mass spectrometry, perhaps never to rise again (2).

Parallel to the activities surrounding adaptations to higher liquid flow rates was the excitement surrounding the new-found capability to determine proteins, peptides, and other biopolymers by mass spectrometry (MS) (3,4). This was heralded as a scientific breakthrough although, in these pre-proteome years, the commercial opportunities were dwarfed by an immediate demand for small-molecule HPLC/MS. Activities blossomed primarily in university and public research institutions where ideas to solve the particular issues surrounding trace level protein characterization were incubating and eventually gave rise to the majority of the work described in this chapter. These issues centered on the problem of obtaining sequence information from femtomole levels of proteins in sub-microliter sample sizes, typically extracted from two-dimensional (2-D) gel electrophoresis spots. Of particular note were the seminal works of Wilm and Mann (5–8), who developed the fundamental physics of nanoflow electrospray (5) and demonstrated the practicality of the technique, referred to as nanoES, for protein identification after 2-D gel electrophoretic separations (6,7). Their contributions went far beyond the proof-of-principle stage and rapidly evolved into an entire analytical methodology including the development of the necessary software (9) and sample preparation tools required to unleash its full potential (10,11). A series of week-long courses, held at the European Molecular Biology Institute (EMBL) in Heidelberg, taught the practical implementation of this methodology and were instrumental in popularizing the technique throughout the second half of the 1990s. It is a tribute to their system, coupled with the unique readability of the peptide sequence information obtained from tandem mass spectrometry (MS/MS) of doubly charged tryptic peptides (12–14), that very similar if not identical systems for protein identification are still in use today.

From one point of view, operating electrospray at high flow rates is forcing the process into an unnatural state, where stabilization of the Taylor cone and formation of an aerosol of droplets are practically impossible with electric fields alone. To generate stable ion currents one must provide additional energy input, in the form of pneumatic nebulization (15) and heat (1,16,17), to force droplet formation, leaving the task of droplet charging to the electric field. Proper implementation of this additional energy is of overriding concern in the design of high flow rate systems, far overshadowing in importance other details of the electrospray process such as Taylor cone formation and stabilization. For the nanoflow techniques, the opposite is true; factors affecting the formation and stabilization of the Taylor cone are of paramount concern. Other forms of external energy input to generate charged droplets are not required because the electric field is sufficient to charge the liquid and simultaneously generate an aerosol.

The essence of the nanoflow methodology is to reduce the flow rate of the sprayed sample liquid by orders of magnitude below the microliter per minute regime. As stable flows are achieved at lower and lower flow rates, the efficiency of the ionization process improves approximately in proportion to the flow rate reduction. Even though sample molecules enter the sprayer at a much lower rate than with the high flow systems, the signal per unit time detected by MS remains constant and can often be seen to improve by factors of 2–3 (6,18). However, the substantial advantage, where gains in sample utilization efficiency and detection limits are measured in orders of magnitude, comes from the length of time one can implement signal-averaging techniques in the tandem MS mode, where chemical noise is of less concern. Small volumes of multicomponent mixtures of peptides from enzymatic digests of proteins can now be conveniently interrogated by MS/MS with sufficient time to obtain high quality spectra for sequencing purposes. Figure 1 illustrates this point. When a 1 µL sample is flow injected into a heated and pneumatically nebulized ion source with a flow of 10 mL/min, the signal is fleeting (Fig. 1A). The same sample produces ions for 25–30 mins when it is introduced through a nanoES electrode at a flow rate of 30 nL/min (Fig. 1B). Thus, one can readily see that the efficiency of sample utilization, defined as the proportion of analyte molecules introduced into the ion source and detected by the mass spectrometer, is improved by roughly 300-fold in this case.

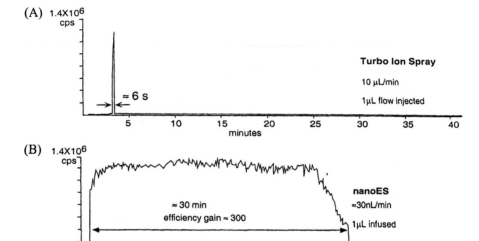

Figure 1 Comparison of the selected ion current traces from two different electrospray sample introduction techniques. A solution of glufibrinopeptide (0.5 pmol/μL) was used to monitor the doubly charged molecule (selected ion monitoring; SIM). (A) A 1 μL aliquot of solution was injected into a 10 μL/min flow of solvent and ionized with a Turbo IonSpray heated pneumatically assisted electrospray nebulizer with a 50 μm i.d. fused silica capillary emitter. (B) One microliter of the same sample as that used in (A) was deposited into a nanoES tube with a drawn tip of approximately 2 μm exit aperture. Gas pressure of 20 psi was applied, and the sample was electrosprayed at a rate of approximately 30 nL/min until the sample was consumed.

Spanning the decade of the 1990s there were significant developments in the area of nanoflow electrode design, and a variety of strategies for implementing these new configurations emerged. In the following pages we take a detailed look at some of these designs, with a particular focus on nanoLC separation systems that take advantage of this low flow technology. On the horizon of our expectations is the implementation of nanofabrication technology for nanoflow ion source design. Discussion of this technology is absent from this discourse because, at least until now, there have been few examples of practical implementation of this approach.

For the sake of this discussion, nanoflow rates are defined as the range from less than 1 nL/min to 1 μL/min. The low nanoflow regime is from less than 1 to 50 nL/min, intermediate from 50 to 150 nL/min, and high from 150 to 1000 nL/min.

II. NANOFLOW ELECTRODES

A. Flow Rate Versus Electrospray Needle Architecture

The early theoretical derivations of nanoES by Wilm and Mann (5) showed a strong correlation between the liquid flow rate and the diameter of the droplets being emitted from the Taylor cone. At nanoliter per minute flows, these droplets were calculated to be small enough to result in almost immediate ion emission, with very little time or distance required for further desolvation (5,19). This insight into the effects of low liquid flow rates led Wilm to devise an electrospray needle that delivered very low flows and could thus be positioned in close proximity to the entrance aperture of the mass spectrometer, maximizing the collection of ions and posing no danger of overwhelming the vacuum system with jets of liquid and particles penetrating through the MS entrance aperture. However, the theory provided little guidance as to what these electrospray needles should look like. Inspired intuition directed Wilm and coworkers and other groups working on the same problem to experiment with capillaries drawn as fine as possible. At the EMBL, where these investigations were taking place, sophisticated equipment for the production of glass micropipette electrodes was readily available. The pipettes are used as channels of communication with the interior of living cells and are relatively commonplace tools for fundamental investigations into the neural sciences and in general cellular communication studies. Making use of this resource, the researchers could rapidly test and optimize a variety of tip architectures. The appearance of many of the nanoES tips used today in laboratories worldwide bears a close resemblance to these earliest empirically optimized prototypes, although many subtle but important improvements have since been implemented. For an in-depth treatise on the science and history of microelectrode technology, see the handbook by Brown and Flaming (20), available from Sutter Instruments (21), one of the leading manufacturers of this type of electrode manufacturing equipment.

In addition to some other enlightened early forays into the development of nanoflow electrospray emitters (22–27), many studies subsequent to the Wilm and Mann publications began to emerge rapidly, optimizing tip configurations for specific applications such as infusion, capillary electrophoresis coupled with mass spectrometry (CE/MS), and capillary HPLC/MS. A common theme among them was the investigation of the various aspects of the physical architecture of these nanoflow electrodes and its effect on their operational behavior, the factor of primary interest being the lowest flow rate an electrode of a particular size and shape could support without compromising the sensitivity or stability of the ion beam (18,28–50). The picture has developed into a complex interrelationship between tip inner diameter, outer diameter, channel taper to the tip, and solvent composition, all exerting some influence on the lowest attainable flow with a particular nanoflow electrode. The various researchers in this area derived their preferred combinations of these parameters to achieve the type of operational results they desired for their application. Reference is given to the work of Covey (28), Emmett and Caprioli (26), Valaskovic and McLafferty (51,52), Geromanos et al. (18), Kelly et al. (36), and Bateman et al. (35), where some tabular data are presented relating flow rate to the geometry of several different tip geometries. Some guidelines are put forth below, based on our experience and the collective considerations of the above-mentioned studies (28–50) regarding the relationship between electrode geometry and flow rate.

The primary determinant of the lowest attainable flow from a nanoflow electrode is the inner diameter of the tip at the exit. Figure 2 presents some results from experiments showing the relationship between tip inner diameter and lowest sustainable flow (28). These experiments were done by mounting 3 cm lengths of fused silica tubing of various inner diameters in a metal fitting. The outside diameter (o.d.) of the tubing was 150 μm, and the inner diameters (i.d.) ranged from 1 to 50 μm along the entire length of the tube; tubing of these dimensions is commercially available (53). The exteriors were metallized

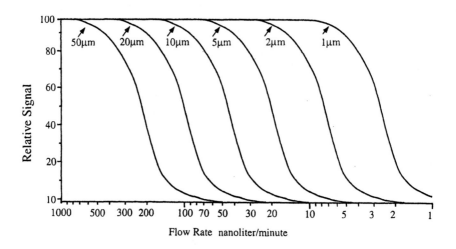

Figure 2 Flow profiles for fused silica tips of different inner diameters. The arrows identify the inner diameter of the electrode and the point of optimum flow for that electrode. For experimental details see text.

according to the method of Kriger et al. (30), and no attempt was made to shape the tips other than preparing as square a cut as possible. A sample containing a peptide mixture dissolved in 50/50 water–acetonitrile and 0.1% aqueous formic acid was delivered at ever-decreasing flows with a syringe pump until the ion current became unstable.

An example of the experimental data used to construct the curves in Fig. 2 is shown in Fig. 3. Below a certain minimum flow, referred to here as the "optimum flow," the ion current signal begins to decrease with flow, exhibiting a mass-flux-sensitive response (i.e., the ion abundance varies with the rate at which sample molecules are introduced into the ion source). This is what would be typically expected of detection with a mass spectrometer, which measures the rate at which ions impinge on a detector (here a channel electron multiplier). But this will occur only if the efficiency of conversion of liquid-phase ions to gas-phase ions in the ion source (ionization efficiency) remains constant as the flow changes. Indeed, in the region below 200 nL/min for this electrode, this must be the case.

In practice, this behavior is seldom observed or reported with electrospray sample introduction. More typically, flow rates above the optimum flow for a particular electrode geometry are used. In Fig. 3, where the 20 μm aperture

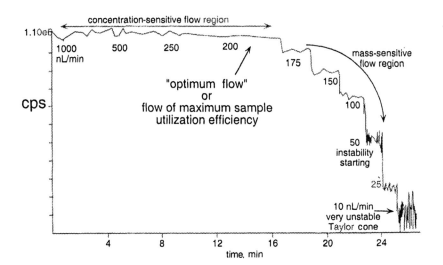

Figure 3 Flow rate profile for a 20 μm i.d. fused silica tip. The ion current trace was obtained by monitoring the molecular ion of glufibrinopeptide (selected ion monitoring) at a concentration of 0.5 pmol/μL on a quadrupole mass spectrometer.

was used, one observes that the ion abundance remains constant with changing flow rate in the region above 200 nL/min. The response resembles that of a concentration-sensitive detector (i.e., the signal is constant and independent of sample introduction rate and varies only as a function of the concentration of the sample in the solution). This can occur only if the ionization efficiency is dropping as the flow increases, just as predicted by Wilm and Mann's model (5).

Although rarely described, others have noticed the region of mass-sensitive response with nanoflow electrodes. Results very similar to those in Fig. 3 were published by Bateman et al. (35) and commented on by others (24,27,39). Bateman et al. (35) used an electrode of the same inner diameter as the one used to obtain the data in Fig. 3 but of considerably different architecture in all other aspects. Their electrode was produced by flame drawing 365 μm o.d. × 50 μm i.d. fused silica to an inner diameter of 20 μm and an outer diameter of approximately 100 μm, then depositing gold on the tip for electrical contact. That very similar results were obtained from two drastically different electrodes with the same inner diameter is an indication of the strong relationship between emitter inner diameter and flow rate characteristics and also quantifies this relationship for the 20 μm i.d. emitter.

As a point of reference, it is noted that a key distinguishing feature between nanoflow operation and operation in the high flow regime (>1 μL/min) is the general absence of observations of a region of mass-sensitive response at the higher flows (54). This is because electrodes used for higher flow operations typically have inner diameters of 50–100 μm and are operated far above the optimal flow for electrodes of their particular inner diameter, well into the concentration-sensitive zone. Explanations for this observed relationship between emitter geometry and optimum flow rate are proposed at the end of this section.

The experiment depicted in Fig. 3 for the 20 μm i.d. capillary was duplicated for capillaries ranging from 1 to 50 μm (the results shown in Fig. 2) (28). Figure 2 can be used as a rough estimate of the optimum flow from electrodes of known internal diameter whether they are fabricated from pulled, shaped, or unmodified fused silica capillaries of 150–360 μm o.d or from 1 mm o.d. borosilicate glass tubes. However, it should be kept in mind that there are other electrode architectural features involved in establishing the optimum flow rate. Among these is the outer diameter of the tip at the exit of the electrode. Most important, this dimension establishes the minimum voltage required to produce sufficient electric field strength to initiate the electrospray process. As such, sharper tips can generally be operated closer to the entrance aperture of the mass spectrometer. The taper of the channel leading up to the exit aperture and the restriction to flow it imposes have also been noted to have an effect. A long narrow channel results in flows somewhat lower than expected for a particular inner diameter (Fig. 3). Resistance to flow in narrow channels is also increased by solvent viscosity, which superimposes a solvent composition effect.

Reference to the specific literature cited will give a more accurate estimate than the one provided in Fig. 2 for the optimum flow for a specific electrode geometry. However, details of all aspects of the geometry are often omitted from many articles owing to the difficulty of making accurate measurements at these dimensions. Characterization of the optimum flow for a particular electrode, as defined above, is often also overlooked. What is frequently discussed is the ability to establish a stable ion beam at a particular flow, which, as can be seen from Figs. 2 and 3 and the data presented by Bateman et al. (35), can potentially be suboptimum from a sensitivity point of view if one is operating in the mass-flow-sensitive region. Large discrepancies from the relationships shown in Fig. 3 (factors of 10 in flow) can often be accounted for by such an explanation. It should be pointed out that the vast majority of the nanoflows cited in the literature for a particular electrode exit inner diameter are within a factor of 2 of those determined in the experiments summarized in Fig. 2.

Several possibilities can be proposed to explain this observed relationship between aperture size and optimum flow rate and the observation of decreasing signal intensity at flows below the optimum. Central to this understanding is the quantification of the factors affecting the size, shape, and stability of the Taylor cone, which represents a significant fluid dynamic challenge. The large disparities of geometrical shape, solvent viscosity, and conductivity as encountered in the nanoflow electrospray field have added sufficient complications to the modeling efforts to result in active debate. In 1999 an entire issue of the *Journal of Aerosol Science* (55) was devoted to the consideration of these many deliberations regarding the formation and stabilization of the Taylor cone.

One possible model is a simple hydrodynamic one based on Taylor's original derivation of a constant cone semiangle α at all flow rates. Using this model as a basis, Wilm and Mann (5) derived that the diameter of the droplets being emitted from the tip of the cone will decrease with decreasing flow. With this reduction in the initial droplet size, an increase in ionization efficiency will be observed (i.e., more of the sample ions in solution will be released to the gas phase and transferred into the vacuum of the mass spectrometer because of the minimal amount of solvent evaporation required). Fewer ions are lost in undesolvated droplets. As the mass flux or flow rate drops, no net loss in absolute signal will be observed because the ionization concomitant efficiency is increased (the so-called concentration-sensitive region in Fig. 3). As the flow is lowered and the cone semiangle remains constant, as predicted by Taylor, both the radius of the Taylor cone base and the length of the cone must decrease in order for smaller droplets to be formed, as diagramatically portrayed in Fig. 4. A point must eventually be reached, as dictated by the inner diameter of the electrode tip at the exit, where the base of the Taylor cone can collapse no further. This is the flow of optimum efficiency, and a lowering of the flow beyond this point will not result in smaller, more efficient emission droplets, because the Taylor cone can no longer shrink. In response to the lowered flow, a lower rate of droplet emission occurs, thus decreasing the signal. A mass-sensitive response is observed as a proportional decrease in signal with decreasing flow, as shown in Figs. 2 and 3, until the flow is so low that the Taylor cone collapses into the lumen of the electrode and an unstable droplet emission regime is entered.

As a corollary to the above, the diameter of the droplets emitted from the tip will increase with increasing flow, which, if the cone semiangle were to remain constant, could occur only if the cone were to increase in volume by broadening at the base and lengthening, also portrayed in Fig. 4. Eventually the base of the cone will begin to envelop the outer diameter of the electrode, generally without adverse effects on signal stability but eventually requiring pneumatic nebulization to sustain stable droplet formation. Systems designed for higher flow rates (>1 µL/min) typically have electrode inner diameters of between 50 and 100 µm and are therefore always operated in this flooded or overflow condition where the diameter of the base of the Taylor cone never approaches the inner diameter of the electrode. For this reason a mass-sensitive response is not observed with higher flow rate systems unless electrodes of larger inner diameters are used.

In contrast, De La Mora (56) argues that the constant cone semiangle assumption may not be entirely valid, that α changes with flow rate and the

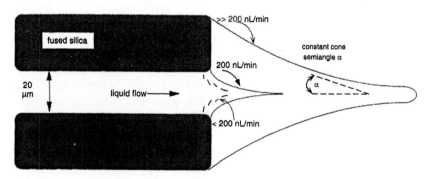

Figure 4 Diagramatic portrayal of the effect of flow on the Taylor cone from a 20 µm i.d. fused silica electrode.

flow rate–Taylor cone stability relationship is primarily driven by electrostatic considerations. Flows in excess of the optimum for a given electrode geometry will produce a Taylor cone angle with insufficient fields at the tip to produce droplets of adequate charge density for optimal ion emission rates. Regardless of the model, and there are undoubtedly many others that would fit the observations, the situation is not unlike that encountered in the dispute over the mechanism of ion formation [i.e., the ion evaporation model (57), the charged residue model (58,59), the electrohydrodynamic model (19,60), or mixtures of the three (61–64)]. Despite the differences in the fundamental premises of what occurs on the microscopic level, these models can be used to predict the important macro behavior or analytical characteristics that are commonly observed (1). Limitations of dynamic range, ionization suppression, and matrix effects are all examples of phenomena that can be predicted by all of the models proposed for ion formation.

An important development with regard to the effect of tip architecture on flow rate is the recent inclusion of frits or packing materials near the exit aperture (52) (see also Section III of this chapter). This appears to have the net effect of reducing the effective inner diameter and also helps to throttle the flow in a fashion similar to a narrow channel leading up to the tip. Thus, tips of this nature can achieve lower flows than would otherwise be expected and add an element of robustness. Multiple small channels are less likely to occlude than a single one. These are discussed further in the following section with reference to their utility for chromatographic separations.

One final practical note regarding tip architecture and the essential requirement to draw or taper tips to achieve the low nanoliter per minute flow regime: The experiments that produced the results in Figs. 2 and 3 used short lengths (3 cm) of fused silica of 1–50 μm i.d. as an alternative to drawing tips and as a means of implementing reproducible tip apertures to quantify the relationships studied. However, the backpressure generated by channels of small inner diameters in relatively long lengths of fused silica such as those used here is proportional to the fourth power of the channel i.d. As a consequence the backpressure rapidly escalates, as the i.d. decreases, to values difficult to work with from a practical point of view. Using glass syringes and infusion pumps to deliver these low flows often results in the splitting of the glass barrel unless extreme care is taken. The fracture is barely perceptible to the eye unless viewed at the correct light angle. At the flows delivered in these experiments (nL/min range), the liquid leak through the barrel is also imperceptible owing to the fact that the evaporation rate is as fast as the leak. Difficulties maintaining a spray were frequently observed due to this minor but difficult-to-troubleshoot detail. This method of producing nanoflow tips of <20 μm i.d. is not practical for routine use, necessitating the drawing or shaping of tips to produce short regions of small inner diameter in fused silica. In tubing of 20 μm i.d. or greater, these backpressure issues become much less of a concern.

B. Examples of Some Nanoflow Sample Introduction Systems

The following discussion focuses on the area of nanoflow electrode and systems design. The first part of this discussion deals with the bulk flow or discrete sample introduction systems (i.e., those not involving separations). Tips constructed for purposes of nano liquid chromatography (nanoLC) and capillary electrophoresis (CE) are also addressed here, but they are covered from the point of view of the whole system in more detail in Section III. The idea of this chapter is not to provide a description of every foray into the nanoflow electrospray area but to provide information about systems that have either been widely implemented or present promising new interesting twists on the technology.

1. Drawn Glass Tubes

Figure 5 presents four photographs at different magnifications of the type of electrode drawn from a borosilicate glass tube, similar to the original concept of the early Wilm and Mann device (5,6) referred to as a nanoES electrode.

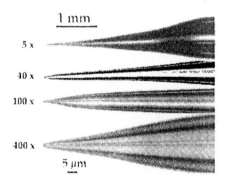

Figure 5 Four magnifications of a nanoES tip drawn from borosilicate glass tubing of 1.2 mm i.d. This tip was produced by New Objective, with the exit aperture opened by HF etching. (Photograph courtesy of New Objective.)

Electrodes of this general configuration, drawn from 1–1.2 mm o.d. × 0.8 mm i.d. borosilicate glass capillaries, are available from New Objective (52), Protana (65), and World Precision Instruments (66). New Objective added some features to make operation and handling of these tips easier. After the drawing process, the capillaries have extremely small or sealed exit apertures, typically requiring a slight touch to a metal surface in the ion source to fracture the tip and open a 1–2 μm sprayer hole. The electrode shown in Fig. 5 circumvents this problem because it is opened with a hydrogen fluoride (HF) etching procedure, which also allows for careful dimensional control of the aperture. Figure 6 is an electron micrograph of a 5 μm tip produced in this manner by New Objective. These calibrated tips have highly predictable flow rates for a given delivery pressure and solvent composition.

As for establishing electrical contact to the tip, the original method of Wilm and Mann was to deposit from vacuum vapor a thin film of gold or platinum onto the surface and exit aperture of the tips. This procedure is exceedingly simple and inexpensive provided one has access to a small vapor deposition instrument, a common accessory in an electron microscopy laboratory and readily available from suppliers of microscopy accessories. Although the metal layer can easily be brushed off, these tips were designed for single-use application. The electrode designed by Valaskovic, shown in Fig. 5, has an additional vapor-deposited coating of SiO_x over the gold monolayer to mechanically secure it in place, providing a more robust and reliable electrical contact (31,51). A nonmetallized version has been reported that uses a tungsten wire inserted into the

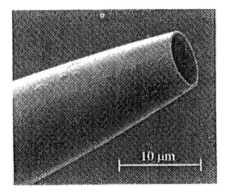

Figure 6 An electron micrograph of a 5 μm tip from a nanoES electrode etched with HF. This tip and photograph were produced by New Objective. (Photograph courtesy of New Objective.)

back of the capillary (43), and the electrical properties of this approach have been extensively studied (67,68).

The general method of operation is depicted in Fig. 7. Micromanipulators are used to position the capillaries to within 1–2 mm of the entrance aperture with the aid of a microscope of 16× or greater magnification. Flow rates delivered through these electrodes tend to vary between 10 and 50 nL/min; variations are due to the tip aperture size, the length and diameter of the restricting channel, solvent composition, and applied pressure. If it is necessary to open the tip by touching it to a metal surface and fracturing it, the resulting flow is somewhat unpredictable but will, with operator experience, generally fall within this flow rate range and provide an optimal flow provided excessive pressure is not applied. Tips whose aperture openings are precalibrated by the HF procedure will provide more predictable flows and circumvent the fracturing procedure. Tips, mounting hardware, and visualization equipment are commercially available from both New Objective (52) and Protana (65) for most commercial atmospheric pressure ionization mass spectrometers.

The practical implementation of the above-described nanoES device marked the beginning of what was to become the widespread utilization of nanoflow techniques in protein sequencing. A flurry of important developments followed to extend and improve the utility of this technology for protein

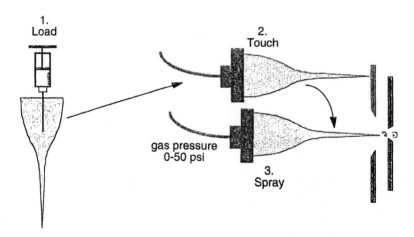

Figure 7 General sequence for nanoES operation. *1*, Sample is put into the capillary with gel-loading Eppendorf tips. *2*, If the tips are sealed, a touch/fracture step with the aid of microscopic inspection is required to initiate flow. *3*, Slight gas pressure is used to tune the flow.

sequencing. Programs for rapidly correlating MS-derived peptide molecular mass (69,70) and MS/MS-derived peptide sequence information (9,69,70) with databases containing known sequence information provided a major leap forward in the speed at which known proteins could be identified. Sample preparation protocols for the low level sequencing of proteins from silver-stained polyacrylamide gels were detailed (10,11) and specifically tailored for the nanoES system of sample introduction. Refined sample preconcentration strategies emerged (73), and techniques for rapid proteolyic digestion (74), again tailored to dovetail with the nanoES system. Instrumentation to automate this extraction process began to emerge (75–78), and a mass spectrometric method to distinguish chemical noise from low level peptide signals was published (79). Finally, a series of seminal papers in *Nature* (7,80) on the use of this technique for the solution of some difficult protein sequencing problems solidified the legitimacy of the nanoES methodology and spurred further advances by a multitude of investigators in this field.

2. Mechanically Ground and Chemically Etched Fused Silica Electrodes

In conjunction with the introduction of the nanoES system, methods to provide more robust electrical contact to the nanoflow electrodes and to develop electrodes inherently longer lived and amenable to on-line separations or continuous-flow operation began to emerge. Nanoflow electrodes constructed of fused silica tubing with narrow inner bores were the obvious substitute for the borosilicate tubes of large inner diameters described for nanoES. The primary motivation was the desire to integrate the nanoflow electrode tip in the CE or HPLC column.

Systems described by Emmett and Caprioli, (26) Kriger et al. (30), and Gale and Smith (23) used lengths of fused silica, commercially available in dimensions ranging from 1 to 100 µm i.d., to provide small exit apertures. However, the high backpressures and plugging problems encountered with even short lengths (1 cm) of the very narrow inner diameter material (<20 µm) made them problematic for practical implementation. Fused silica tubes with inner diameters of 15–20 µm and greater have generally been accepted as a reasonable compromise. To achieve as low a flow as possible with electrodes having exit diameters of 15–20 µm or more, the tips were shaped by different means to reduce the outer diameters. Emmett and Caprioli (26) and Gale and Smith (23) took advantage of the ability of HF to etch silica and thereby reduced the wall thickness at the tip. Voltage contact to the liquid was made through a metal fitting in which the 2 cm long electrode was inserted. The negative side of this approach is that the exit aperture inner diameter was simultaneously being increased as the wall thickness was decreased, but this was eventually overcome by continuously purging the tube with water during the etching process; purging prevents the inner diameter from being eroded.

Kriger and Cook (30) used a mechanical system whereby the outer diameter (365 μm) was ground back with 600 grit sandpaper by turning the tip in a minilathe, producing shapes that maintained the original inner diameter of roughly 20 μm, like the one shown in Fig. 8. The metallization procedure involved the use of an organosilane coupling agent to form a strongly adhering intermediate layer between the silica surface and the gold. In this case, the bifunctional organosilane (3-mercaptopropyl) trimethoxysilane was used. The silane moieties covalently bind to the oxide surface of the silica through siloxane bonds, while the thiol groups engage in strong binding to the vapor-deposited gold coating. Although the chemical treatment steps are rather laborious, involving several iterations of surface pretreatment, baking, and chemical bonding, the result is a highly durable gold-plated electrode resistant to chemical and physical abrasion. We have reproduced the work of Kriger and Cook in our laboratory and can attest to the remarkable sturdiness of these electrodes, stemming from the very high affinity of the noble metals for thiol groups. Several weeks of stable operation as electrodes for CE/MS applications were reported (50). It must be stressed that the effectiveness of the mercaptosiloxane

Figure 8 Example of a mechanically ground tip with an overlay of vapor-deposited gold chemically bonded to the silica surface. Fused silica dimensions are 365 μm o.d. × 20 μm i.d. The exit aperture i.d. is 20 μm. (From Ref. 30, reproduced with permission.)

layer requires strict adherence to the procedure described in the literature. The labor-intensive aspect of this procedure can be ameliorated by processing large batches of electrodes at the same time, which is simple to do with common distillation equipment and laboratory ovens.

The above-mentioned mechanically or chemically shaped tips of 20 µm i.d. or greater have been reported to give operational flow rates of roughly 100–1000 nL/min. In 1995, a commercial device was introduced, called Micro IonSpray (28), that provided a 20 nL dead volume mount for 5 cm lengths of shaped or unshaped fused silica (see Fig. 9). Its purpose is to serve as an electrospray interface for packed capillary HPLC or CE operating in the 10–10,000 nL/min flow rate range, having pneumatic nebulization as an option for operation at flows in excess of 1000 nL/min. This is typically used with 20 µm i.d. fused silica without further shaping of the tip, a process that limits the low flow operation to 200 nL/min. Drawn tips provide lower flows. This device employed the concept of applying the high voltage to a metal union upstream of the emitter tip to avoid requirements to metallize individual tips. Many others have reported this general approach of applying high voltage with considerable success (23,26,40–42,81), and a similar system, without the pneumatic nebulization option, was studied in great detail (34,39) for capillary LC/MS applications.

A mechanical grinding procedure similar to that used by Kriger and Cook (30) was adapted by Barnidge et al. (44) for tubing of nearly equivalent

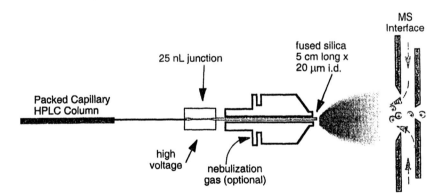

Figure 9 Diagram of a Micro IonSpray nebulizer for coupling packed capillary LC and CE to electrospray. Operational flow range with unshaped, square-cut 20 µm i.d. fused silica is 200–10,000 nL/min. Flow can be reduced with the use of tips drawn to smaller inner diameters.

dimensions, 25 μm i.d. x 365 μm o.d. One interesting development from this group is their most recent metallization procedure, called the "fairy dust" technique (45) because of the shimmering appearance of these electrodes (see Fig. 10). Glass or fused silica can be coated with a thin film of polyimide resin by simply drawing it through a piece of lens paper wetted with the resin. Gold dust is then sprinkled on the surface, the electrode is cured in an oven, and an extremely stable layer of gold particles fused into the polyimide matrix provides a conductive coating resilient to physical and chemical insults. This approach has been applied to tips that had either been mechanically ground or pulled down to fine apertures with heat from a torch (45). This metallization procedure probably represents the ultimate in simplicity for those who want to construct their own electrodes.

Another approach to providing electrical contact to electrodes shaped by grinding down the outer diameter, and applicable to drawn tips as well, is to deposit a transition metal undercoating to establish a better contact layer for the electrochemically resistant noble metal. Materials such as chromium, titanium, and tungsten have been employed for this purpose (44). The method involves depositing by electron beam emission a layer of chromium on the exposed silica tip as well as on a significant portion of the polyimide coating, followed by vapor deposition of gold. Mechanical stability is imparted to this tip owing to the fact that most of the polyimide coating is left intact. The use of a transition metal undercoating was noted to present problems relating to the diffusion of the transition metal along grain boundaries to the noble metal surface, presenting sites for electrochemical attack. The problem of electrochemical attack of the undercoating has been severe with its application to the production of elec-

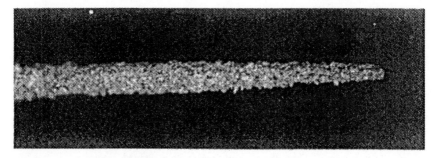

Figure 10 Photograph of the "fairy dust" tips. Metallization is achieved by adhering micrometer-sized gold flakes to the fused silica with a polyimide bonding resin. (Reproduced with permission from Ref. 45.)

trochemical sensors but has not been shown to be a deficiency during a lifetime of use for nanoflow electrospray electrodes.

3. Drawn Fused Silica Electrodes

The following discussion concentrates on continuous flow systems using fused silica with tips drawn to reduced apertures for lower flow operation. For methods involving on-line nanoscale separations or systems endeavoring to achieve some degree of automated sample introduction in the nanoliter per minute range, tips prepared on the emitter end of lengths of fused silica tubing are an important development. Apertures of 10 μm and greater can be reproducibly made by melting the tubing with a torch and pulling with a weight (35,41,45,46). Below 10 μm this method tends to lack reproducibility. In the mid-1990s, Sutter Instruments (21) developed a laser-powered puller, Sutter Model P-2000, that was based on the successful resistive filament–heated puller but could deliver sufficient heat to melt fused silica. This represented a significant development for the routine and reproducible production of aperture dimensions of 0.5–10 μm, and many research groups obtained this equipment to experiment with the production of tips with various sizes and shapes.

One of the first attempts to improve upon the gold vapor deposition method for providing electrical contact to the tip involved painting the tip with a conductive epoxy adhesive (27). This technique has not been generally adapted owing to lack of durability, the high chemical background, and, in the case of silver colloids, chemical and electrochemical degradation, leading to significant silver and silver adduct background. The popular methods in use today employ gold affinite underlayers such as those described above (30,44), the gold capturing over layers of Valaskovic and McLafferty (31) and New Objective (5), "fairy dusts" (45), electroplating (36), and metal/liquid junctions, which were discussed earlier and are treated in more detail below (23,26,28,40–42,81).

Gold electroplating of drawn tips has been used to enhance significantly the durability of a vapor-deposited gold underlayer (35,36). An overcoating 1–2 μm thick was applied by electroplating on top of the typical 100 nm thick vapor deposition layer. In this work, a study of aperture sizes was undertaken to determine the most robust, trouble-free dimensions. Apertures of 20 μm drawn from 350 μm × 50 μm i.d. tubing provided flows of 100–300 nL/min and were settled upon for their reliability and freedom from plugging. Several days of stable CE/MS operation could be obtained with this approach. It was also noted that apertures of this size were readily prepared with a simple butane minitorch and did not need the sophisticated laser pullers required for electrodes of much smaller apertures. A very similar procedure for making integrated CE column/nanoflow electrodes of 16 μm aperture size was reported by Barroso and de Jong (46). Corroborating the work of Kelley, the 10–20 μm aperture sizes were found to reliably spray the electro-osmotic flows produced with 50 μm CE capillaries flows of approximately 150 nL/min.

Valaskovic and McLafferty (31,51) carried out extensive investigations to develop reproducible nanoflow electrodes for CE and other continuous flow applications having submicrometer- to micrometer-sized tips for ultralow flow operation. The methodologies developed eventually led to the establishment of the company New Objective, which offers the most diverse assortment of fused silica and borosilicate glass electrodes for nanoflow operations commercially available today (52). With these electrodes, as with their glass tube electrode counterparts shown in Figs. 5 and 6, the vapor-deposited gold is protected with a robust electrically insulating overcoating of SiO_x and the aperture openings are carefully quality controlled with HF etching.

The use of metallized "liquid junctions" to apply voltage to low flow systems was originally implemented by Lee et al. (82) as one means for CE coupling. In this instance the junction was used both for electrical contact and as a means to "T" in an additional liquid to raise the flow in the sprayer to the microliters per minute range. Variations on this theme involve the use of metallized unions to apply the voltage upstream of a drawn fused silica electrode (18,23,26,28,40–42,81). As mentioned earlier, the electrical properties of these devices, in contrast to those of metallized tips, have been extensively investigated (67,68). Other methods have also been devised to deliver the voltage upstream of the tip. A microdialysis membrane sleeved over a section of open tubing was utilized by Severs et al. (83) for CE/MS, and Cao and Moini (37) did another variation on this theme by inserting a platinum wire through a hole drilled in the side of the fused silica; this latter approach has been used in CE/MS applications.

Efforts to implement automated nanoflow systems based on the continuous flow, pulled fused silica concept are highly active today primarily in support of the high throughput proteomic laboratory initiatives. This automation and high throughput endeavor has required efforts on several fronts: sample handling and preparation, hardware, and software to produce integrated systems. This includes efforts to develop microautosamplers and microextractors (75–78,84,85), robust continuous nanoflow systems (18,40,71), and customized bioinformatics software to speed up the data acquisition and protein identification process (9,69,70,86). The current goal is to achieve protein identification with throughputs of 100–200 2-D spots per day per instrument, roughly one order of magnitude higher than what is routinely achieved today.

Nanospray ESI Development

Sloan Kettering Research Institute has invested considerable time, talent, and financial resources to develop an automated system to support its proteomic efforts (18,40,86). This approach, described below, is based on the principle of maximizing sample utilization efficiency to achieve limits of identification as low as possible without compromising sample throughput. Generally speaking, speed and sensitivity are juxtaposed; one is gained at the detriment of the other. Their system uses variable flows—high flows to gain speed on the sample delivery and purge steps and very low flows for the analysis step to achieve high efficiency of sample utilization. The software then ensures that the minimum amount of time is consumed to acquire the requisite protein sequence information to obtain a database match before the sample is purged at a higher flow and the analyzer moves on to the next sample.

Figure 11 is a schematic of the Sloan Kettering autosampling and ionization system referred to as the JaFIS system (86). Control of the delivery pressure

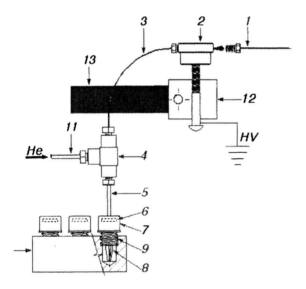

Figure 11 Diagram of the JaFIS nanoelectrospray autosampling system. *1*, Fused silica capillary 5 cm long × 20 μm i.d. drawn to a 0.8 μm i.d. tip. *2*, Twenty nanoliter metal union to which the high voltage is applied. *3*, Fused silica transfer line, 20 μm i.d. *4*, "Tee" fitting, 1.6 mm. *5*, Fused silica guide and septum-puncturing tube, sharpened at vial end. *6*, Septum. *7*, Cap for PCR tube container. *8*, PCR tube sample reservoir. *9*, Container to hold PCR tube. *10*, Sliding sample rack to advance vials. *11*, Helium gas pressure 0–50 psi and vent line. *12*, PEEK source mount. *13*, Standoff to micromanipulator mechanism. (Reprinted with permission from Ref. 18.)

allows for dynamic regulation of the flow rate during the analysis Flow rates on the order of 20–100 nL/min are used to rapidly deliver the sample from the PCR tube reservoir to the sprayer, at which time the flow is reduced, by dropping the pressure, to the high sensitivity, low flow regime (1–4 nL/min). At this point customized software takes control to guide the data acquisition. Peptide molecular ions are chosen for MS/MS sequencing either on the basis of the signal-to-noise ratio or by correlation with data obtained by matrix-assisted laser desorption ionization (MALDI) on the same sample. Spectra are interpreted, databases are searched on the fly, and when a positive identification is made the remaining ions expected from that protein are eliminated from further analyses. Peptide ions from unidentified proteins in that sample are then sought out, this process being referred to as "subtractive reiteration." A considerable time saving is realized in the decision-making process on what and what not to determine. When the sample is sufficiently interrogated, the high flow resumes and the sample is purged.

The electrodes they use are produced from pulled fused silica capillaries, drawn by the Sutter P-2000 to tip inner diameters of 0.8 µm (Fig. 12). No metallization is utilized; instead, voltage is applied to the metal fitting in which the 5.5 cm long tip is inserted. Sample utilization efficiency is high at these very low flows (0.5–5 nL/min)—reported to be as high as one sample molecule in solution out of 26 reaching the MS detector (18). Practically speaking this amounts to the ability to obtain readable sequence data from standard solutions

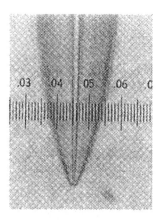

Figure 12 Magnified view of an electrode from the low flow JaFIS system. The smallest division equals 0.8 µm, approximating the exit diameter. (Reprinted with permission from Ref. 18.)

of protein tryptic digests at concentrations of 1–5 femtomoles per microliter (fmol/μL). Extensive efforts have been exerted by this group, as well as by Valaskovic and coworkers (31–33,51,52), to optimize all parameters of tip architecture to achieve the highest efficiencies and lowest flow rates, pushing the limits into the sub to low nanoliter per minute flow regime.

4. Fritted and Packed Fused Silica Electrodes

Some recent innovations involve the development of processes to produce pulled tips with integral sintered silica frits to capture reverse-phase packing materials and offer an element of robustness. Figure 13A is a magnified view of a tip with a 15 μm i.d. exit aperture containing a sintered frit produced by New Objective. Figure 13B shows the column packed with reverse-phase material. An interesting operational behavior of tips with sintered frits or HPLC packing material lodged inside near the exit is that they have the capacity to achieve lower flows than would normally be expected from that large an exit diameter by effectively reducing the exit diameter with internal material. The utility of this characteristic is discussed further in the following paragraph.

Nanoflow electrode tips designed for coupling to packed capillary HPLC have generally had apertures of 10–20 μm i.d. for robust operation with flows of several hundred nanolitere per minute. Publications by Gatlin et al. (38) and by Davis and coworkers (87–89) emphasized the use of 2 μm pulled fused silica apertures to achieve higher efficiencies. Davis et al. (88) demonstrated that with

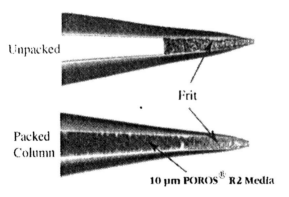

Figure 13 Photographs of a pulled fused silica tip from 75 μm i.d. tubing with an exit aperture of 15 μm produced by New Objective. A sintered silica frit is fixed in place. The lower view shows the column packed with reverse-phase stationary phase material. (Photograph courtesy of New Objective.)

these tips one could operate in the chromatographically more advantageous higher flow regime (>100 nL/min) but also drop the flow during sample elution, the "peak-parking" technique, to take advantage of signal averaging in the MS/MS mode. Since then, interest has arisen to produce tips that would operate in the 100–1000 nL/min flow range with the robustness of the large aperture sprayers, yet at the same time be able to rapidly drop the flow to around 50 nL/min or lower while "parking" on an eluting chromatographic peak and maintaining efficiency of ionization. Recently the fritted tips described above (available from New Objective) and "flame-pulled/hand-packed" tips discussed in more detail in Section III have demonstrated these desirable characteristics. Figure 14A is a scanning electron micrograph of a flame- and gravity-pulled tip with an exit aperture of approximately 15 μm. When 5 μm reverse-phase material is slurry packed into these tips, a "log jam" occurs at the end, forming a virtual frit (Fig. 14B). Lower flows of approximately 50 nL/min can be accessed with these, as well as with the fritted tips, for use in the peak-parking mode (89) during MS/MS acquisition. The separation step occurs in the 100–400 nL/min flow range in the single MS mode of data acquisition. This approach is an attempt to combine the best of both worlds: relatively fast chromatographic sample concentration, separation, and cleanup with slow sample introduction for the MS/MS step, allowing long periods of time and the signal-averaging advantage to obtain

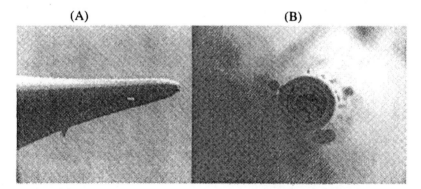

Figure 14 Side view (A) and end view (B) of a flame-pulled tip of approximately 15 μm exit diameter to be packed with reverse-phase material without a frit in place. The tip was intentionally cut very close to the end, in an attempt to expose the log jam. In the process some of the packing material was dislodged. The capillary dimensions are 160 μm o.d. × 100 μm i.d. × 5 cm length. The capillary volume is 1.6 μL before packing; after packing the void volume is 200 nL. Typical injection volumes are 0.5–10 μL. (Reprinted with permission from Ref. 108.)

low level peptide product ion spectra for sequence interpretation. More details regarding this technique are provided in the next section.

III. NANOFLOW LIQUID CHROMATOGRAPHY MASS SPECTROMETRY

Liquid chromatography (LC) is often the method of choice for the analysis of biological samples such as proteins, especially for components present at low levels. LC is a convenient method for performing sample cleanup steps such as desalting and detergent removal. In addition, the use of solvent conditions that allow for concentration of the sample at the head of the column provides exceptionally high sensitivity. The constant pursuit of even higher sensitivity drives efforts to decrease the size of the LC column. An added benefit of decreased column size is the concomitant decrease in the optimal flow rate, an important consideration for coupling LC to electrospray ionization mass spectrometry. This section outlines some of the recent advances in miniaturization of the analytical LC column. The applications reviewed here again focus on LC/MS applications in proteomics, a field that is ideally suited to miniaturized separation systems.

A. Miniaturization of the Analytical LC Column

For a given mass of sample injected, the analyte concentration is inversely proportional to the square of the column internal diameter (i.d.). As the column inner diameter is reduced, the optimum flow rate also lowers by the same exponential function. Similarly, the ionization efficiency increases with lowered flows. Therefore, there is an ongoing effort to maximize sensitivity by reducing the i.d. of the LC column. However, a decrease in capacity accompanies the decrease in column i.d.; in practice, only when the sample volume is limited do small i.d. columns provide increased sample concentration detection limits. This is the case for proteins extracted from SDS-PAGE gels, where typical sample volumes are 2–5 µL, with as little as a few femtomoles of digested protein. These types of samples dictate the use of nanoscale columns. This section focuses on some of the initial developments in nanoscale columns, their coupling to nanoflow electrodes, and the eventual merging of the two components into integrated nanoLC/nanoflow electrodes.

The optimal flow rate of the nanoscale columns (104,105), defined as columns with an internal diameter of 10–100 µm, falls within the intermediate nanoflow regime of 100–400 nL/min. Davis et al. (88) were one of the first groups to describe a nanoscale LC/MS system. They used various low dead volume unions to couple a nanoflow electrode to a 100 μm (i.d.) × 5.0 cm column. The electrode was prepared from 360 μm × 150 μm fused silica that was pulled by using a flame and a weight to produce a tapered tip with a 1–5 μm orifice. A platinum sheath glued into the union supplied the electrospray voltage. A narrow fused silica (20 μm i.d.) capillary served as a transfer line between the end of the LC-UV detection system and the union. This arrangement allowed for the analysis of 100 fmol of a protein standard that had been digested in the liquid phase.

The construction of the nanoscale LC/MS systems described above requires a high degree of technical expertise. In particular, the use of nanoscale columns places strict restrictions on the use of extracolumn devices; all dead volumes must be minimized. Therefore, an important development in nanoscale LC/MS was the elimination of any components between the column and the nanospray (nanoES) electrode, which was soon followed by the integration of columns and electrodes into a single device. Four groups (38,106–108) developed similar methods for the construction and operation of integrated nanoLC electrodes. These systems are described below and summarized in Table 1.

B. Integration of Nanoflow Electrodes and Columns

In the work of Gatlin et al. (38), the authors used a laser puller to produce columns of 100 or 180 μm i.d. with a tip i.d. of ~2 μm and packed them with 10 μm beads using a relatively high pressure of 8000 psi. Alternatively, the tip could be enlarged to ~15 μm by breaking it against a hard surface, allowing for packing at 400 psi in 2–3 min. Although these authors stress that reproducible tips can be fabricated only by using a laser puller, not a microtorch, in our laboratory both methods are successful, and the microtorch method is the method of choice. The 10–12 cm columns were mounted in a PEEK tee or cross, and the electrospray voltage was applied via a gold rod inserted into the fitting. A conventional LC solvent delivery system with a precolumn split supplied mobile phase at 400 nL/min. This system provided efficiencies of 4000 theoretical plates. The authors were able to identify as little as 76 fmol of bovine serum albumin from an in-gel digest of a one-dimensional band.

Using a different approach, Davis and Lee (106) prepared columns from 150 μm i.d. capillaries pulled to a tip i.d. of 5–10 μm using a laser puller. With the aid of a microscope, a short piece of 150 μm o.d. × 25 μm i.d. capillary and

Table 1 Compilation of Some Recent References Regarding the Integration of Nanocolumns and Electrodes into a Single Device

Authors (Ref)	Capillary dimensions (i.d. × o.d. in μm)	Tip i.d. (μm) and pulling method	Frit	Packing pressure (psi)	ESI voltage contact
Davis Lee (106)	150 × 360	5–10 μm, laser puller	Capillary support plus membrane	4000	Gold wire
Gatlin et al. (38)	100 or 180 × 360	~2 μm, laser puller; ~15 μm after breaking	None	8000 400	PEEK tee plus gold rod
Moore et al. (107)	150 × 360	5–10 μm, laser puller	None	NA (polymer-filled)	Gold wire
Pinto and Figeys (108)	25–100 × 360	~10–15 μm, torch	None	100–500	Stainless steel union

a membrane were inserted into the column to serve as a frit. The column was then packed with 5 μm beads at 4000 psi and inserted into a PEEK tee. The electrospray voltage was applied via a gold rod inserted into the fitting. Davis and Lee report that a column and tip endured more than 40 injections if the samples were filtered prior to injection. Periodic trimming of the column head extends the lifetime of the column to roughly 100 injections.

Moore et al. (107) devised a fritless method whereby needles were prepared from 360 μm o.d. × 150 μm i.d. capillaries pulled to a tip i.d. of 5–10 μm using a laser puller. The needles were filled with a mixture of reagents necessary for the formation of a porous monolith of poly(styrene divinylbenzene). After the polymerization was complete, the needles were inspected and trimmed to 3–4 cm. Moore et al. report a success rate of 80% compared to a success rate of "greater than half the time" for their needles packed with 5 μm beads. The polymeric columns provide efficiencies of 20,000 plates and are suitable for the analysis of in-gel digested protein samples at the 200–400 fmol level.

The systems described above provide several benefits. First, combining the columns and nanoelectrospray source into a single unit eliminates the possibility of extracolumn broadening. Second, the short columns are amenable to rapid analysis, a critical element for high throughput applications such as screening. Third, the small footprint of the column–source unit makes multiplexing practical. To this end, our laboratory designed a rugged, simple interface for multiplexed nanoLC/MS analysis. The remainder of this section describes this system.

C. Example of a Multiplexed NanoLC System

1. General Description

The goal of a multiplexed nanoLC system is to provide faster and more sensitive analysis than traditional HPLC at a cost that makes the technique affordable for multiplexing. Figure 15 depicts the basic, one-column system. The pneumatic solvent delivery system (A) uses a high pressure (500psi) for sample injection, desalting, and column conditioning and a low pressure (30–50 psi) for extended LC/MS/MS analysis. Electrically actuated valves allow for pressure

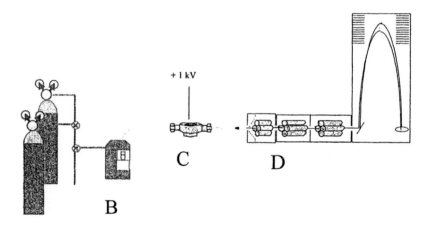

Figure 15 Diagram of a single-column integrated nanoLC system with solvent delivery by pneumatic means. (A) Gas pressure source. (B) Pressure bomb containing sample and mobile phases. (C) Microunion (or tee for gradient operation) for mounting the nanoLC column/sprayer. (D) Tandem quadrupole TOF mass spectrometer. (Reprinted with permission from Ref. 108.)

selection and venting. The solvents and sample are contained in a manually operated pressure bomb (B). Venting, switching solvents, and repressurizing takes approximately 15 s. A fused silica transfer line and a stainless steel union connect the bomb to the nanoLC column (C). The data presented were collected by using a prototype tandem quadrupole time-of-flight mass spectrometer (D) (109). The small dimensions of the nanoLC column permit the installation of multiple columns in the source. In Fig. 16, 12 columns are arrayed according to the dimensions of a single row of a 96-well plate and mounted on a removable strip, which is driven into position by a stepper motor, essentially making the ion source the autosampler. A strip of 12 samples is simultaneously loaded and washed off-line, gaining the significant time advantages of parallel batch processing while the samples are run in the conventional serial manner at a rate of approximately 2–8 min per analysis. The off-line process involves mounting the strip of 12 on top of a chamber containing the samples in a 96-well plate. Samples are loaded and washed in parallel by pressurizing the chamber. Extending this principle to larger arrays (8 × 12) appears very feasible.

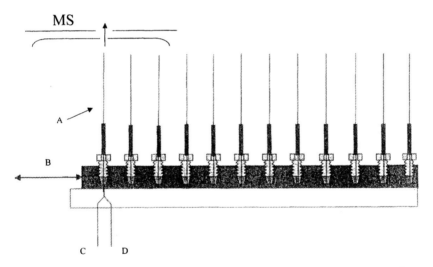

Figure 16 A combined nanoLC ion source/autosampler comprising a 12-column array. (A) A linear row of packed nanoLC columns spaced according to the dimensions of a 96-well sample plate mounted on a removable PEEK strip. (B) Stepper motor drive to serially position each column in front of the MS entrance aperture. (C, D) Fluid lines from pressure-driven solvent reservoir or conventional pumps. Samples loaded in a parallel batch fashion (see text). (Reprinted with permission from Ref. 108.)

2. Flow Characteristics

In traditional LC/MS systems, the use of postcolumn splitting decouples the chromatographic and electrospray flow rates. The column operates at a high flow rate to provide optimal resolution, while the ESI source operates at a lower flow that is compatible with electrospray or pneumatically assisted electrospray. However, the integration of the electrospray electrode with the column narrows the flow range that can be used. Thus, it becomes desirable to produce electrodes with as broad a flow range as possible. Figure 17 is a plot of intensity as a function of flow rate for a flame-pulled 20 μm i.d. electrode packed with 5 μm stationary phase to the tip, similar to the one depicted in Fig. 14. These data exhibit trends similar to those presented in Fig. 3: a concentration dependence at high flow rates and a mass flow response at low flow rates, with the exception that a lower optimal flow rate for a given emitter i.d. appears to be achievable with these packed or fritted

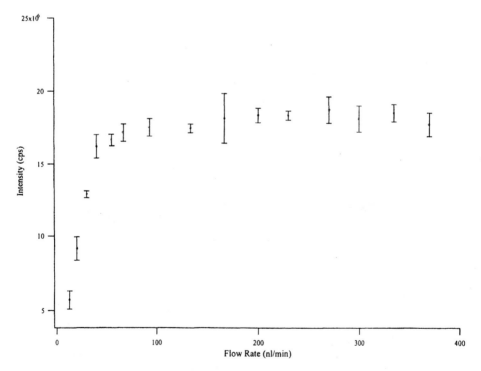

Figure 17 Intensity versus flow rate relationship for the nanoLC columns described in the text. (Reprinted with permission from Ref. 108.)

electrodes. It is desirable to maximize this type of behavior. From the chromatographic perspective, optimum separation efficiency, analysis speed, and durability can be obtained in the flow range of hundreds of nanoliters per minute with electrodes of fairly large inner diameters. From the MS/MS sensitivity perspective, the use of these types of electrodes allows operation at a considerably lower flow rate and makes the peak-parking technique, originally developed by Davis and Lee (89), a very practical method of operation.

3. Separation of Complex Samples

In protein identification, the sample complexity is often high, requiring detailed MS/MS analysis to successfuly identify the protein. Even if the protein sample is isolated by using 2-D gel electrophoresis, which is capable of separating thousands of proteins, each sample often contains more than one protein. The samples are then digested with trypsin to produce a mixture of dozens of peptides. To further complicate matters, samples are often contaminated with keratin, a ubiquitous protein present in skin and hair, as well as fragments from trypsin, which is itself, a protein. Despite this sample complexity, the standard method for protein analysis is continuous infusion of the sample, without any separation, by nanoelectrospray ionization (nanoESI) (6).

NanoESI, although successful for routine protein identification, is often unsuitable for analyzing complex mixtures of proteins or for performing detailed analysis such as identification of post-translational modifications. Figure 18 contains an example of a nanoelectrospray spectrum from a simple digest of a pure protein standard. Without separation, peptides with the same nominal mass are unresolved, rendering the acquisition of independent production spectra for both peptides difficult or even impossible. In our experience, the occurrence of isobaric peptides in protein digests is a common situation. However, when performing nanoLC/MS analysis, peptides of this type are easily resolved (Fig. 19), permitting detailed MS/MS analysis.

4. Sensitivity

As mentioned earlier, nanoscale columns provide high sensitivity in cases where sample volume is limited owing to the increase in concentration of the analytes compared to traditional large or microbore columns. Figure 20 contains data from the injection of 30 fmol of a solution digest of a protein. An extracted ion chromatogram and the corresponding mass spectrum for one of the proteolytic fragments are given to demonstrate the high sensitivity that

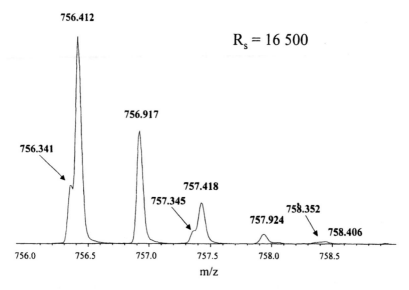

Figure 18 Expansion of a portion of the mass spectrum from a continuous-introduction nanoESI analysis of a tryptic digest of a protein standard with no chromatographic separation. A resolving power of 16,000 obtained from the TOF analyzer was insufficient to resolve isobaric peptides.

nanoscale columns provide. Using this system, we regularly identify faint protein spots from silver-stained 2-D gels.

5. Gradient Operation

The system described here provides the ability to perform separations in an isocratic mode or a step-gradient mode. Figure 21 contains a display of separations obtained using these two modes. For routine protein identification, isocratic elution is preferable over step-gradient or gradient elution because of its simplicity of operation and because of the small time delay between the start of the run and the elution of the first component. However, in isocratic elution, early-eluting components are often poorly resolved, whereas late-eluting components can go undetected because the peak width, and consequently the dilution of the analyte by the mobile phase, increases rapidly with elution time. Nevertheless, we use isocratic elution for routine identification of abundant proteins. The analysis of samples containing low levels of analyte or the identification of post-translational modifications, which requires very high coverage, dictates the use of

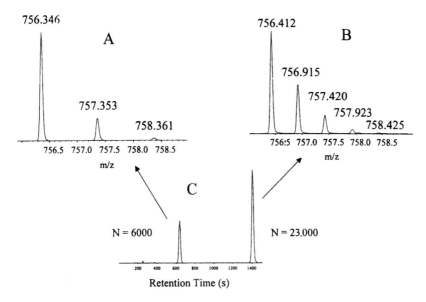

Figure 19 The same tryptic digest sample as the one described in Fig. 18 subjected to nanoLC separation. (A, B) Expansions of the molecular ion regions of the spectra obtained from the chromatographic peaks depicted in (C). (C) The extracted ion current chromatogram for m/z 756.4 of this sample. The second peak from this gradient elution showed 23,000 theoretical plates. (Reprinted with permission from Ref. 108.)

gradient elution. In Fig. 21, the separation performed using a step gradient exhibits far greater sensitivity for late-eluting components as well as increased resolution of early-eluting components.

D. Future Directions

Efforts to further decrease the dimensions of nanoscale columns may be impeded by our inability to construct tubing for columns with internal diameters in the micrometer and submicrometer ranges, not to mention the challenge of constructing uniform submicrometer packing materials. A potential solution to this problem is the use of microfabricated columns, packing materials, and electrospray emitters. He et al. (110) developed a microfabrication method for producing open tubular columns. These chips contained up to 4096 diamond-shaped posts. Each post measured 5 μm × 5 μm and was electrostatically coated with a poly(styrene sulfate) stationary phase. Solvents were delivered

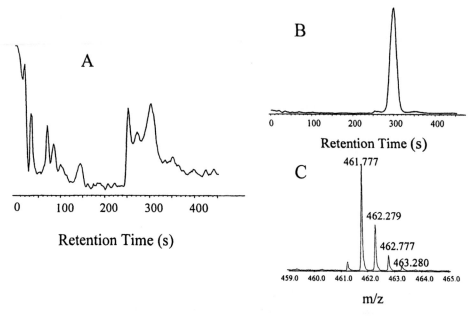

Figure 20 NanoLC analysis of a solution tryptic digest of a 30 fmol sample. (A) Total ion current trace. (B) Extracted ion current trace for *m/z* 461.7. (C) Expansion of the molecular ion region of the mass spectrum derived from the component in (B). (Reprinted with permission from Ref. 108.)

by electro-osmotic pumping at flow rates in the range of a few tens of nanoliters per minute. Such a system is readily multiplexed and could be interfaced with a microfabricated nanoES device such as the one reported by Corso et al. (111). They developed a circular chip with 96 nanoES electrodes, each of which had a volume of 200 pL, making them suitable for coupling to microfabricated nanocolumns. The diameter of the chip was a minuscule 1 mm. Therefore, prototypes of multiple chip-based LC columns, each with a nanoES electrode at its terminus, will undoubtedly emerge in the near future.

IV. CONCLUSIONS

Nanoflow techniques, for all the reasons discussed in this chapter, have become the method of choice for the characterization of low levels of proteins in biological sample. Suffice it to say that one can expect developments to con-

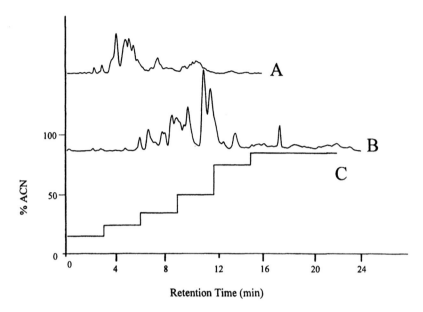

Figure 21 Total ion current traces for (A) isocratic elution and (B) gradient elution analysis of the same protein digest. (C) The step-gradient profile used for the analysis. (Reprinted with permission from Ref. 108.)

tinue to enhance the throughput and convenience of performing these analyses. However, with the sample utilization efficiency gains of several orders of magnitude that the nanoflow techniques have to offer, why are they not rapidly supplanting the high flow approaches in all areas of application? The answer to this question indicates an important direction for future developments in the nanoflow arena.

Referring back to the flow injection comparison in Fig. 1, we see that in this case the difference in the detection limits, defined by the signal-to-noise ratio (S/N) of the ion current trace, is not significantly different between the high and low flow sample introduction systems. In a vast number of application areas, the quantification of small molecules is all that counts. If the efficiency of sample utilization of the nanoflow methods can be manifested only by signal averaging over long time periods, with little change in limits of detection as measured by S/N or ion current traces, then the scope of applications will be limited.

However, if the sample in Figs. 1A and 1B could be introduced into the two sprayer systems over the same amount of time, then, in principle, the difference in absolute intensity (and presumably detection limit) would be directly proportional to the difference in sample utilization efficiency, in this case a factor of roughly 300. Chromatography or electrophoresis should take care of this problem. That is to say, when the column inner diameter is properly matched to the flow rate so as to provide the optimum linear velocity for efficient separations, the chromatographic peak widths over time should be nearly the same for high flow and low flow systems. For the same absolute amount of material injected, the nanoflow technique should give signals several orders of magnitude higher.

The trend toward improved chromatographic detection limits with nanoflow techniques has been demonstrated (34,36, 39,48,89), but in general practice few have taken advantage of the several orders of magnitude enhancement expected from these techniques. Part of the problem lies on the chromatography side. Pump, column, autosampler, and sample preparation technology simply do not support high efficiency separations in the nanoflow regime. It is exceedingly difficult to maintain the same concentration factors (chromatographic peak widths) that are routinely obtained with the high flow systems; thus, much of what is gained with ionization efficiency is lost by sample dilution in broadened peaks.

Other reasons besides the sensitivity issue also impede wider acceptance of the nanoflow techniques. Although the robustness and reliability of sprayer technology have come a long way, major improvements are still required. The small electrode apertures are prone to plugging from sample particles, precipitates, and particle bleed from chromatographic media. The trend for sample preparation techniques to be minimized to enhance throughput causes increasing degradation of the ion source. For the determination of small molecules extracted from blood plasma, the concentrations of both dissolved and undissolved solids far exceed those commonly encountered in a typical protein digest extract from a 2-D gel separation. Throughputs of thousands of samples per day must be accommodated without issues of plugging and fouling slowing the process.

Speed of sample transfer into the source is increasingly becoming an important issue. When faced with daily sample loads in the hundreds to several thousands per day, sample transfers must be fast. Hydraulic limitations come into play with the nanoflow techniques in this regard; the transfer speeds of the high flow techniques cannot be matched by the nanoflow techniques. In the future, microfabrication of short, small diameter channels in monolithic, connection-free separation/sprayer systems may help in this regard, but they have yet to be demonstrated to be a solution to this problem.

Finally, there is the consideration of sample concentration detection limits. When all is said and done, will there be any improvements? In a large number of high flow application areas, the concentrations of analytes to be measured are exceedingly low, roughly three orders of magnitude lower than a typical protein digest sample submitted for nanoESI analysis, which is typically 10–100 fmol/μL. Plasma drug level assays at the 1 pg/mL plasma level are not uncommon today. For a compound of 300 Da molecular mass, this translates into roughly 15 attomoles per microliter (after a typical extraction concentration procedure) in a very complex sample submitted for quantitative analysis by a targeted (MRM) multiple reaction monitoring technique. The difference is that in this popular application area involving the quantification of small molecules in biological fluids, approximately three orders of magnitude more sample volume is available for injection than is typical of a protein analysis from a 2-D gel separation. Because the sample loading capacities of the high flow techniques are so much higher than those of the nanoflow systems, and because with these applications much larger volumes of low concentration samples are available for injection, the ionization efficiency losses are negated by the higher loading capacities, resulting in no substantial difference in the concentration detection limits between the high flow and nanoflow techniques, unless special measures are taken to increase column capacities (34,46,90–93). The challenge is again a chromatographic one; not only must loading capacities be raised for the nanoflow separation techniques, but that must be done by a means amenable to high throughput applications. This represents a substantial barrier for the future torchbearers of the nanoflow techniques to overcome. Developments in microfabricated devices are beginning to show some promise (94–104) to reduce these hurdles, but miniaturization alone will not be the panacea. The true breakthroughs will lie in the more difficult task of developing new chemical and electrophoretic techniques for rapid sample preconcentration and separation.

REFERENCES

1. TR Covey. Analytical characteristics of the electrospray ionization process. In: P Snyder, ed. Biochemical and Biotechnological Applications of Electrospray Ionization Mass Spectrometry. Am Chem Soc, ACS Symp Ser 619. Washington, DC.: 1995, pp 21–59.
2. TR Covey, ED Lee, AP Bruins, JD Henion. Liquid chromatography/mass spectrometry. Anal Chem 58(14): 1451A–1461A, 1986.
3. JB Fenn, M Mann, CK Meng, SK Wong, C Whitehouse. Electrospray ionization for mass spectrometry of large biomolecules. Science 246: 64–71, 1989.

4. TR Covey, RF Bonner, B Shushan, J Henion. The determination of protein, oligonucleotide and peptide molecular weights by ion-spray mass spectrometry. Rapid Commun Mass Spectrom 2(11):249–256, 1988.
5. M Wilm, M Mann. Electrospray and Taylor-cone theory, Dole's beam of macromolecules at last? Int J Mass Spectrom Ion Process 136:167–180, 1994.
6. M Wilm, M Mann. Analytical properties of the nanoelectrospray ion source. Anal Chem 68:1–8, 1996.
7. M Wilm, A Shevchenko, T Houtaeve, S Breit, L Schweigerer, T Fotsis, M Mann. Femtomole sequencing of proteins from polyacrylamide gels by nano-electrospray mass spectrometry. Nature 379:466–469, 1996.
8. M Mann, M Wilm. Electrospray mass spectrometry for protein characterization. Trends Biochem Sci 20:219–230, 1995.
9. M Mann, M Wilm. Sequence tags obtained by mass spectrometry/mass spectrometry for the error tolerant identification of peptides. Anal Chem 136:167–180, 1995.
10. A Shevchenko, M Wilm, O Vorm, M Mann. Mass spectrometric sequencing of proteins from silver-stained polyacrylamide gels. Anal Chem 68(5):850–858, 1996.
11. A Shevchenko, M Wilm, O Vorm, ON Jensen, AV Podtelejnikov, G Neubaurer, P Mortensen, M Mann. A strategy for identifying gel-separated proteins in sequence databases by MS alone. Biochem Mass Spectrom 24:893–896, 1996.
12. TR Covey, EC Huang, JD Henion. Structural characterization of protein tryptic peptide via liquid chromatography/mass spectrometry and collision-induced dissociation of their doubly charged molecular ions. Anal Chem 63:1193–1200, 1991.
13. TR Covey. Liquid chromatography/mass spectrometry for the analysis of protein digests. In: JR Chapman, ed. Methods in Molecular Biology, Vol. 61: Protein and Peptide Analysis by Mass Spectrometry. Totowa, NJ: Humana Press, 1995, pp 83–99.
14. TR Covey, E Huang, J Henion. Protein sequencing by mass spectrometry. US Patent 5,952,653, 1999.
15. AP Bruins, TR Covey, JD Henion. Ion spray interface for combined liquid chromatography/atmospheric pressure ionization mass spectrometry. Anal Chem 59:2642–2646, 1987.
16. TR Covey. Dynamic gas focusing for achieving high flow rate electrospray ionization. Proc 42nd ASMS Conf on Mass Spectrometry and Applied Topics, Chicago, IL, 1994, p 872.
17. TR Covey, JF Anacleto. Ion spray with intersecting flow. US Patent 5,412,208, 1995.
18. S Geromanos, G Freckleton, P Tempst. Tuning of an electrospray ionization source for maximum peptide-ion transmission into a mass spectrometer. Anal Chem 72:777–790, 2000.

19. R Juraschek, T Dulcks, M Karas. Nanoelectrospray—More than just a minimized-flow electrospray ionization source. J Am Soc Mass Spectrom 10(4): 300–308, 1999.
20. KT Brown, DG Flaming. Advanced micropipette techniques for cell physiology. In: A. D. Smith, ed. Methods in the Neurosciences, Vol. 9. IBRO Handbook Ser. New York: Wiley, 1992.
21. Sutter Instrument Company, 51 Digital Drive, Novato, CA 94949. Tel (415) 883–0128. www.sutter.com
22. SK Chowdhury. Method for the electrospray ionization of highly conductive aqueous solutions. Anal Chem 63:1660–1662, 1991.
23. DC Gale, RD Smith. Small volume and low flow-rate electrospray ionization mass spectrometry of aqueous samples. Rapid Commun Mass Spectrom 7:1017–1020, 1993.
24. RD Smith, JH Wahl, DR Goodlett, SA Hofstadler. Capillary electrophoresis/mass spectrometry. Anal Chem 65(13):574A–584A, 1993.
25. PE Andren, MR Emmett, RM Caprioli. Micro-electrospray: Zeptomole/attomole per microliter sensitivity for peptides. J Am Soc Mass Spectrom 5:867–869, 1994.
26. MR Emmett, RM Caprioli. Micro-electrospray mass spectrometry: Ultra-high sensitivity analysis of peptides and proteins. J Am Soc Mass Spectrom 5: 605–613, 1994.
27. JH Wahl, DC Gale, RD Smith. Sheathless capillary electrophoresis—Electrospray ionization mass spectrometry using 10 μm I.D. capillaries: Analyses of tryptic digests of cytochrome c. J Chromatogr A 659:217–222, 1994.
28. TR Covey. Characterization and implementation of micro-flow electrospray nebulizers. Proc 43rd ASMS Conf, Atlanta, 1995, p 669.
29. GA Valaskovic, FW McLafferty. Attomole-sensitivity electrospray source for large-molecule mass spectrometry. Anal Chem 67:3802–3805, 1995.
30. MS Kriger, KD Cook, RS Ramsey. Durable gold coated fused silica capillaries for use in electrospray mass spectrometry. Anal Chem 67:385–389, 1995.
31. GA Valaskovic, FW McLafferty. Long-lived metallized tips for nanoliter electrospray mass spectrometry. J Am Soc Mass Spectrom 7:1270–1272, 1996.
32. GA Valaskovic, FW McLafferty. Sampling error in small-bore sheathless capillary electrophoresis/electrospray ionization mass spectrometry. Rapid Commun Mass Spectrom 10:825–828, 1996.
33. GA Valaskovic, NL Kelleher, FW McLafferty. Attomole protein characterization by capillary electrophoresis-mass spectrometry. Science 372:1199–1202, 1996.
34. K Vanhoutte, W Van Dongen, I Hoes, F Lemiere, E Esmans, H Van Onckelen, E Eeckhout, REJ Van Soest, AJ Hudson. Development of a nanoscale liquid chromatography/electrospray mass spectrometry methodology for the detection and identification of DNA adducts. Anal Chem 69:3161–3168, 1997.
35. KB Bateman, RL White, P Thibault. Disposable emitters for on-line capillary zone electrophoresis/nanoelectrospray mass spectrometry. Rapid Commun Mass Spectrom 11:307–315, 1997.

36. JF Kelly, L Ramaley, P Thibault. Capillary zone electrophoresis-electrospray mass spectrometry at submicroliter flow rates: Practical considerations and analytical performance. Anal Chem 69:51–60, 1997.
37. P Cao, M Moini. A novel sheathless interface for capillary electrophoresis/electrospray ionization mass spectrometry using an incapillary electrode. J Am Soc Mass Spectrom 8:561–564, 1997.
38. CL Gatlin, GR Kleemann, LG Hays, AJ Link, JR Yates. Protein identification at the low femtomole level from silver-stained gels using a new fritless electrospray interface for liquid chromatography-microspray and nanospray mass spectrometry. Anal Biochem 263:93–101, 1998.
39. K Vanhoutte, W Van Dongen, EL Esmans. On-line nanoscale liquid chromatography nano-electrospray mass spectrometry: Effect of the mobile phase composition and the electrospray tip design on the performance of a Nanoflow™ electrospray probe. Rapid Commun Mass Spectrom 12:15–24, 1998.
40. S Geromanos, J Philip, G Frekleton, P Tempst. Injection adaptable fine ionization source (JaFIS) for continuous flow nano-electrospray. Rapid Commun Mass Spectrom 12:551–556, 1998.
41. JC Hannis, DC Muddiman. Nanoelectrospray mass spectrometry using nonmetalized tapered fused-silica capillaries. Rapid Commun Mass Spectrom 12:443–448, 1998.
42. H Wang, M Hackett. Ionization within a cylindrical capacitor: Electrospray without an externally applied high voltage. Anal Chem 70:205–212, 1998.
43. KWY Fong, TWD Chan. A novel nonmetallized tip for electrospray mass spectrometry at nanoliter flow rate. J Am Soc Mass Spectrom 10:72–75, 1999.
44. DR Barnidge, S Nilsson, KE Markides, H Rapp, K Hjort. Metallized sheathless electrospray emitters for use in capillary electrophoresis orthogonal time-of-flight mass spectrometry. Rapid Commun Mass Spectrom 13:994–1002, 1999.
45. DR Barnidge, S Nilsson, KE Markides. A design for low-flow sheathless electrospray emitters. Anal Chem 71:4115–4118, 1999.
46. MB Barroso, AP de Jong. Sheathless preconcentration-capillary zone electrophoresis-mass spectrometry applied to peptide analysis. J Am Soc Mass Spectrom 10:1271–1278, 1999.
47. F Hsich, E Baronas, C Muir, SA Martin. A novel nanospray capillary zone electrophoresis/mass spectrometry interface. Rapid Commun Mass Spectrom 13:67–72, 1999.
48. CL Gatlin, JK Eng, ST Cross, JC Detter, JR Yates. Automated identification of amino acid sequence variations in proteins by HPLC/microspray tandem mass spectrometry. Anal Chem 72:757–763, 2000.
49. B Feng, RD Smith. A simple nanoelectrospray arrangement with controllable flow rate for mass analysis of submicroliter protein samples. J Am Soc Mass Spectrom 11:94–99, 2000.
50. D Figeys, I Oostveen, A Ducret, R Abersold. Protein identification by capillary zone electrophoresis/microelectrospray ionization-tandem mass spectrometry at the sub-femtomole level. Anal Chem. 68(11):1822–1828, 1996.

51. GA Valaskovic, FW McLafferty. Electrospray ionization source and method of using the same. US Patent 5,788,166, Aug. 4, 1998.
52. New Objective, Inc. 763 D Concord Avenue, Cambridge, MA 02138-1044. Tel (617) 576-2255. new obj@ix.netcom.com
53. Polymicro Technologies 18019 North 25th Ave. Phoenix, AZ. Tel (602) 375-4100. www.polymicro.com
54. G Hopfgartner, K Bean, J Henion, R Henry. Ion spray mass spectrometric detection for liquid chromatography—a concentration-flow-sensitive or a mass-flow sensitive device? J Chromatogr 647(1):51–61, 1993.
55. J Aerosol Sci 30(7): entire issue, 1999.
56. JF De La Mora. The effect of charge emission from electrified liquid cones. J Fluid Mech 243:561–574, 1992.
57. BA Thomson, JV Iribarne. Field induced ion evaporation from liquid surfaces at atmospheric pressure. J Chem Phys 71(11):4451, 1979.
58. M Dole, LL Mack, RL Hines. Molecular beams of macroions. J Chem Phys 49(5):2240–2249, 1968.
59. M Gamero-Castano, J Fernandez de la Mora. Mechanisms of electrospray ionization of singly and multiply charged salt clusters. Anal Chim Acta 406(1): 93–104, 2000.
60. FW Rollgen, E Bramer-Weger, L Butfering. Field ion emission from liquid solutions: Ion evaporation against electro hydrodynamic disintegration. J Phys C6(11):48, 1987.
61. P Kebarle, Y Ho. On the mechanism of electrospray mass spectrometry. In: RB Cole, ed. Electrospray Ionization Mass Spectrometry. New York: Wiley, 1997, pp 3–63.
62. P Kebarle, M Peschke. On the mechanisms by which the charged droplets produced by electrospray lead to gas phase ions. Anal Chim Acta 406(1):11–35, 2000.
63. M Labowsky, JB Fenn, J Fernandez de la Mora. A continuum model for ion evaporation from a drop: Effect of curvature and charge on ion solvation energy. Anal Chim Acta 406(1):105–118, 2000.
64. TL Constantopoulos, GS Jackson, CG Enke. Challenges in achieving a fundamental model for ESI. Anal Chim Acta 406(1):37–52, 2000.
65. Protana, The Protein Analysis Company, Odense, Denmark. Tel 45 63 15 20 30. www.protana.com
66. World Precision Instructions, International Trade Center, 175 Sarasota Center Blvd, Sarasota, FL. Tel (941) 371-1003. www.piinc.com
67. GJ Van Berkel. The electrolytic nature of electrospray. In: RB Cole, ed. Electrospray Ionization Mass Spectrometry. New York: Wiley, 1997, pp 65–105.
68. GS Jackson, CG Enke. Electrical equivalence of electrospray ionization with conducting and nonconducting needles. Anal Chem 71:3777–3784, 1999.
69. WJ Henzel, TM Billeci, JT Stultz, SC Wong, C Grimley, C Watanbe. Identifying proteins from two-dimensional gels by molecular mass searching of peptide fragments in protein sequence databases. Proc Natl Acad Sci USA 90:5011–5015, 1993.

70. P James, M Quadroni, R Carafoli, G Gonnet. Protein identification by mass profile fingerprinting. Biochem Biophys Res Commun 195:58–61, 1993.
71. AL McCormack, DM Schieltz, B Goode, S Yang, G Barnes, D Drubin, JR Yates. Direct analysis and identification of proteins in mixtures by LC/MS/MS and database searching at the low-femtomole level. Anal Chem 69:767–776, 1997.
72. DN Perkins, DJ Pappin, DM Creasy, JS Cottrell. Probability-based protein identification by searching sequence databases using mass spectrometry data. Electrophoresis 20(18):3551–3567, 1999.
73. H Erdjument-Bromage, M Lui, L Lacomis, A Grewal, RS Annan, SA Carr, P Tempst. Examination of micro-tip reversed-phase liquid chromatographic extraction of peptide pools for mass spectrometric analysis. J Chromatogr A 826:167–181, 1998.
74. PK Blackburn, RJ Anderegg. Characterization of femtomole levels of proteins in solution using rapid proteolysis and nanoelectrospray ionization mass spectrometry. J Am Soc Mass Spectrom 8:482–484, 1997.
75. T Houthaeve, H Gausephohl, K Ashman, T Nillson, M Mann. Automated protein preparation techniques using a digest robot. J Protein Chem 16:343–348, 1997.
76. T Houthaeve, H Gausepohl, M Mann, K Ashman. Automation of micropreparation and enzymatic cleavage of gel electrophoretically separated proteins. FEBS Lett 376:91–94, 1995.
77. K Ashamn, T Houthaeve, H Gausepohl, ON Jensen, M Mann. A commercial robot workstation for the in-gel digestion of proteins and its application to the rapid identification of yeast proteins. Proteomics: Integrating Protein-Based Tools and Application for Drug Discovery. IBC Library Ser. 1998.
78. Genomic Solutions, Ann Arbor, MI. Tel (734)975-4800. www.genomicsolutions.com
79. M Wilm, G Neubauer, M Mann. Parent ion scans of unseparated peptide mixtures. Anal Chem 68(3):527–533, 1996.
80. I Just, J Seizer, M Wilm, C von Eichel-Streiber, M Mann, K Aktories. Glucosylation of Rho proteins by Clostridium difficile toxin B. Nature 375:500–503, 1995.
81. S Konig, HM Fales. Comment on the cylindrical capacitor electrospray interface. Anal Chem 70:4453–4455, 1998.
82. ED Lee, W Muck, JD Henion, TR Covey. Liquid junction coupling for capillary zone electrophoresis/ion spray mass spectrometry. Biomed Environ Mass Spectrom 18:844–850, 1989.
83. JC Severs, AC Harms, RD Smith. A new high-performance interface for capillary electrophoresis/electrospray ionization mass spectrometry. Rapid Commun Mass Spectrom 10:1175–1178, 1996.
84. LC Packings, Amsterdam, Netherlands. www.lcpackings.nl
85. MT Davis, DC Stahl, KM Swiderek, TD Lee. Capillary chromatography/mass spectrometry for peptide and protein characterization. Methods 6:304–314, 1994.

86. S Geromanos, J Philip, G Freckleton, P Tempst. An integrated micro sampler and LIMS system for data dependent unattended analysis of low µL sample volumes for sequential nano-electrospray mass spectrometric analysis. Proc 47th ASMS Conf on Mass Spectrometry and Allied Topics, Dallas, TX, June 1999.
87. MT Davis, DC Stahl, TD Lee. Low-flow high performance liquid chromatography solvent delivery system design for tandem capillary liquid chromatography-mass spectrometry. J Am Soc Mass Spectrom 6:571–577, 1995.
88. MT Davis, DC Stahl, SA Hefta, TD Lee. A microscale electrospray interface for on-line, capillary liquid chromatography/tandem mass spectrometry of complex protein mixtures. Anal Chem 67:4549–4566, 1995.
89. MT Davis, TD Lee. Variable flow liquid chromatography-tandem mass spectrometry and the comprehensive analysis of complex protein digest mixtures. J Am Soc Mass Spectrom 8:1059–1069, 1997.
90. AJ Tomlinson, LM Benson, S Jameson, DH Johnson, S Naylor. Utility of membrane preconcentration–capillary electrophoresis-mass spectrometry in overcoming limited sample loading for analysis of biologically derived drug metabolites, peptides, and proteins. J Am Soc Mass Spectrom 8(1):15–24, 1997.
91. KP Bateman, RL White, P Thibault. Evaluation of adsorption preconcentration/capillary zone electrophoresis/nanoelectrospray mass spectrometry for peptide and glycoprotein analyses. J Mass Spectrom 33:1109–1123, 1998.
92. E van der Heeft, GJ ten Hove, CA Herberts, HD Meiring, CACM van Els, APJM de Jong. A microcapillary column switching HPLC-electrospray ionization MS systems for the direct identification of peptides presented by major histocompatibility complex class 1 molecules. Anal Chem 70:3742–3752, 1998.
93. HD Meiring, BM Barroso, E van der Heft, GJ ten Hove, APJM de Jong. Sheathless nanoflow HPLC-ESI/MS in proteomics research and MHC bound peptide identification. Proc 47th ASMS Conf on Mass Spectrometry and Applied Topics, Dallas, TX, 1999.
94. J Li, J F Kelly, I Chernushevich, JD Harrison, P Thibault. Rapid separation of peptides from gel-isolated membrane proteins using a microfabricated device coupled to a high performance quadrupole/time-of-flight mass spectrometer. Anal Chem 20:599–609, 2000.
95. J Li, C Wang, JF Kelly, JD Harrison, P Thibault. Rapid and sensitive separation of trace level protein digests using microfabricated device coupled to a high performance quadrupole/time-of-flight mass spectrometer. Electrophoresis 21:198–210, 2000.
96. J Li, P Thibault, NH Bings, CD Skinner, C Wang, K Colyer, DJ Harrison. Integration of microfabricated devices to capillary electrophoresis-electrospray mass spectrometry using a low dead volume connection: Application to rapid analyses of proteolytic digests. Anal Chem 71:3036–3045, 1999.
97. NH Bings, C Wang, CD Skinner, KL Colyer, JD Harrison, J Li, P Thibault. Microfluidic devices connected to glass capillaries with minimal dead volume. Anal Chem 71:3292–3296, 1999.

98. B He, N Tait, F Regner. Fabrication of nanocolumns for liquid chromatography. Anal Chem 70:3790–3797, 1998.
99. Q Xue, F Foret, YM Dunayevskiy, PM Zavracky, NE McGruer, BL Karger. Multichannel microchip electrospray mass spectrometry. Anal Chem 69:426–430, 1997.
100. D Figeys, R Aebersold. Nanoflow solvent gradient delivery from a microfabricated device for protein identifications by electrospray ionization mass spectrometry. Anal Chem 70:3721–3727, 1998.
101. D Figeys, Y Ning, R Aebersold. A micro fabricated device for rapid protein identification by microelectrospray ion trap mass spectrometry. Anal Chem 69:3153–3160, 1997.
102. D Figeys, C Lock, L Taylor, R Aebersold. Micro fabricated device coupled with an electrospray ionization quadrupole time-of-flight mass spectrometer: Protein identifications based on enhanced-resolution mass spectrometry and tandem mass spectrometry data. Rapid Commun Mass Spectrom 12:1435–1444, 1998.
103. S Ekstrom, P Onnerfjord, J Nilsoon, M Bengtsson, T Laurell, G Marko-Varga. Integrated micro analytical technology enabling rapid and automated protein identification. Anal Chem 72:286–293, 2000.
104. KB Tomer, A Mosely, LJ Detarding, CE Parker. Capillary liquid chromatography mass spectrometry. Mass Spectrom Rev 13:431–457, 1994.
105. JPC Vissers, HA Classens, CA Cramers. Microcolumn liquid chromatography: Instrumentation, detection and applications J Chromatogr A 779:1–28, 1997.
106. MT Davis, TD Lee. Rapid protein identification using a microscale electrospray LC/MS system on an ion trap mass spectrometer. J Am Soc Mass Spectrom 9:194–201, 1998.
107. RE Moore, L Licklider, D Schumann, TD Lee. A microscale electrospray interface incorporating a monolithic, poly(styrene-divinylbenzene) support for on-line liquid chromatography/tandem mass spectrometry analysis of peptides and proteins. Anal Chem 70(23):4879–4884, 1998.
108. DM Pinto, D Figeys. Miniaturized separation systems for high sensitivity protein identification. Presented at 47th Conf on Mass Spectrometry and Allied Topics, Dallas, TX, 1999.
109. A Shevchenko, IV Chernushevich, W Ens, KG Standing, B Thomson, M Wilm, M Mann. Rapid de novo peptide sequencing by a combination of nanoelectrospray isotopic labeling and a quadrupole/time-of-flight mass spectrometer. Rapid Commun Mass Spectrom 11:1015–1020, 1997.
110. B He, N Tait, F Regnier. Fabrication of nanocolumns for liquid chromatography. Anal Chem 70:3790–3797, 1998.
111. TN Corso, GA Schultz, S Prosser. A fully integrated monolithic microchip-based electrospray device for microfluidics systems. IBC 4th Annual Conf on Microfabrication and Microfluidics Technologies, San Francisco, CA, 1999.

3
Characterization of Pharmaceuticals and Natural Products by Electrospray Ionization Mass Spectrometry

A. K. Ganguly
Stevens Institute of Technology, Hoboken, New Jersey

Birendra N. Pramanik, Guodong Chen, and Petia A. Shipkova
Schering-Plough Research Institute, Kenilworth, New Jersey

I. INTRODUCTION

A typical drug discovery process involves the identification of a biochemical target and screening of libraries of synthetic compounds or compounds obtained from natural sources (i.e., plants, microorganisms, etc.). Collections of synthetic compounds in the pharmaceutical industry consist of novel structures designed around structural leads by medicinal chemists. More recently, compounds synthesized using combinatorial chemistry have been added to this collection. Once a suitable lead is identified, its activity, selectivity, and pharmacokinetic properties are optimized before the compound is recommended for further development as a clinical candidate. A large number of samples are generated during this process, and they require rapid and accurate determination.

In this chapter, we review the application of electrospray ionization mass spectrometry (MS) for the characterization of pharmaceuticals and natural products. Mass spectrometry is a powerful analytical technique that is used for the qualitative and quantitative identification of small organic molecules, peptides, proteins, and nucleic acids (1,2). The MS technique offers speed, accuracy, and high sensitivity. The development of ionization techniques and mass analyzers over the past two decades has enabled MS to solve a wide variety of structural identification problems. The introduction of electrospray ionization (ESI) (3) greatly expanded the role of MS in pharmaceutical analysis (4–14). One of the characteristic features of ESI is the generation of multiply charged ions for large molecular weight compounds (i.e., peptides, proteins, etc.); these differently charged molecules enable accurate determination of the molecular weight of these compounds and their analysis in complex biological media. The application of ESI in peptides and proteins is the subject of several chapters in this book and is not discussed here.

Important features of ESI are the simplicity of its source design and its capability to operate with solutions at atmospheric pressure; these allow the coupling of ESI to high performance liquid chromatography (HPLC) for analysis of complex mixtures (15,16). The LC/MS combination utilizes both the separation capability of HPLC and the detection and characterization ability of MS. Thus, it offers a useful method for qualitative and quantitative analysis of mixtures of components present in pharmaceutical samples. In addition, ESI is one of the most sensitive ionization techniques; it can achieve detection at the attomole (10^{-18} mole) level for large biomolecules (17,18). For pharmaceutical applications, micro- to nanomoles of synthetic samples are usually subjected to mass spectral analysis. As a nearly general method of ionization, ESI can be successfully applied to over 90% of organic compounds in pharmaceutical research, and that immediately makes it the method of choice for characterization of drug substances. Additionally, ESI is a soft ionization technique that yields a simple, easy-to-interpret mass spectrum in which the protonated $[M + H]^+$ or cationized ($[M + Na]^+$, $[M + K]^+$, etc.) molecules typically correspond to the base peak.

The degree of fragmentation in ESI-MS can be readily controlled in a selective fashion (where a particular ion is subjected to fragmentation) by collision-induced dissociation (CID) in tandem mass spectrometry (MS/MS), or nonselectively (where all ions are subjected to fragmentation) by varying the cone (orifice) voltage (in-source CID). Characterization of structures by MS/MS is an important aspect of pharmaceutical analysis. This technique correlates the product ions to preselected precursor ions (19). In tandem MS experiments, the precursor ions of interest are selected in the first mass analyzer according to their mass-to-charge ratios. Then these ions are allowed to dissociate by CID or other methods of activation (photodissociation, etc.), and the resulting fragment ions are characterized by a second mass measurement, usually using another mass analyzer. The advantage of MS/MS experiments is the enhanced specificity, which is very useful in analyzing mixtures.

Common MS/MS instruments in the pharmaceutical laboratory include triple quadrupole mass spectrometers (20) and quadrupole ion-trap mass spectrometers (21,22). To obtain a product ion spectrum with a triple quadrupole mass spectrometer, the precursor ions are mass-selected by the first quadrupole and then made to collide with a collision gas, such as argon, in the second quadrupole. The resulting fragment ions are then analyzed by the third quadrupole. In some cases, a single MS/MS experiment may not be sufficient for structural elucidation. The quadrupole ion trap can be used to perform multiple-stage MS/MS (MS^n) experiments in a timed sequence using only a single mass analyzer. Nearly complete fragmentation pathways can be readily established by this strategy. This capability is invaluable for structural elucidation of complex molecules.

II. CHARACTERIZATION OF SMALL MOLECULES
A. Determination of Molecular Weight

A molecular weight (MW) determination is often the first and most important step in characterizing an organic compound. The ESI process is quite suitable for providing this information, and thus this technique has become an indispensable tool for medicinal chemists for monitoring the progress of chemical reactions and detecting reaction intermediates, targeted compounds, and by-products. These tasks can be accomplished by operating an ESI-MS instrument manually or through automation. With the reliable high throughput technology developed over the past decade, ESI-MS started to play a major role in pharmaceutical research with flow injection analysis (FIA) and LC/MS operations.

Typically, an automated HPLC/ESI-MS system is used in the FIA mode to analyze samples, although a manual direct infusion analysis can also be employed for sample introductions. The flow rate for the direct infusion analysis of samples can vary between 1 and 20 μL/min, and the solvents that are most commonly used include water, acetonitrile, and methanol. Figure 1 shows an example of the FIA-ESI-MS analysis of an organic reaction product. The peak ranging from 0.5 min to 1.0 min in the total ion chromatogram represents the ESI signal of the analyte (Fig. 1A). The corresponding mass spectrum is illustrated in Fig. 1B. The base peak in the spectrum is the protonated molecule $[M + H]^+$ at m/z 517.5, indicating a molecular mass of 516.5 daltons (Da). The spectrum also displays a proton-bound dimer $[2M + H]^+$ of the reaction product at m/z 1034.2, which further confirms the assignment of the molecular weight. The formation of this type of dimeric ions is quite common in ESI-MS; the formation depends on the nature of the compound as well as on the experimental conditions. Proton-bound dimers can

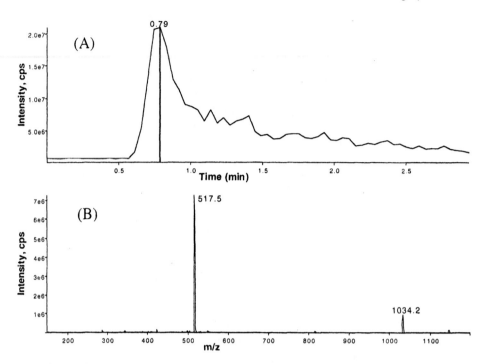

Figure 1 (A) Total ion chromatogram of an organic reaction product determined by FIA-ESI-MS. The experiment was performed on a Sciex API 150 single quadrupole mass spectrometer (cone voltage was at 7 V). (B) The ESI mass spectrum derived from (A) at a retention time of 0.79 min. (Reproduced with permission from Ref. 4.)

be easily distinguished from their covalent analogs on the basis of the observed mass-to-charge ratios. In the example illustrated in Fig. 1, if a covalent noncyclic dimer had been present, it would have been observed at m/z 1032.2 or two mass units lower than the proton-bound analog. Furthermore, noncovalent dimers dissociate readily when the experiment is conducted at a higher cone voltage (the voltage applied between the cone and skimmer of the ESI source); this voltage serves to accelerate precursor ions. It should be noted that very few fragment ions are observed in the spectrum. However, if different ESI conditions are employed, such as a higher cone voltage (orifice voltage), the appearance of the mass spectrum may be different owing to the enhanced fragmentation as a result of larger internal energy deposition in the ions. This is evident by comparing Fig. 1B (7 V cone voltage) and Fig. 2B (60 V cone voltage) for the same sample. The low mass fragment ions are observed at higher abundance with a higher cone voltage,

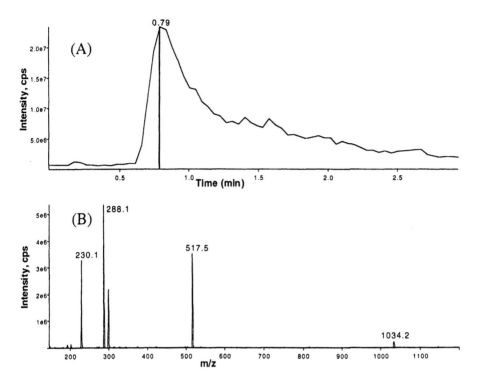

Figure 2 (A) Total ion chromatogram of an organic reaction product determined by FIA-ESI-MS. The experiment was performed on a Sciex API 150 single quadrupole mass spectrometer (cone voltage was at 60 V). (B) The ESI mass spectrum derived from (A) at a retention time of 0.79 min. (Reproduced with permission from Ref. 4.)

whereas the abundances of both the protonated molecule and the proton-bound dimeric ion are substantially decreased (Fig. 2B). This "cone voltage" fragmentation strategy is also useful in obtaining fragment ions that provide useful structural information. However, it should be noted that the use of a lower cone voltage is preferable for the determination of molecular weights of compounds.

B. Multiple Charging

When there are more than one basic site in the analyte, the formation of multiply charged ions can be detected in ESI mass spectra. At a lower cone voltage, the multiply charged ions are present at higher relative abundances. Figure 3 illustrates the relationship between cone voltage and relative abundance of

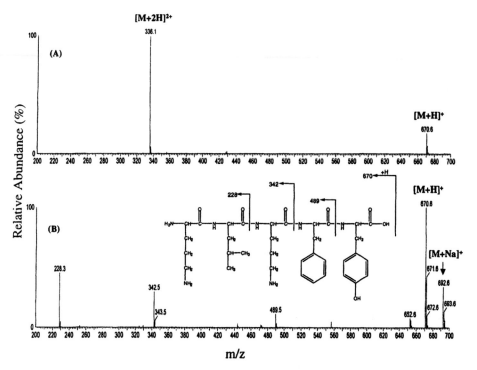

Figure 3 (A) Mass spectrum of an ESI-produced pentapeptide at a cone voltage of 20 V. (B) Mass spectrum of the same pentapeptide at a cone voltage of 80 V.

singly and doubly charged ions for a small peptide. Doubly charged ions of small molecules are intrinsically less stable than their singly charged analogs, and they can easily fragment to form singly charged ions (23). For example, at a cone voltage of 20 V, the singly charged molecular ion $[M + H]^+$ of a pentapeptide is observed at m/z 670.6 at approximately 20% relative abundance, and the doubly charged molecular ion (m/z 336.1) is the most abundant ion in the spectrum (Fig. 3A). When the cone voltage is set at 80 V, the doubly charged ion has disappeared completely, whereas the singly charged molecular ion is most abundant, with some fragment ions at m/z 228, m/z 342, and m/z 489 providing structural information (Fig. 3B). Multiply charged ions (charge state $z \geq 2$) can be readily distinguished from their singly charged analogs ($z = 1$) by the observed mass difference n ($n = 1/z$) between the isotopic peaks, i.e., $n = 1$ Da for $z = 1$, $n = 0.5$ Da for $z = 2$, etc.; however, for small molecules (mass < 2000 Da), z is typically 3 or less.

There are two key factors that affect the ESI-MS response: cone voltage, as discussed above, and choice of solvent. The ionization of molecules tends to be suppressed in the presence of nonvolatile salts (phosphates, sodium dodecylsulfate, etc.), surfactants, and inorganic acids. Ideal solvents are water, acetonitrile, and methanol. Suitable additives for promoting ionization include acetic acid (<1%), formic acid (<1%), ammonium acetate (<50 mM), and trifluoroacetic acid (<0.1%) in the positive ion mode, and ammonium hydroxide (<0.1%) and ammonium acetate (<50 mM) in the negative ion mode. The presence of solvent, additives, and impurities in samples complicates the ESI-MS spectrum, which also often contains adduct ions, such as dimeric ions as discussed above (Fig. 1). Other common adduct ions are complex cations ($[M + NH_4]^+$, $[M + Na]^+$, $[M + K]^+$, etc.), solvent cluster ions ($[M + H_2O + H]^+$, $[M + CH_3CN + H]^+$, $[M + CH_3OH + H]^+$, $[M + Cl]^-$, $[M + CH_3COO]^-$, etc.) and multimer ions ($[2M + H]^+$, $[3M + H]^+$, $[2M + Na]^+$, etc.). For example, Fig. 4 displays the FIA-ESI-MS spectrum of a pharmaceutical compound. Clearly, the spectrum shows the protonated molecule $[M + H]^+$ at m/z 503, along with several other adduct ions, including $[M + Na]^+$ at m/z 525, $[2M + H]^+$ at m/z 1005, and

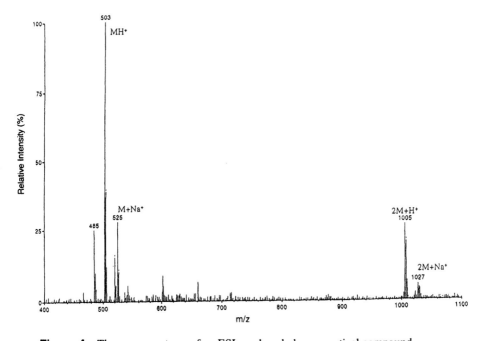

Figure 4 The mass spectrum of an ESI-produced pharmaceutical compound.

[2M + Na]⁺ at m/z 1027. The presence of these adduct ions may complicate the mass spectrum; however, they do provide additional information for the correct assignment of the molecular weight.

C. Salts

For compounds that exist in the form of salts, it may be necessary to perform both positive ion and negative ion ESI-MS experiments to obtain the molecular weight information. Recently, a potent CCR5-specific antagonist was discovered to inhibit R5 HIV-1 replication even at concentrations less than 10 nM (24). This quaternary ammonium salt was analyzed by both positive and negative ion FIA-ESI-MS (Fig. 5). The molecular mass of the cation portion of the molecule was found to be 495 Da (Fig. 5A), and the anion portion of the molecule was determined to contain chlorine atom(s) (Fig. 5B) as suggested by its isotopic pattern (^{35}Cl/^{37}Cl). Further structural confirmation was obtained by MS/MS experiments.

The product ion mass spectrum of the cation portion of this ionic compound (Fig. 6) shows the cleavage of the amide bond, leading to the dominant fragment ion at m/z 261. The other minor product ions observed are also

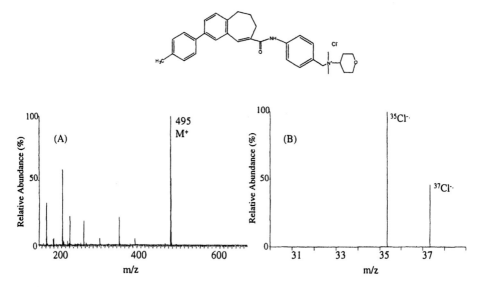

Figure 5 (A) Positive ion mass spectrum of a CCR5-specific antagonist. (B) Negative ion mass spectrum of a CCR5-specific antagonist. Ions were produced by ESI.

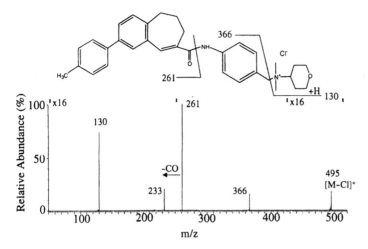

Figure 6 The positive product ion mass spectrum of the cation at m/z 495 produced by ESI.

consistent with the structure (the insert illustrates the fragmentation pattern). It should be noted that the relative abundance for the mass range from 0 to 250 Da has been magnified by a factor of 16 to display the low abundance product ions at m/z 130 and 233. This MS/MS experiment was performed on a triple quadrupole mass spectrometer, which is known for its simplicity and ease of operation as an MS/MS instrument.

D. Liquid Chromatography/Mass Spectrometry

In many cases, the samples of interest are complicated mixtures. The FIA-ESI-MS technique may not be sufficient for the identification of individual components. The preferred analytical approach is to employ HPLC/ESI-MS to identify components based on their retention times in chromatographic separation and molecular weights measured by ESI-MS. For HPLC/ESI-MS analysis with a low flow rate (100–200 μL/min), the sample solution can be sprayed directly into the ESI source. However, most samples from the analytical development and chemical development groups in the pharmaceutical industry require HPLC separations at high flow rates (0.5–2 mL/min). In such cases, a postcolumn split is often used to reduce actual flow rates to the ESI source at 40–200 μL/min. For most samples with low concentrations of organic compounds and biomolecules, HPLC columns with smaller diameters are employed; they have typical flow rates of 1–40 μL/min. Alternatively, a nanoflow device (capillary LC) can be used to deliver the sample solution to a nanospray source for analysis.

Figure 7A illustrates a total ion chromatogram for a 35 min HPLC/ESI-MS analysis of a nine-component mixture. Clearly, all nine components are well separated. The molecular weight information can be obtained from the mass spectrum of each individually selected chromatographic peak. The detected protonated molecules for the main components are indicated in Fig. 7A. When two or more peaks are partially overlapping, molecular ions corresponding to both peaks might be observed. For example, the mass spectrum corresponding to peak 2 shows a protonated molecule at m/z 415, suggesting a molecular mass of 414 Da (Fig. 7B). The presence of its proton-bound dimeric ion at m/z 829 confirms the assignment of the molecular mass. In addition, a lower intensity peak at m/z 329, corresponding to component 3, is also observed. These two

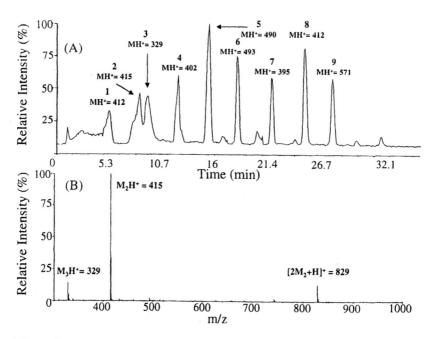

Figure 7 (A) Total ion chromatogram of a nine-component mixture obtained from HPLC/ESI-MS. The experiment was performed on a Sciex API 100 single quadrupole mass spectrometer. The gradient was run from 10% CH_3CN to 90% CH_3CN in 35 min on a C18 column. (B) The mass spectrum corresponding to peak 2 ionized by ESI. (Reproduced with permission from Ref. 4.)

Pharmaceuticals and Natural Products

peaks can be easily distinguished by displaying a selected ion chromatogram. The ability to successfully separate and characterize individual components in complex mixtures has made ESI-MS effective in determining most pharmaceutical compounds. The recent introduction of high throughput analysis by ESI-MS has also made a great impact on combinatorial chemistry (see Chapter 4).

When analyzing pharmaceuticals that exist as salts, such as sodium or potassium phosphates or sulfates, by infusion ESI and HPLC/ESI-MS, one should be aware of the possible proton–metal ion exchange during the chromatographic elution, which can somewhat complicate the molecular weight determination. An example is illustrated in the case of the corticosteroid celestone disodium phosphate. When the sample is analyzed by direct infusion ESI, a protonated molecule $[M+H]^+$ of the drug component is detected at m/z 517 along with a sodiated adduct $[M+Na]^+$ at m/z 539, as shown in Fig. 8A. This mass spectrum is obtained directly from a formulation mixture, which explains

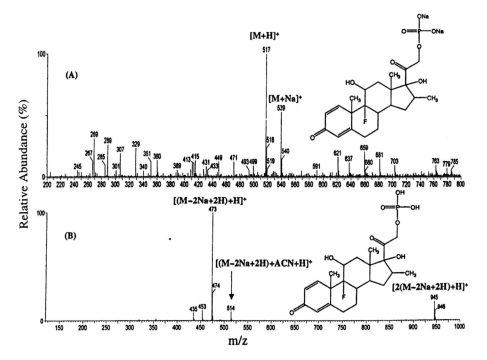

Figure 8 (A) Direct infusion ESI-MS of celestone phosphate. (B) Mass spectrum of celestone phosphate separated by HPLC and ionized by ESI.

the presence of additional peaks at both lower and higher masses; many of them correspond to sodium adduct ions. For example, the ions at m/z 329 and 681 are sodium adducts for ions at m/z 307 and 659, respectively. Such infusion analysis quickly confirms the presence of the main drug substance when it is not necessary to isolate off-line the component of interest. For a more detailed analysis, the main component can be easily separated from the formulation additives by using an HPLC column. When the sample is analyzed by HPLC/ESI-MS under acidic pH conditions, the two sodium cations associated with the phosphate group are readily exchanged with two protons during the chromatographic separation, generating an $[M - 2Na + 2H + H]^+$ ion at m/z 473 (Fig. 8B). Further confirmation of the molecular weight assignment is obtained from the detection of an acetonitrile adduct (m/z 514) and the proton-bound dimer (m/z 945).

E. Derivatization and Atmospheric Pressure Chemical Ionization

One of the drawbacks in ESI-MS is that it cannot be directly applied for the analysis of nonpolar compounds. This is due to an intrinsic characteristic of ESI-MS: Ionization is most effective for compounds that exist as ions in solution (4). This limitation, however, can be overcome by simple chemical derivatization, where an ionizable group is covalently attached to the nonpolar compound of interest. Nonpolar or moderately polar species such as cholesterol, lanosterol, and ergosterol do not produce molecular ions under standard ESI conditions (positive or negative ion mode) and therefore cannot be detected. Such nonpolar sterols play a major role in biochemistry. The traditional analytical methods, such as sterol color tests and thin-layer chromatography, do not readily distinguish among different sterols and often lack specificity and sensitivity. In such cases, mass spectrometric analysis becomes very important and can offer superior accuracy and specificity. One approach for analyzing cholesterol by ESI-MS involves a simple chemical derivatization performed by reacting cholesterol with a pyridine sulfur trioxide complex (25). The obtained sulfate derivative exhibits high ionization efficiency in the negative ion mode and can be easily determined (25,26). Figure 9 shows the ESI-MS spectrum of a derivatized cholesterol (cholesterol 3-sulfate) with an intense negative ion signal $[M - H]^-$ at m/z 465 (26). Other ionization techniques have also been used to obtain similar results with nonpolar compounds. In addition to the traditionally used electron ionization (EI) and chemical ionization (CI), atmospheric pressure chemical ionization (APCI) (27,28) responds well to most moderately polar or nonpolar compounds and generates CI-like spectra. Unlike ESI, the key components in APCI include a heated nebulizer and a corona discharge needle that

Pharmaceuticals and Natural Products

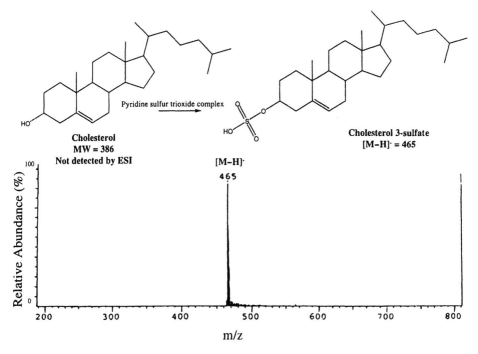

Figure 9 Negative ion mass spectrum of a derivatized cholesterol ionized by ESI. (Reproduced with permission from Ref. 26.)

is maintained at a higher voltage (5–6 kV). Ion–molecule reactions often play important roles in ionization processes in APCI.

F. Tandem Mass Spectrometry

Structural characterization of complex mixtures in pharmaceutical analysis often relies on HPLC/ESI-MS/MS techniques that combine the separation capability of HPLC with the mass fragmentation ability of tandem mass spectrometry. Some specific applications include identification of impurities and degradation products in bulk drug substances and determination of metabolites in drug metabolism and pharmacokinetcs studies. The discussion on drug metabolism and pharmacokinetics is a subject of Chapter 5 in this book.

The identification of impurities and degradation products in bulk drug substances is a requirement in submission of new drug applications (NDAs) to the regulatory agencies. Typically, all components present in >0.1% of the parent drug substance must be fully characterized. The advantage of using LC/MS and LC/MS/MS for structural characterization of components in a mixture is the elimination of time-consuming isolation and purification steps. Accurate structural information for minor components in complex mixtures can be quickly obtained. Figure 10A displays a total ion chromatogram of a steroid, mometasone furoate, with peak D representing the main drug component. This steroid is the active ingredient in a drug substance used for treating seasonal allergies. The mass spectrum of the minor component peak C is illustrated in Fig. 10B. The spectrum exhibits two coeluting components, displaying their protonated molecules at m/z 535 and 581, respectively. The presence of ammoniated adduct (m/z 552, 598) and sodiated adduct (m/z 557, 603) in the mass spectrum further confirms their molecular weights as 534 Da and 580 Da, respectively. The LC/MS/MS analysis on the protonated molecule at m/z 535 shows an abundant product ion at m/z 135 (Fig. 11A), whereas the product ion spectrum of the protonated mometasone furoate (m/z 521) obtained by

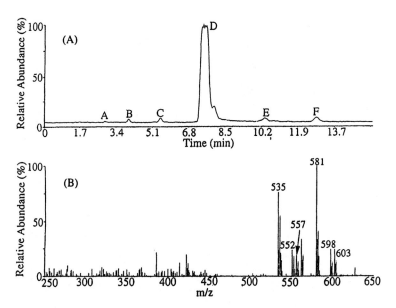

Figure 10 (A) Total ion chromatogram of the steroid drug mometasone furoate (Schering-Plough) obtained from HPLC/ESI-MS analysis. Peak D corresponds to molecular ions of the parent drug compound. (B) Mass spectrum of the component corresponding to peak C, indicating two coeluting components with molecular ions $[M + H]^+$ at m/z 535 and 581, respectively. (Reproduced with permission from Ref. 4.)

Pharmaceuticals and Natural Products

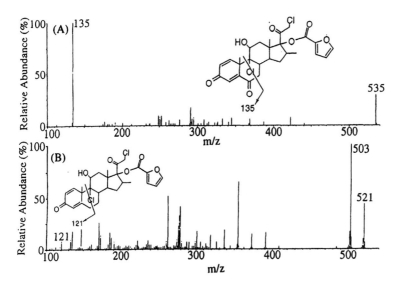

Figure 11 (A) The product ion mass spectrum obtained from dissociation of molecular ions at m/z 535 (impurity component), indicating an abundant product ion at m/z 135. (B) The product ion mass spectrum obtained from dissociation of molecular ions at m/z 535 for momentasone furoate. The components were separated by on-line HPLC and ionized by ESI. (Reproduced with permission from Ref. 4.)

LC/MS/MS displays a less abundant product ion at m/z 121 (Fig. 11B). A comparison of the fragmentation pattern of the mometasone furoate and the impurity indicates the presence of a 6-keto function in the structure of the impurity. This example demonstrates the power of LC/MS/MS in providing rapid structural characterization of low level, coeluting components in bulk drug preparations.

III. STRUCTURAL ELUCIDATION OF ANTIBIOTICS AND NATURAL PRODUCTS

A. Antibiotics

A major part of the drug discovery process is focused on natural products, and many of the currently available drugs are based on such compounds. For instance, over 60% of all approved and pre-NDA candidates are either natural products or synthetic molecules based on known natural products. Most natural product extracts are mixtures that require chromatographic separation prior to MS analysis, and ESI-MS is especially suitable for this task because of its compatibility with HPLC. In this section we discuss the applications of ESI-MS for the discovery and development of these compounds as therapeutic agents; we include major classes of clinically used antibiotics such as quinolones, β-lactams, macrolides, and aminoglycosides.

Quinolones are a group of highly potent broad-spectrum synthetic antibiotics. The common structural feature among them is the 4-oxo-1,4-dihydroquinoline group. For example, sarafloxacin is used to treat respiratory and enteric bacterial infections in animals (29). Its structure and mass spectrum are shown in Fig. 12A. The protonated molecule $[M + H]^+$ at m/z 386 readily fragments to give structurally informative product ions in the MS/MS experiment of a triple quadrupole mass spectrometer (Fig. 12A). The loss of H_2O and CO_2 from the ion at m/z 386 leads to the ions at m/z 368 and 342, respectively. The further loss of HF (20 Da) from the ion at m/z 342 generates the ion at m/z 322. The ring opening and expulsion of C_2H_5N from the ion at m/z 342 yields the ion at m/z 299. The fragment ion at m/z 368 undergoes further fragmentation to yield more structural information, and this information is found by employing a quasi-MS^3 technique in which a higher cone voltage (95 V in this case) is used (Fig. 12B) (30). For example, the loss of the dehydrated diethylenediamine ring ($-C_4H_2N_2$) from the ion at m/z 348 gives the product ion at m/z 270. The data demonstrate the use of multiple-stage MS/MS (MS^n) for the structural elucidation of complex molecules.

β-Lactams such as penicillins and cephalosporins are perhaps the most widely used antibiotics against both gram-positive and gram-negative bacteria (31). As a representative example of the application of ESI-MS in β-lactams, the structures and the negative ion fragmentation pathway of dicloxacillin are shown in Scheme 1. This β-lactam is present in the form of a salt, and negative ion ESI-MS shows product ions of an intense negative ion signal at m/z 468. This ion can readily lose CO_2 to give the ion at m/z 424. Further loss of C_5H_7NO (97 Da) via opening of the lactam ring leads to the ion at m/z 327. The rearrangement of the ion at m/z 327 with the loss of HCl generates the ion at m/z 291. The sequential MS^n studies on an ion trap mass spectrometer are summarized in Fig. 13 (32). The MS^n studies provide evidence for the proposed fragmentation pathways.

Another example of the application of MS in antibiotics involves the macrolide tylosin. It is composed of a 16-membered oxygenated lactone ring attached to three carbohydrate residues. The structure of tylosin A is shown in Scheme 2. Collisional activation of the protonated tylosin A molecule at m/z

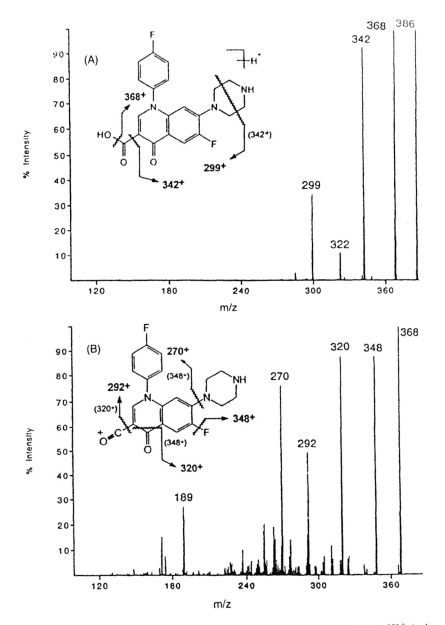

Figure 12 Product ion spectra of sarafloxacin. (A) Product ions of $[M+H]^+$ (m/z 386) at a collision energy of 30 V and a cone voltage of 35 V. (B) Product ions of $[M+H-H_2O]^+$ (m/z 368) at a collision energy of 40 V and a cone voltage of 95 V. The elevated cone voltage induces the first stage of fragmentation. (Reproduced with permission from Ref. 30.)

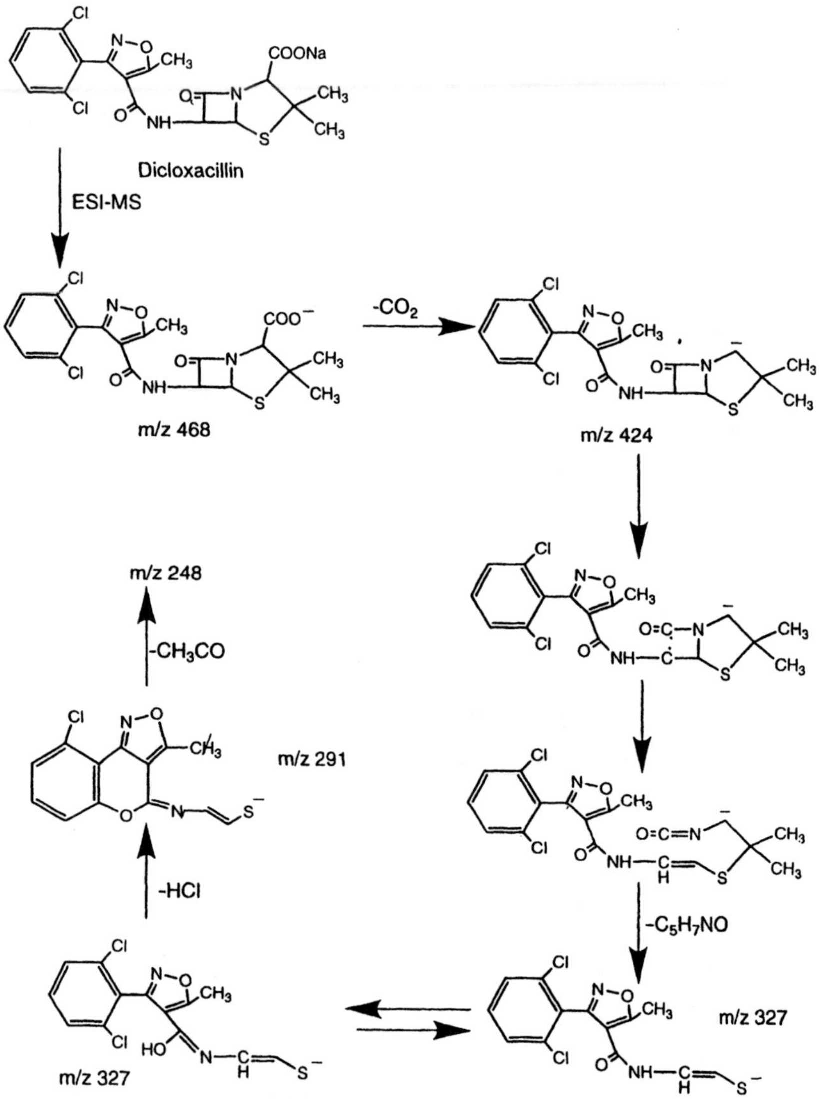

Scheme 1 Fragmentation pathways of dicloxacillin. (Reproduced with permission from Ref. 32.)

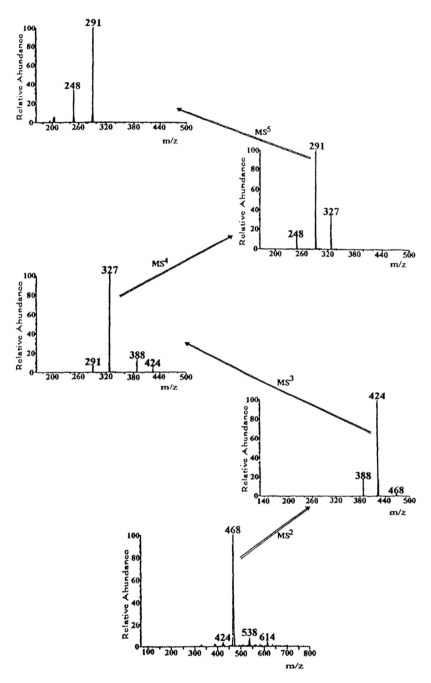

Figure 13 The product ion spectra by MS^n obtained for dicloxacillin in the negative ion mode. (Reproduced with permission from Ref. 32.)

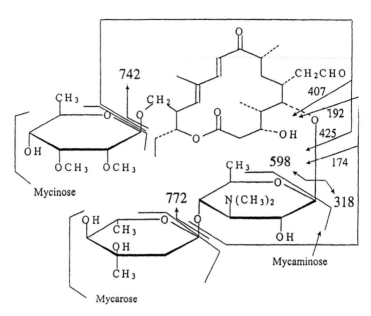

Scheme 2 Fragmentation pathway of tylosin A.

916.5 leads to a product ion at m/z 772 from the loss of mycarose (Fig. 14A), along with other product ions at m/z 898 (loss of H_2O), 754 (loss of H_2O from ions at m/z 772), 742 (loss of mycinose), 598/318 (cleavage of aglycon–desoamine glycosidic bond), 407/425 (charged aglycone part), 389 (loss of H_2O from ions at m/z 407), and 371 (loss of $2H_2O$ from ions at m/z 407). Further fragmentation of the ion at m/z 772 results in a number of product ions (Fig. 14B), including two new ions at m/z 174 and 192. They are likely to derive from cleavage of the glycosidic bond between the aglycone and amino sugar at both sides of glycosidic oxygen (Scheme 2). The observed fragmentation schemes are similar to an early study by fast atom bombardment (FAB) MS and MS/MS (33).

B. Use of [M + H]⁺ and [M + Na]⁺ for Structural Elucidation

As discussed in previous sections, the formation of metal complex ions in ESI-MS is often observed for some classes of compounds, such as aminoglycoside antibiotics (32–36). In early studies of aminoglycosides, EI and CI

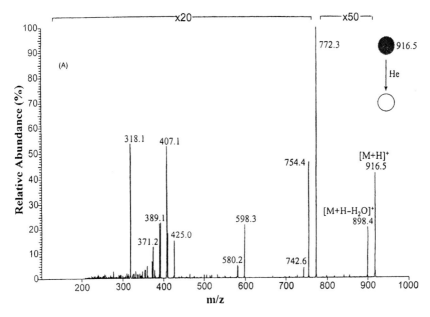

Figure 14 (A) Product ion mass spectrum of ions at m/z 916.5 in an ion trap mass spectrometer.

techniques were used to characterize various members of this class of compounds (35). Although the mass spectra displayed low abundance molecular ions, they often contained extensive fragment ions that are useful for structural determination (35). Early studies by Gross, Pramanik, and their coworkers also showed informative fragmentations for alkali metal ions of macrolides by FAB-MS/MS (33). Figure 15A exhibits the ESI-MS spectrum of the aminoglycoside antibiotic gentamicin B (dissolved in methanol), showing abundant molecular ions with very limited or no fragmentation. The dominant ion in this spectrum is the sodiated molecule $[M + Na]^+$ at m/z 505. The major fragmentation patterns for both protonated molecules (Fig. 15B) and sodiated molecules (Fig. 15C) involve cleavage of sugar rings A or C, or both. For example, the product ion at m/z 324 from protonated gentamicin B is a result of the loss of ring C (Fig. 15B), whereas the loss of ring C from sodiated gentamicin B gives ions at m/z 346 (Fig. 15C). In the case of the protonated molecule, an apparent cleavage within sugar ring C is also evident (m/z 366). The fragmentation patterns for the sodiated molecule suggest that the sodium cation is most likely located in the region of ring B. Coordination of metal ions with organic compounds alters the

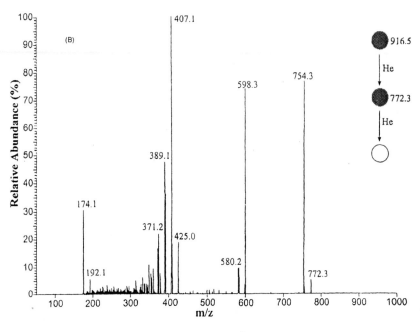

Figure 14 (B) Product ion mass spectrum (MS^3) of ions at *m/z* 772 derived from ions at *m/z* 916.5.

stability of product ions. Thus, different fragmentation pathways are likely to result from such metal complexes. This could be a potentially useful technique for probing structures of such compounds.

Another example showing different fragmentation patterns of protonated and sodiated molecules is demonstrated for actinonin, a natural product that acts as an inhibitor of aminopeptidase M and leucine aminopeptidase. The ESI mass spectrum of actinonin displays a protonated molecule [M + H]$^+$ at *m/z* 386 and a sodiated adduct [M + Na]$^+$ at *m/z* 408 (Fig. 16A). The MS/MS fragmentation patterns for the protonated and sodiated molecules are illustrated in Figs. 16B and 16C, respectively. These two product ion spectra differ significantly. As discussed above, the location of the charge carrier (H$^+$ or Na$^+$) affects the fragmentation pattern. The major cleavage site for the protonated ion is the amide bond between the proline (Pro) and valine (Val) amino acid residues, generating the ions at *m/z* 102 and 285, respectively. The relatively high abundance of the product ion at *m/z* 102 suggests that the proton is most likely located at the piperidine nitrogen atom. For the sodiated molecule at *m/z* 408, the cleavages occur at amide bonds (Fig. 16C).

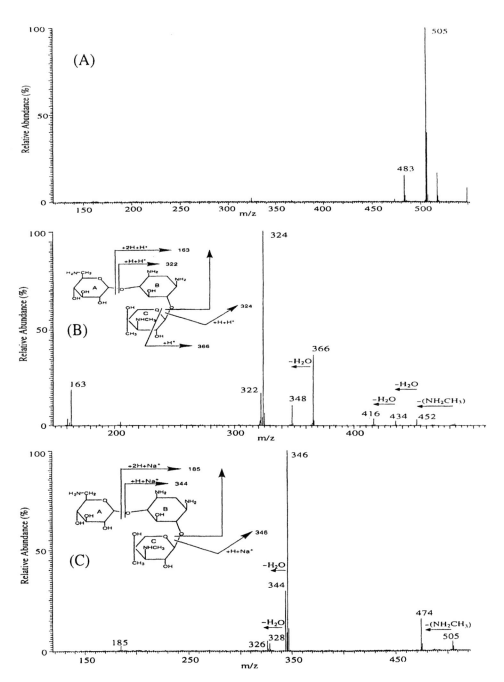

Figure 15 (A) Mass spectrum of gentamicin B dissolved in methanol. (B) Product ion mass spectrum of protonated gentamicin B molecule at m/z 483 in an ion trap mass spectrometer. (C) Product ion mass spectrum of sodiated gentamicin B molecule at m/z 505 in an ion trap mass spectrometer. All precursor ions were produced by ESI.

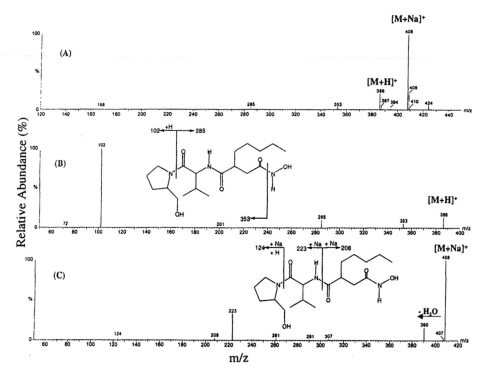

Figure 16 (A) Mass spectrum of actinonin. (B) Product ion mass spectrum of the protonated molecule at *m/z* 386. (C) Product ion mass spectrum of the sodiated molecule at *m/z* 408. All precursor ions were produced by ESI.

When the sodium cation is associated with the Pro-Val fragment, the ion at *m/z* 223 is generated, and the rest of the molecule is lost. Alternatively, when the sodium ion is attached to the right-hand side of the molecule, the ion at *m/z* 208 is generated, and the Pro-Val residues are lost as neutrals. The different fragmentation patterns resulting from different precursor ions provide complementary structural information, which can be valuable for structural elucidation of unknown compounds.

C. Mass Spectra of Analogs

Most natural products exhibit complex structures, and their exact stereochemistry is usually determined by using NMR, MS, and X-ray crystallography. However, once the structure of a member of the group of natural products is determined, the use of MS becomes the method of choice in determining the structures of related members of the group. This is very clearly demonstrated in the case of everninomicins.

Everninomicin belongs to a new class of oligosaccharide antibiotics obtained by the fermentation of *Micromonospora carbonacea* (37). They are highly active against gram-positive bacteria, which are sensitive to and resistant to methicillin and vancomycin. The structure of Sch 27899 contains eight sugar backbones, two ortho ester functionalities (rings C and H), a methylene dioxy group (ring J), a nitro sugar (ring K), and two substituted aromatic ester residues (terminal groups 1 and 2) (Scheme 3). Early structural characterization work of other members of the everninomicin class of antibiotics was based on FAB-MS (38–41), establishing fragmentation patterns. However, when only small amounts of material are available or separation is required, the ESI-MS and ESI-MS/MS techniques are better suited for providing detailed structural information.

The positive ion ESI-MS of Sch 27899 shows an abundant precursor ion at m/z 1652, corresponding to the sodiated molecule $[M + Na]^+$ (Fig. 17). The peak at m/z 1670 corresponds to the sodium adduct of the Sch 27899 hydrolysis product $[M + Na + H_2O]^+$. The abundant ions at m/z 935 and 957 are generated by the cleavage of ortho ester ring C. Although many fragment ions are observed in the mass spectrum, most of them are not sequence-specific. As a result, the interpretation of the mass spectrum becomes complicated. The issue about the location of sodium cation in the metal complex ion is fundamentally important in understanding the fragmentation mechanism of this sodiated complex ion. The ESI-MS^n data obtained with an ion trap mass spectrometer on the sodiated Sch 27899 molecule offer detailed fragmentation patterns. One particular fragmentation pathway is displayed in Fig. 18, where the sodiated Sch 27899 complex ions undergo sequential CID under multiple collision conditions in the ion trap. This six-stage mass analysis (MS^6) suggests that the location of sodium cation in the complex ion is between rings F and G, as indicated by the appearance of frag-

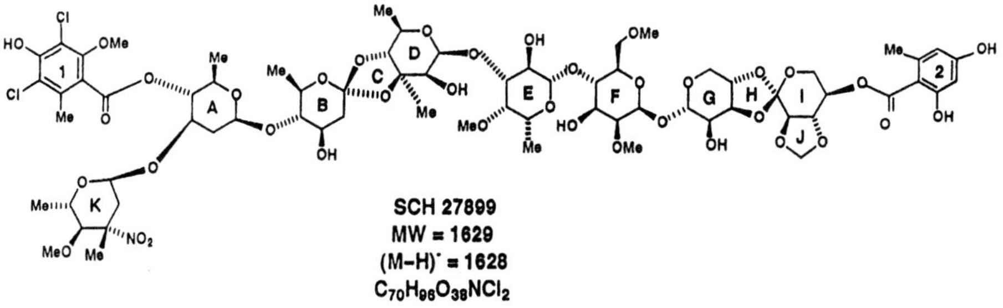

SCH 27899
MW = 1629
$(M-H)^- = 1628$
$C_{70}H_{98}O_{38}NCl_2$

Scheme 3 Structure of Sch 27899.

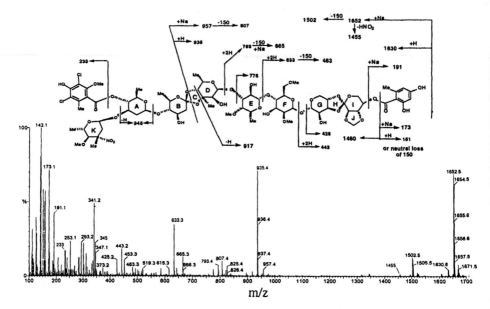

Figure 17 The positive ion mass spectrum of Sch 27899 dissolved in acetonitrile and ionized by ESI.

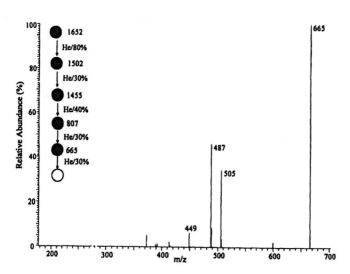

Figure 18 The positive product-ion mass spectrum of ions at *m/z* 665 derived from the sodiated Sch 27899 molecule in an MS^6 experiment.

Pharmaceuticals and Natural Products

ment ions at m/z 505 (rings E, F, and G and partial ring H), 487 (rings F, G, H, I, J), and 449 (rings E and F and partial ring G). This information is useful for the identification of structures of other everninomicin compounds.

D. Negative Ions

In many cases, negative ion MS displays a different pattern of fragmentation, and it can be complementary to the positive ion data. This becomes especially useful when the analyte is of unknown structure and its response to positive ion or negative ion ionization cannot be predicted. In the case of Sch 27899, negative ion ESI-MS exhibits a simpler and more informative mass spectrum (Fig. 19). The deprotonated molecule $[M - H]^-$ at m/z 1628 fragments to yield the ion at m/z 1478. Several ions at m/z 1646, 1496, 1291, and 712 result from fragmentation of the hydrolysis product of Sch 27899. The cleavages of the ester and ether sugar linkages generate the ions at m/z 1186, 1014, 854, 694, 548, and 249. The loss of an HNO_2 unit from the fragment ions at m/z 1478, 694, and

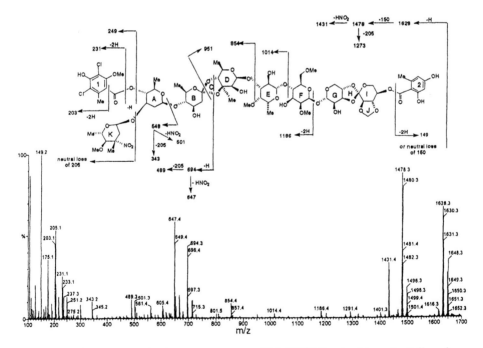

Figure 19 The negative ion mass spectrum of Sch 27899 dissolved in acetonitrile and ionized by ESI.

548 generates the ions at m/z 1431, 647, and 501, respectively. The terminal aromatic group 2 is defined by the ion at m/z 149. The comparison between positive and negative ion ESI-MS for Sch 27899 illustrates that the negative ion spectrum provides more structurally informative information for the elucidation of the carbohydrate residues in this complex molecule. In general, the negative ion ESI-MS spectrum shows much less background interference (i.e., alkali metal adducts).

E. Exact-Mass Measurement

With appropriate MS instrumentation (sector, time-of-flight), high resolving power accurate mass measurements of components separated on-line with HPLC and ionized by ESI can provide elemental compositions of the analytes. This information is used to assist the structural identification process. Figure 20 shows a negative ion high resolution HPLC/ESI-MS reconstructed ion chromatogram (RIC) of a mixture containing the degradation products of Sch 27899 and impurities present in the fermentation broth

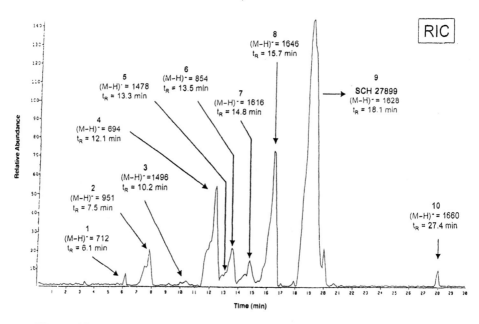

Figure 20 Reconstructed ion chromatogram on Sch 27899 mixtures separated by HPLC and monitored under exact-mass measurement conditions.

(42). High resolution HPLC/ESI-MS becomes especially important when minor components can be identified directly from the mixture without prior separation and isolation. The data shown in Fig. 20 were obtained on a double-focusing sector instrument (JEOL MStation) with a resolving power of 5000 (10% valley). The calibration was done with PEG sulfates as internal standards. The proposed elemental compositions were measured with an experimental error of less than 2 ppm, as shown in Table 1. The structures of 10 mixture components (degradation products) were identified based on the empirical formulas and established fragmentation pattern of Sch 27899 (Scheme 4).

As illustrated above, exact-mass measurements in ESI-MS are invaluable tools for the identification of structures of unknown compounds. This is exemplified in the elucidation of the structure of a new cerebroside isolated from the methanol extract of the red alga *Amphiroa fragilissima*. A sample of this natural product isolated from the natural sources was shown to have strong antiviral activities. Cerebrosides belong to a class of compounds called sphingolipids, in which the terminal hydroxy group (C-1) of sphingosine is linked to a single sugar residue, either glucose or galactose (43). Sphingolipids usually have a long-chain fatty acid group attached to the amino group at C-2, to form an amide. The most common sphingolipids are formed from sphingosine, an amino glycol with a C_{18} aliphatic backbone that includes a trans double bond

Table 1 Exact-Mass Measurements for Mixture Components 1–10[a]

No.	t_R (min)	$(M-H)^-$	Elemental composition	Theoretical value	Experimental value	Δ (ppm)
1	6.1	712[b]	$C_{29}H_{40}O_{15}NCl_2$	714.1733(^{37}Cl)	714.1746	1.8
2	7.5	951	$C_{41}H_{59}O_{25}$	951.3345	951.3353	1.1
3	10.2	1496	$C_{62}H_{92}O_{36}NCl_2$	1496.4776	1496.4781	0.4
4	12.1	694	$C_{29}H_{38}O_{14}NCl_2$	694.1669	694.1678	1.3
5	13.3	1478	$C_{62}H_{90}O_{35}NCl_2$	1478.4670	1478.4689	1.2
6	13.5	854	$C_{36}H_{50}O_{18}NCl_2$	854.2405	854.2414	1.1
7	14.3	1616	$C_{69}H_{96}O_{38}NCl_2$	1616.4957	1616.4987	1.9
8	15.7	1646	$C_{70}H_{98}O_{39}NCl_2$	1646.5093	1646.5082	0.7
9	18.1	1628	$C_{70}H_{96}O_{38}NCl_2$	1628.4987	1628.4999	0.7
10	27.4	1660	$C_{71}H_{100}O_{39}NCl_2$	1660.5250	1660.5262	0.8

[a] See Scheme 4.
[b] The most abundant ion was observed at m/z 712 (^{35}Cl). However, due to reference interference at m/z 712, the exact-mass measurement was performed on the peak at m/z 714 (^{37}Cl).

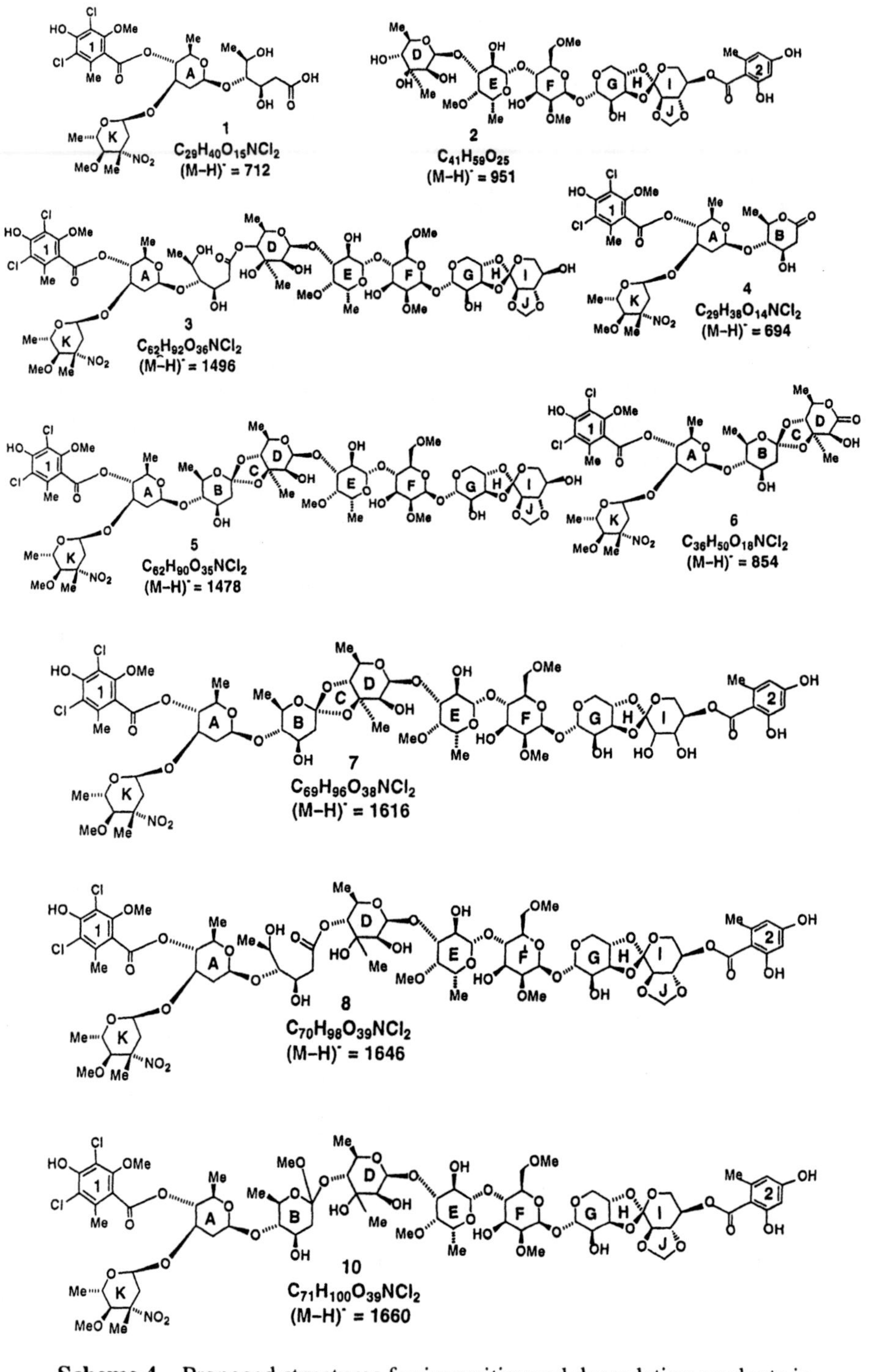

Scheme 4 Proposed structures for impurities and degradation products in Sch 27899 mixtures.

between C-4 and C-5. Structural characterizations of another related class of compounds, namely sphingosine long-chain based phospholipid (sphingomyelin), have been shown in the literature by tandem mass spectrometry with ESI (44).

The structure of this novel cerebroside is illustrated in Scheme 5. The positive ion ESI-MS spectrum of this extract shows an abundant sodiated molecule $[M+Na]^+$ at m/z 879 (Fig. 21A). Exact-mass measurement under high resolving power of this sodiated molecule established the elemental composition to be $C_{49}H_{93}NO_{10}Na$ (within 1 ppm error of calculated value). The collisional activation of the sodiated molecule at m/z 879 generates several product ions (Fig. 21B). The loss of H_2O from the sodiated molecule yields the product ion at m/z 861, whereas the loss of a sugar group ($C_6H_{10}O_5$) leads to product ions at m/z 717 and 699 (further loss of H_2O from ions at m/z 717). The fragment ion at m/z 510 is a result of the loss of H_2O from the ion at m/z 528,

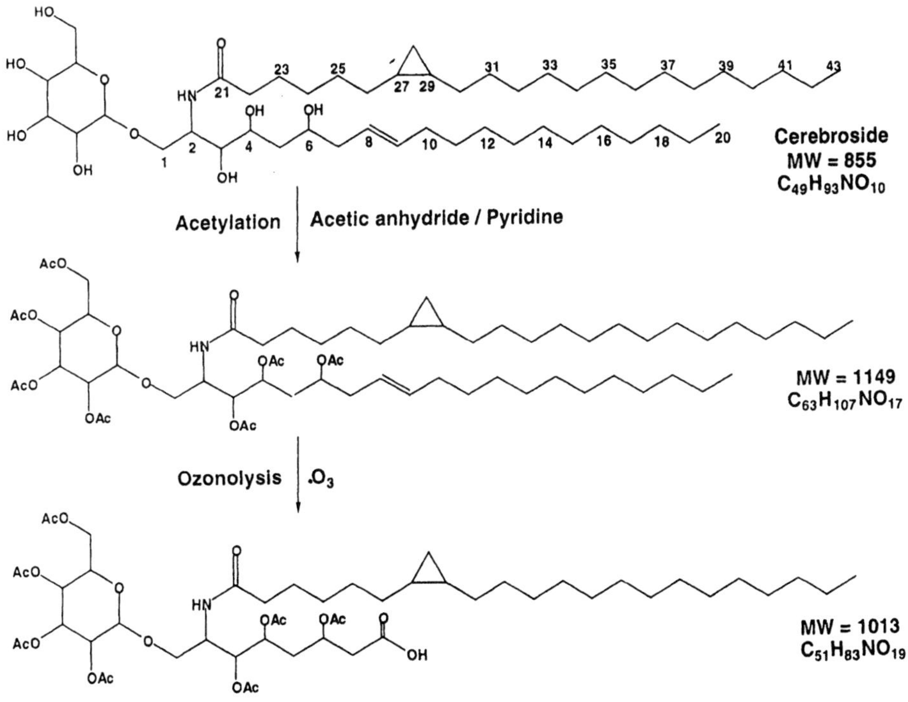

Scheme 5 Structure of cerebroside and chemical modification of cerebroside.

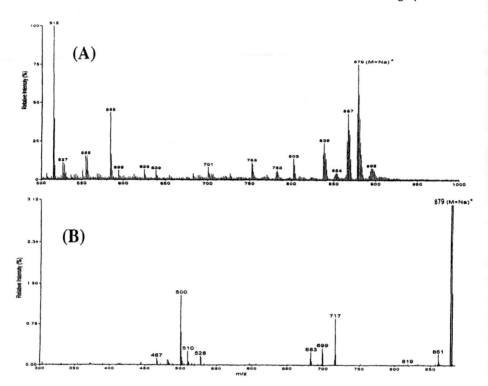

Figure 21 (A) The positive ion mass spectrum of cerebroside. (B) The positive product ion mass spectrum of the sodiated molecule at m/z 879 from cerebroside. Ionization was by ESI.

which in turn is derived from the loss of the N-acyl group ($C_{23}H_{43}NO$) of the precursor sodiated molecule. The loss of a C_2H_4 unit from the ion at m/z 528 likely results in the abundant ion at m/z 500.

To firmly establish the structural assignment of this novel cerebroside, chemical modifications were carried out, as shown in Scheme 5. The extract was first acetylated to form the heptaacetate, the composition of which was derived by exact-mass measurements. The heptaacetate was further treated with ozone to identify the position of the double bond. The ozonolysis reaction led to cleavage of the double bond and formation of an acidic group at C-8. Exact-mass measurements supported this proposed structure by giving an elemental composition of $C_{51}H_{83}NO_{19}Na$ (m/z 1036). The sodiated molecule [M + Na]$^+$ at m/z 1036 fragments under MS/MS conditions to yield several product ions as results of the loss of acetic acid and acetyl groups (Fig. 22). A series of ions at m/z 976, 916, and 856 are produced via consecutive losses of acetic acid molecules from the sodiated molecule. These ions further lose CO_2 to yield ions at m/z 932, 872, and 812. The cleavage of the C–O bond connecting the sugar ring to the rest of the molecule leads to product ions at m/z 371 (sugar ring) and 688. The further loss of acetic acid molecules from ions at m/z 371 gives ions at m/z 311 and 251. The MS data in conjunction with chemical modifications clearly indicate that the double bond is located between C-8 and C-9. From MS data and further NMR studies, the structure of this cerebroside extract is determined to contain a cyclopropane ring with a unique hydroxylation pattern (positions 1,3,4, and 6) and a trans double bond (positions 8 and 9) (Scheme 5) (45).

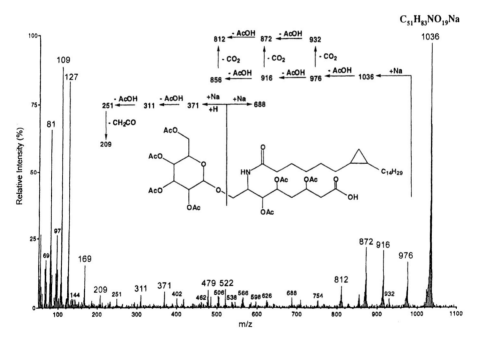

Figure 22 The positive product ion mass spectrum of the sodiated molecule at m/z 1036 from the ozonolysis product of the acetylated cerebroside. Ionization was by ESI

IV. CONCLUSIONS

The structural characterization of biologically active molecules as well as the identification and quantification of impurities and degradation products present in bulk drug substances are integral parts of pharmaceutical research and development. ESI has become the method of choice in mass analysis in the pharmaceutical industry because it has high sensitivity, soft ionization, and compatibility with separation techniques. The ESI-MS technique can rapidly provide accurate molecular weight information on the major components in samples, and this information can be used for monitoring chemical reactions in organic synthesis for confirming final reaction products, for determining intermediates, and for identifying major impurities and by-products. The HPLC/ESI-MS and MS/MS techniques are valuable tools for detection and identification of minor components in complex mixtures without the need of prior separation or isolation. In this chapter we have illustrated the power of ESI-MS using examples based on analytical research on various classes of drug molecules. It is our expectation that ESI-MS will continue to play an important role in the pharmaceutical industry.

References

1. RG Cooks, G Chen, P Wong, H Wollnik. Mass spectrometers. In: GL Trigg, Ed. Encyclopedia of Applied Physics. New York: VCH, 1997, Vol 19, pp 289–330.
2. JT Watson. Introduction to Mass Spectrometry. 2nd ed. New York: Raven Press, 1985.
3. JB Fenn, M Mann, CK Meng, SF Wong, CM Whitehouse. Electrospray ionization for mass spectrometry of large biomolecules. Science 246:64–71, 1989.
4. BN Pramanik, PL Bartner, G Chen. The role of mass spectrometry in the drug discovery process. Curr Opin Drug Discov Dev 2:401–417, 1999.
5. Y Hua, RB Cole. Electrospray tandem mass spectrometry for structural elucidation of protonated brevetoxins in red tide algae. Anal Chem 72:376–383, 2000.
6. K Vishwanathan, MG Bartlett, JT Stewart. Determination of antimigraine compounds rizatriptan, zolmitriptan, naratriptan and sumatriptan in human serum by liquid chromatography/electrospray tandem mass spectrometry. Rapid Commun Mass Spectrom 14:168–172, 2000.
7. JP Antignac, B Le Bizec, F Monteau, F Poulain, F Andre. Collision-induced dissociation of corticosteroids in electrospray tandem mass spectrometry and development of a screening method by high performance liquid chromatography/tandem mass spectrometry. Rapid Commun Mass Spectrom 14:33–39, 2000.
8. N Van Eckhout, JC Perez, J Claereboudt, R Vandeputte, C Van Peteghem. Determination of tetracyclines in bovine kidney by liquid chromatography/

tandem mass spectrometry with on-line extraction and clean-up. Rapid Commun Mass Spectrom 14:280–285, 2000.
9. BP Lau, PM Scott, DA Lewis, SR Kanhere. Quantitative determination of ochratoxin A by liquid chromatography/electrospray tandem mass spectrometry. J Mass Spectrom 35:23–32, 2000.
10. JB Lee, T Hayashi, K Hayashi, U Sankawa. Structural analysis of calcium spirulan (Ca-SP)-derived oligosaccharides using electrospray ionization mass spectrometry. J Nat Prod 63:136–138, 2000.
11. W Lang, J Mao, Q Wang, C Niu, TW Doyle, B Almassian. Isolation and identification of metabolites of porfiromycin formed in the presence of a rat liver preparation. J Pharm Sci 89:191–198, 2000.
12. K Yamaguchi, M Takashima, T Uchimura, S Kobayashi. Development of a sensitive liquid chromatography-electrospray ionization mass spectrometry method for the measurement of KW-5139 in rat plasma. Biomed Chromatogr 14:77–81, 2000.
13. K Matsuura, K Murai, Y Fukano, H Takashina. Simultaneous determination of tiopronin and its metabolites in rat blood by LC-ESI-MS-MS using methyl acrylate for stabilization of thiol group. J Pharm Biomed Anal 22:101–109, 2000.
14. DR Doerge, HC Chang, MI Churchwell, CL Holder. Analysis of soy isoflavone conjugation in vitro and in human blood using liquid chromatography-mass spectrometry. Drug Metab Dispos 28:298–307, 2000.
15. WMA Niessen. Advances in instrumentation in liquid chromatography-mass spectrometry and related liquid-introduction techniques. J Chromatogr A 794:407–435, 1998.
16. WMA Niessen. Analysis of antibiotics by liquid chromatography-mass spectrometry. J Chromatogr A 812:53–75, 1998.
17. GA Valaskovic, NL Kelleher, DP Little, DJ Aaserud, FW McLafferty. Attomole-sensitivity electrospray source for large-molecule mass spectrometry. Anal Chem 67:3802–3805, 1995.
18. GA Valaskovic, NL Kelleher, FW McLafferty. Attomole protein characterization by capillary electrophoresis-mass spectrometry. Science 273:1199–1202, 1996.
19. KL Busch, GL Glish, SA McLuckey. Mass Spectrometry/Mass Spectrometry: Techniques and Applications of Tandem Mass Spectrometry. New York: VCH, 1988.
20. RA Yost, CG Enke. Selected ion fragmentation with a tandem quadrupole mass spectrometer. J Am Chem Soc 100:2274–2275, 1978.
21. RE March, JFJ Todd. Practical Aspects of Ion Trap Mass Spectrometry. New York: CRC Press, 1995.
22. RG Cooks, G Chen, C Weil. Quadrupole mass filters and quadrupole ion traps. In: RM Caprioli, A Malorni, G Sindona, eds. Selected Topics in Mass Spectrometry in the Biomolecular Sciences. Dordrecht: Kluwer Academic, 1997, Ser C, Vol 504, pp 213–238.

23. AR Katritzky, PA Shipkova, RD Burton, SM Allin, CH Watson, JR Eyler. Investigation of doubly charged organic cations by electrospray ion cyclotron resonance mass spectrometry. J Mass Spectrom 30:1581–1587, 1995.
24. M Baba, O Nishimura, N Kanzaki, M Okamoto, H Sawada, Y Iizawa, M Shiraishi, Y Aramaki, K Okonogi, Y Ogawa, K Meguro, M Fujino. A small-molecule, non-peptide CCR5 antagonist with highly potent and selective anti-HIV-1 activity. Proc Natl Acad Sci USA 96:5698–5703, 1999.
25. R Sandhoff, B Brügger, D Jeckel, WD Lahmann, FT Wieland. Determination of cholesterol at the low picomole level by nano-spray ionization tandem mass spectrometry. J Lipid Res 40:126–132, 1999.
26. K Metzger, PA Rehberger, G Erben, WD Lehmann. Identification and quantification of lipid sulfate esters by electrospray ionization MS/MS techniques: Cholesterol sulfate. Anal Chem 67:4178–4183, 1995.
27. EC Horning, MG Horning, DI Carroll, I Dzidic, RN Stillwell. New picogram detection system based on a mass spectrometer with an external ionization source at atmospheric pressure. Anal Chem 45:936–943, 1973.
28. A Raffaelli. Atmospheric pressure ionization (ESI and APCI) theory and application. In: RM Caprioli, A Malorni, G Sindona, eds. Selected Topics in Mass Spectrometry in the Biomolecular Sciences. Dordrecht: Kluwer Academic, 1997, Ser C, Vol 504, pp 17–31.
29. I Kempf, F Gesbert, M Guiltel, G Bennejean, AC Cooper. Efficacy of danofloxacin in the therapy of experimental mycoplasmosis in chicks. Res Vet Sci 53:257–259, 1992.
30. DA Volmer, B Mansoori, SJ Locke. Study of 4-quinolone antibiotics in biological samples by short-column liquid chromatography coupled with electrospray ionization tandem mass spectrometry. Anal Chem 69:4143–4155, 1997.
31. A Goodman-Gilman, LS Goodman, TW Rav, F Murad. The Pharmacological Basis of Therapeutics. 7th ed. New York: Macmillan, 1979.
32. S Rabbolini, E Verardo, MD Col, AM Gioacchini, P Traldi. Negative ion electrospray ionization tandem mass spectrometry in the structural characterization of penicillins. Rapid Commun Mass Spectrom 12:1820–1826, 1998.
33. RL Cerny, DK MacMillan, ML Gross, AK Mallams, BN Pramanik. Fast-atom bombardment and tandem mass spectrometry of macrolide antibiotics. J Am Soc Mass Spectrom 5:151–158, 1994.
34. TL Nagabhushan, GH Miller, MJ Weinstein. In: A Whelton, HC Neu, eds. The Aminoglycosides: Microbiology, Clinical Use and Toxicology. New York: Marcel Dekker, 1982, pp 3–27.
35. PJL Daniels, AK Mallams, J Weinstein, JJ Wright, GWA Milne. Mass spectral studies on aminocyclitol-aminoglycoside antibiotics. J Chem Soc Perkin Trans 1:1078–1088, 1976.
36. P Hu, EK Chess, S Brynjelsen, G Jakubowski, J Melchert, RB Hammond, TD Wilson. Collisionally activated dissociations of aminocyclitol-aminoglycoside

antibiotics and their application in the identification of a new compound in tobramycin samples. J Am Soc Mass Spectrom 11:200–209, 2000.
37. MJ Weinstein, GH Wagman, EM Oden, GM Luedemann, P Sloane, A Murawski, J Marquez. Purification and biological studies of everninomicin B. Antimicrob Agents Chemother 5:821–827, 1965.
38. AK Ganguly, BN Pramanik, VM Girijavallabhan, O Sarre, PL Bartner. The use of fast atom bombardment mass spectrometry for the determination of structures of everninomicins. J Antibiotics 38:808–812, 1985.
39. BN Pramanik, AK Ganguly. Fast atom bombardment mass spectrometry: A powerful technique for study of oligosaccharide antibiotics. Indian J Chem 25B:1105–1111, 1986.
40. BN Pramanik, PR Das, AK Bose. Molecular ion enhancement using salts in FAB matrices for studies on complex natural products. J Nat Prod 52:534–546, 1989.
41. PL Bartner, BN Pramanik, AK Saksena, YH Liu, PR Das, O Sarre, AK Ganguly. Structure elucidation of everninomicin-6, a new oligosaccharide antibiotic, by chemical degradation and FAB-MS methods. J Am Soc Mass Spectrom 8:1134–1140, 1997.
42. PA Shipkova, L Heimark, PL Bartner, G Chen, BN Pramanik, AK Ganguly, RB Cody, A Kusai. High resolution LC/MS for analysis of minor components in complex mixtures: Negative ion ESI for identification of impurities and degradation products of a novel oligosaccharide antibiotic. J Mass Spectrom 35:1252–1258, 2000.
43. G Odham, E Stenhagen. Complex lipids. In: GR Waller, ed. Biochemical Applications of Mass Spectrometry. New York: Wiley-Interscience, 1972, p 240.
44. FF Hsu, J Turk. Structural determination of sphingomyelin by tandem mass spectrometry with electrospray ionization. J Am Soc Mass Spectrom 11:437–449, 2000.
45. PS Parameswaran, AK Bose, PL Bartner, TM Chan, P Shipkova, YH Ing, BN Pramanik. A novel cyclopropane containing cerebroside from the red alga *Amphiroa fragilissima*. 47th ASMS Conf on Mass Spectrometry and Allied Topics, Dallas, TX, June 13–17, 1999.

4
Electrospray Mass Spectrometry Applications in Combinatorial Chemistry

Mike S. Lee
Milestone Development Services, Newtown, Pennsylvania

I. INTRODUCTION

Approaches by combinatorial chemistry for the synthesis of large sets of compounds or libraries have played a significant role in the pharmaceutical industry. Combinatorial chemistry techniques (1–6) have been integrated into drug discovery strategies to provide an automated and systematic approach for the generation of compound diversity in conjunction with biological screens. Innovations in automated synthesis continue to stimulate new paradigms for drug discovery (7,8). The combination of vision and depth of knowledge has helped to promote a greater understanding of technology and the development of new, industrially based strategies for discovering novel lead candidates.

Current industry trends in pharmaceutical analysis emphasize high volume approaches to accelerate lead candidate generation and evaluation. Combinatorial chemistry aims at the high throughput preparation of collections of compounds for screening. The designed libraries are targeted at specific tasks, either to find a "hit" compound or to optimize a good "lead" compound. The preparation of libraries can be achieved either by solid-phase methodology or by parallel synthesis. In solid-phase synthesis, the desired structures are built to the polymer support through a linker that covalently connects the growing molecules and the support. The release of the compounds from the support is obtained either by repeated washing of the support with an acid solution or by sample irradiation at a required wavelength. The library generated by solid-phase synthesis contains a large number of components. On the other hand, the parallel synthesis strategy is designed to load each resin bead with one unique compound, and the beads are pooled at the end of synthesis. Thus, spatially isolated and traceable compounds can be prepared. This strategy enables producing collections of single unique molecules.

These two combinatorial chemistry approaches have led the way, resulting in an unprecedented rate of sample generation. Prior to the advent of combinatorial chemistry technologies, a single bench chemist was capable of synthesizing approximately 50 final compounds per year, depending on the synthesis. Chemists today are capable of generating well over 2000 compounds per year, using a variety of automated synthesis technologies. With the recent integration of combinatorial chemistry workstations, distinctly new requisites for analysis have been created. To meet the challenges of determining constituents in a library, rapid, high throughput, sensitive, and selective methods are essential for pharmaceutical analysis.

Liquid chromatography/mass spectrometry (LC/MS)-based techniques provide unique capabilities for analysis of combinatorial chemistry derived samples. LC/MS methods are applicable to a wide range of compounds of pharmaceutical interest, and they feature powerful analytical figures of merit (sensitivity, selectivity, speed of analysis, and cost effectiveness). These analytical features have continually improved, resulting in easier-to-use and more reliable instruments. These developments coincided with the pharmaceutical industry's focus on describing the collective properties of novel compounds in a rapid, precise, and quantitative way. As a result, pharmaceutical samples shifted from a predominantly non-trace/pure sample type to trace/mixtures. The results of these developments have been significant as LC/MS has become the preferred analytical method for trace mixture analysis (Fig. 1).

Recent reviews and other chapters in this book describe the application of LC/MS throughout the drug development cycle (9) and the role of mass spectrometry in drug discovery (10) and combinatorial chemistry (11,12). This chapter focuses specifically on electrospray ionization (ESI)-LC/MS applications in support of combinatorial chemistry activities. The key elements for recent success are illustrated to include significant advances in instrumentation, methodology, and application. The applications are highlighted with reference to high throughput requirements and the analytical strategies that were implemented.

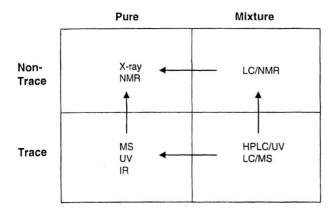

Figure 1 Structural analysis matrix that illustrates pharmaceutical analysis preferences for four specific sample types: non-trace/pure; non-trace/mixture; trace/pure; and trace/mixture. (Courtesy of Milestone Development Services, Newtown, PA.)

II. COMPOUND IDENTIFICATION

A significant benchmark for the acceptance of LC/MS by the pharmaceutical industry occurred when fully automated methods were developed for high throughput open-access instruments in support of conventional synthetic chemistry (13,14). These methods and approaches were developed primarily in response to the significant increase in the rate of sample generation from automated synthesis techniques. Open-access LC/MS systems provide a highly effective approach for maintaining the high throughput molecular weight characterization of synthetic compounds. These systems offer an efficient integration of sample generation and analysis activities from laboratory scale to bench scale. Advances in analytical instrumentation and electronic communication have also played a significant role in the emergence and acceptance of LC/MS as a front-line tool for structural confirmation. Owing to limitations in sample quantities, molecular weight measurement by LC/MS has become a preferred means of structural confirmation over nuclear magnetic resonance (NMR) and infrared (IR) spectroscopy during the early stages of synthetic chemistry activities. Today, open-access LC/MS is used in the same way that chemists previously used thin-layer chromatography (TLC) to monitor reaction mixtures for the desired product and to optimize reaction conditions.

The early procedures described by Pullen et al. (14) used thermospray (TSI) or particle beam (PB) LC/MS interfaces. These systems were successful for about 80–90% of the labile, polar, or higher mass compounds typically determined. A complementary gas chromatography/mass spectrometry (GC/MS) open-access system was used for volatile compounds. The procedures developed by Taylor et al. (13) featured atmospheric pressure chemical ionization (APCI)-LC/MS methods. A total cycle time of 4.5 min per sample was achieved. Abundant molecular ions for a diverse range of compounds were usually observed with little fragmentation. The short analysis time and the ability to yield reliable structural information promoted LC/MS as the first choice for synthetic chemistry applications.

The LC/MS-based methods developed by both Taylor et al. (13) and Pullen et al. (14) formed the foundation for advanced methodologies. Recent reviews (11,15–17) describe the various LC/MS methods that use ESI and that are applied to analyses ranging from complex molecular libraries (18) to open-access formats for drug discovery and development (19). High throughput criteria are central to each application.

An important development in the quest for high throughput combinatorial library analysis was the multiple ESI interface described by Wang et al. (20,21). This novel ESI interface enabled effluent flow streams from an array of four HPLC columns to be sampled independently and sequentially using a quadrupole MS instrument. The interface featured a stepping motor and a rotating plate assembly. The effluent flow from the HPLC columns was connected to a parallel arrangement of electrospray needles coaxial to the mass spectrometer entrance aperture. The individual spray tips were positioned at 90° relative to one another in a circular array. Each spray position was sampled multiple times per second by precise control of the stepping motor assembly.

A parallel sample analysis format using a multiplexed LC/MS interface with an orthogonal time-of-flight (TOF) mass spectrometer was described by Organ et al. (22). This approach demonstrated the high throughput capabilities of a multiplexed ESI interface in combination with an MS format that accommodated fast chromatographic methodologies. The system featured a four-way multiplexed electrospray interface attached directly to the existing source of the TOF-MS instrument (Fig. 2). A rotating aperture driven by a variable-speed stepper motor permitted the sampling of the spray from each electrospray probe tip. The acquisition of data files was synchronized with the corresponding spray. Each spray was sampled for 0.1 s, and mass spectra were acquired from 200 to 1000 Da. The total cycle time for each spray position was 0.8 s.

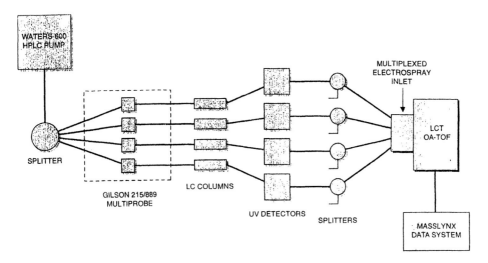

Figure 2 Schematic of the four-way multiplexed electrospray LC/MS interface in combination with a TOF mass spectrometer as reported by Organ et al. (From Ref. 22.)

The need for higher throughput LC/MS systems and increased efficiency via data management and open-access formats resulted in versatile software packages for data manipulation and processing (23–26). These software programs were efficiently implemented with either stand-alone computers or servers that were networked with open-access mass spectrometer data systems (Fig. 3). With this configuration, the data were generated, visualized, processed, and automatically reported to the chemist. The program then compared a template of predicted molecular ions with the actual ions generated by ESI or APCI to permit quick determination of synthetic products, intermediates, reactants, reagents, and contaminants. A list of observed ions and known artifact ions was generated and used to provide a measure of the quality of fit to the predicted product(s).

III. PURIFICATION

Various open-access LC/MS formats were developed to address throughput needs within an industrial laboratory. As the preparation of large libraries for lead discovery has become routine, the burden placed on analytical techniques has focused mainly on throughput and quality (27,28). Biological assays,

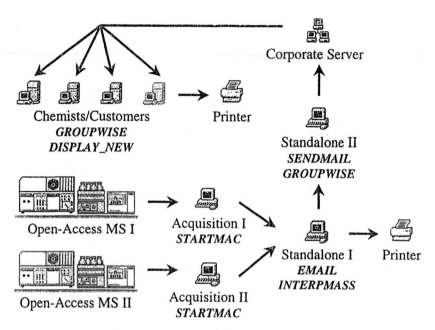

Figure 3 Software packages for data manipulation and processing, using stand-alone computers and servers that are networked with open-access mass spectrometer data systems as described by Tong et al. (From Ref. 24.)

however, normally required pure compounds. Thus, the focus shifted toward the use of automated high throughput purification methods applied to libraries of discrete compounds (29).

Reverse-phase analytical and preparative HPLC methods have been critical for the high throughput purification of components in parallel synthesis libraries. A variety of approaches that featured the use of gradient methods, short columns, and high flow rates were described (30). Highly automated LC/MS approaches for purification at the multimilligram level were described by Zeng et al. (31). These methods involved the use of short columns that were operated at ultrahigh flow rates. Preparative columns were operated at flow rates in excess of 70 mL/min to match the linear velocity of the short analytical columns (4.0 mL/min). Analytical LC/MS analyses of compound libraries were achieved in 5 min for chromatographically well-behaved compounds. Slightly longer preparative LC/MS analysis times (8–10 min/sample) were required for compounds that exhibited poor chromatographic peak shapes and/or for compound mixtures that needed higher resolution separations.

A schematic of the parallel analytical/preparative LC/MS-based system devised by Zeng and coworkers (32,33) is shown in Fig. 4. The fraction collection process is initiated in real time once the reconstructed ion current is observed for an ion of a specific m/z value that corresponds to the compound of interest. This design permits the collection of one sample per fraction. Thus, the need for very large fraction collector beds and postpurification screening and pooling was eliminated. Unattended and automated operation of this system led to the purification of over 100 compounds (milligram quantities) per day.

The increased popularity of LC/MS-based methods combined with limited resources resulted in advances that effectively matched combinatorial chemistry samples (i.e., complexity) with instrument time. Richmond et al. (25) demonstrated the use of an unattended flow injection analysis (FIA)-LC/MS system for rapid purity assessment. This LC/MS-based application addressed two critical bottlenecks: HPLC method development and delivery of results to the synthetic chemists.

The approach developed by Richmond et al. (25) employed an FIA-LC/MS method to generate purity estimates and determine the success of the synthesis. A 96-well sampling plate was used as the standard format for handling LC fractions. Comparability between plates was achieved by using a series of yohimbine standards (installed in row A) (see Fig. 5). The mass spectrometer was operated in the ESI positive and negative ion modes. A software application developed in-house within a Visual Basic graphic interface program was used to provide a color representation of the MS results with numerical purities contained within the background (Fig. 5). The purity results were defined by summing the ion current of the expected compound (including the molecular ions and

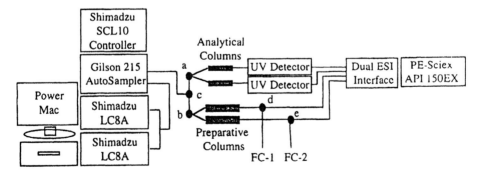

Figure 4 Schematic of the parallel analytical/preparative LC/MS system devised by Zeng and Kassel. (From Ref. 32.)

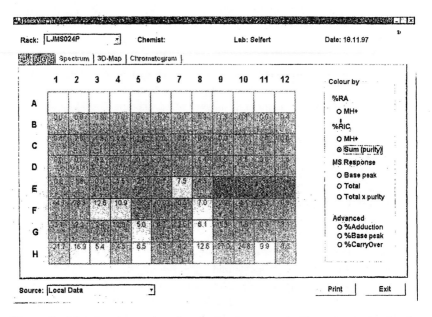

Figure 5 Visual representation of purity for a compound with an expected molecular formula $C_{31}H_{42}N_4O_6$ as reported by Richmond et al. (From Ref. 25.)

adducts) divided by the total ion current. The purity results from each sample plate also can be represented in three-dimensional maps. This representation is a quick approach for surveying and delivering LC fraction data. The facile visualization of data provided a tool both enabling and powerful for judging the success of a chemical synthesis. The data input by the chemist as well as the results reported by the analytical chemist are networked throughout the geographically dispersed research facilities. The data are processed and results are returned to the chemist within 24 h. The chemist can then assess whether to proceed to an LC/MS method in which the LC conditions are optimized.

IV. QUALITY CONTROL

Flow injection analysis (FIA) ESI-LC/MS approaches in combination with a chemiluminescent nitrogen detector (CLND) has been used to provide high throughput quality control of parallel synthesis libraries (34). The FIA ESI-LC/MS method provides proof of structure, and CLND is used for quantification. The method was validated by using a test set of 78 compounds (Table 1).

Table 1 The Test Set of 78 Compounds Used to Validate the FIA ESI-LC/MS and LC/CLND Approach as Described by Shah et al., 2000

	Name		Name
1	FmocSerine(tBu)	42	3-(Dibutylamino)-propylamine
2	FmocTyrosine(tBu)	43	1-(3-Aminopropyl)-2-pyrollidinone
3	FmocProline	44	Octylamine
4	FmocLysine(Boc)	45	Heptylamine
5	FmocPhenylalanine	46	2-Thiophenemethylamine
6	FmocAlanine	47	4-Phenylbutylamine
7	FmocHistamine(Boc)	48	4-Fluorophenethylamine
8	FmocAsparagine(Trt)	49	2-Methoxyphenethylamine
9	FmocAspartic acid(OtBu)	50	N-Phenylenediamine
10	FmocThreonine(tBu)	51	3-Dimethylaminopropylamine
11	FmocGlycine	52	4-Aminomorpholine
12	FmocValine	53	1-(2-Aminoethyl)piperazine
13	Heptylamine	54	1-Indoleacetic acid
14	Isoamylamine	55	Indole-6-carboxylic acid
15	Aminomethylcyclopentane	56	4-Methoxyindole
16	1-Naphthalenemethylamine	57	5-Hydroxyindole
17	4-Trifluoromethylbenzylamine	58	Ethyl 5-hydroxyindole-2-carboxylate
18	3-Phenyl-1-propylamine	59	Ethyl 6-isocyanatohexanoate
19	2-Fluorophenethylamine	60	n-Amyl isocyanate
20	2-Fluorobenzylamine	61	Ethyl isocyanatoacetate
21	3-Fluorobenzylamine	62	Ethyl 3-isocyanatopropionate
22	4-Fluorobenzylamine	63	*Tert*-butyl isocyanate
23	Furfurylamine	64	O-(trimethylsilyl)hydroxylamine
24	2-Methoxyethylamine	65	Methoxylamine hydrochloride
25	3,4-Dimethoxyphenethylamine	66	Hydroxylamine-O-sulfonic acid
26	4-Methoxyphenethylamine	67	O-benzylhydroxylamine hydrochloride
27	N,N-dimethylenediamine	68	O-(2,3,4,5,6-trifluorobenzyl) hydroxylamine
28	1-(2-Minoethyl)pyrrolidine		
29	N-(3-Aminopropyl)imidazole	69	O-ethylhydroxylamine hydrochloride
30	2-(2-Aminoethyl)pyridine	70	O-allylhydroxylamine hydrochloride
31	4-(2-Aminoethyl)morpholine	71	O-tritylhydroxylamine
32	Butylamine	72	5-(Chloro)-2-mercaptoaniline Hydrochloride
33	4-(*tert*-butyl)-aniline		
34	2-Methoxybenzylamine	73	L-penicillamine
35	3,4-Dimethoxybenzylamine	74	2-Aminothiophenol
36	3-Methoxybenzylamine	75	2,5-Diamino-1,4-benzenedithiol dihydrochloride
37	Cyclohexylamine		
38	Aniline	76	2-Amino-4-(trifluoromethyl) benzenethiol hydrochloride
39	Diaminopropane		
40	2-(1-Cyclohexyl)-ethylamine	77	2-Amino-5-nitrobenzophenone
41	2-(1-Cyclohexyl)-methylamine	78	2-Amino-5-chlorobenzophenone

Source: Ref. 34

The compounds represented a diverse range of commercially available building blocks such as amino acids, amines, indoles, isocyanates, hydroxylamines, amides, and carboxylic acids. Standard solutions of each test compound were prepared (1 mM), and relative response factors were derived from the mass concentration divided by the number of nitrogens. The response factors were calibrated versus nitrobenzene. The analysis was completed in approximately 1 min, and 1000 compounds were determined in a single day. The data are summarized using a software routine that generates pass/fail criteria. Quality control using the FIA ESI-LC/MS-based approach provided information equivalent to that given by traditional LC/MS methods. The testing of up to 1000 compounds in a single day was possible. Similar integrated ESI-LC/MS approaches using ultraviolet (UV) and evaporative light scattering detection (ELSD) (35), supercritical fluid chromatography (SFC) (36), and NMR (37) were reported.

V. SCREENING

Combinatorial chemistry initiatives have created a tremendous challenge for screening of these mixtures for activity against a specified target (38). Mass spectrometric approaches that use affinity selection (39), encoding methodologies (40–43), pulsed ultrafiltration (44), and antiaggregatory approaches (45) were reported.

A. Bioaffinity Screening

The recent studies performed by Anderegg et al. (46) describe the use of bioaffinity selection LC/MS methods for the identification of active mixture component(s). This approach features an integrated bioaffinity-based LC/MS screening method to separate and identify compounds from mixtures. This work is an extension of previously described studies performed by Nedved et al. (47) and Dollinger and coworkers (39), where, to achieve "selection," ligands are injected onto chromatography columns that contain target proteins. Compounds that bind to the proteins are selectively retained on the column and are later eluted for identification. Other workers reported on the successful use of ultrafiltration membranes to selectively retain compounds that are bound to target proteins (44,48). Unbound molecules pass through the membrane, and the bound molecules are later released and identified.

In the approach described by Anderegg et al. (46), LC/MS was incorporated as a bioaffinity screening strategy for lead identification in drug discovery. A mixture of compounds was incubated with the target protein, and the compo-

ESI in Combinatorial Chemistry

nents bound to the protein were selected by using a size exclusion chromatography (SEC) "spin column." In this experiment, the unbound compounds are retained on the column. The bound components are then released, eluted, and identified by LC/MS. The spin column enrichment scheme (illustrated in Fig. 6) increases specificity by dissociating the bound compounds and performing a second equilibration incubation with the protein. This procedure preferentially selects for the compounds with higher affinity and results in an enhancement of the quantitative LC/MS response. Iterative stages of incubation, size exclusion, and LC/MS allow the tighter binding components to be enriched relative to more weakly binding components.

In one study, the peroxisome proliferator-activated receptor (PPARγ), which is a target for antidiabetic drugs (construct molecular mass of 32,537 Da), was incubated with 10 ligands that range in molecular mass from 283 to 587 Da. A spin column of 6000 Da cutoff was used for SEC purposes. The retained mixture of components was analyzed by fast perfusive chromatography (49,50), using a standard full-scan LC/MS strategy. This analytical procedure allows for

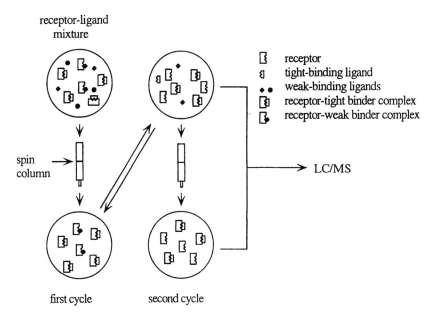

Figure 6 The procedure illustrated by Anderegg et al. for bioaffinity screening of combinatorial libraries with two cycles of iterative size-exclusion chromatography. The receptor and receptor–binder complexes are separated using "spin columns". (From Ref. 46.)

the identification and quantification of the protein and the ligands compared to their levels prior to incubation. The ligand–protein complex that dissociated under the reverse-phase chromatographic conditions is selectively detected. The LC/MS response for the 10-compound mixture is shown in Fig. 7. After one pass through the spin column, several weak-binding ligands (B, F, H) disappeared, and several others (D, E, G) were diminished in relative amount. After a second pass through the spin column, the three weak-binding ligands (D, E, G) were further diminished. Four ligands remained, corresponding to the tight binders (A, C, I, J).

This analysis scheme provided a quick measurement of binding affinity and served as a screening tool for drug candidate selection. Spreadsheets are constructed and used to calculate the binding affinity of the components. In this example, two incubation cycles followed by SEC separation distinguish strong binders from weak binders. This LC/MS-based method provides a unique approach to obtaining information in situations where lower concentrations of tighter binding ligands are present in the same mixture with higher concentrations of weaker binding ligands. This LC/MS approach is more efficient than synthetic deconvolution procedures and does not require the use of radioligands.

B. Accurate Mass with QTOF MS

The use of ESI approaches with MS formats that provide accurate mass capabilities have been recently illustrated for screening combinatorial libraries (51–54). The unambiguous confirmation or identification of combinatorial library components from small quantities of material was illustrated by using a hybrid quadrupole/orthogonal TOF (QTOF) (51,52) and Fourier transform ion cyclotron resonance (FTICR) (53,54) mass spectrometry. Accurate isotope patterns or "isotopic signature" and unique mass differences between isobaric compounds were obtained. Similar ESI-MS approaches (i.e., analysis, database searching) for the dereplication of natural products was recently reviewed (55).

In a novel ESI QTOF-MS approach described by Blom et al. (51), active combinatorial chemistry–derived lead compounds were identified by accurate mass and tandem mass spectrometric (MS/MS) measurements. Mixtures obtained directly from active beads of a dipeptide combinatorial library that was selected by ultrahigh throughput screening protocols (for matrix metalloprotease activity) were analyzed with a single acquisition using an automatic function switching method. Figure 8 illustrates the ESI full scan and MS/MS

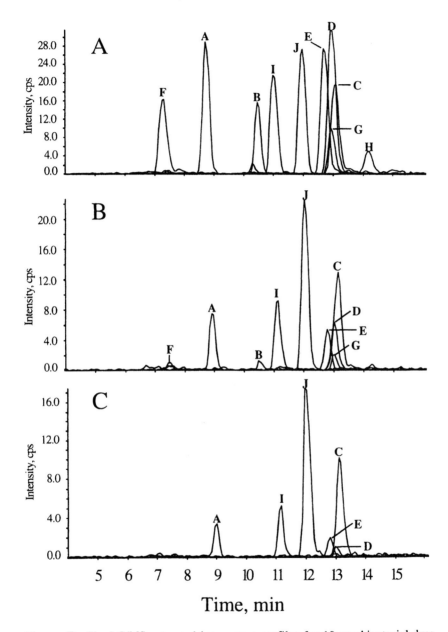

Figure 7 The LC/MS-extracted ion current profiles for 10 combinatorial drug candidate library components using the bioaffinity screening procedure. (A) Before passing through a spin column; (B) after one cycle; (C) after two cycles. (From Ref. 46.)

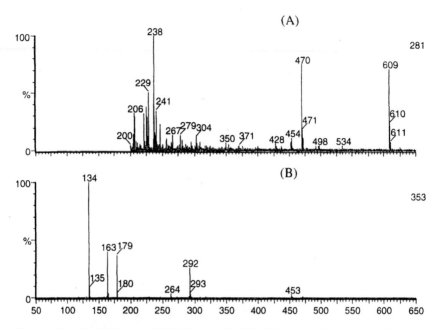

Figure 8 (A) Positive ion ESI-MS and (B) MS/MS product ion spectra of a compound released from a single bead after ultrahigh throughput screening reported by Blom et al. (From Ref. 51.)

product ion spectra for a compound released from a single bead after ultrahigh throughput screening. An EXCEL program was used to generate a list of all components in the library, the exact masses of the $[M+H]^+$ ions, and the masses of the four expected product ions produced in the MS/MS experiment. This ESI-LC/MS methodology was developed to support a discovery leads program that required about 100 analyses per day.

Typical measurements were made on ~20 pmol of compound. Automated data analysis routines provided (a) a quick survey of molecular weight based on $[M+H]^+$ masses and (b) a qualitative measure of best fit based on specific tolerances for the calculated mass and the product ions from MS/MS. This automated approach greatly minimizes the possibility of error. Identification of structures using an automated computer analysis is obtained in about 80% of the cases with a mass measurement accuracy of <5 ppm. Manual interpretation of mass spectra was then carried out to identify the compounds in the remainder of the samples.

The use of generic microbore HPLC methods in conjunction with an accurate mass QTOF-MS configuration was reported by Lane and Pipe (52). In the previous method described above, compounds obtained from single beads were measured directly. In the new approach, identifier tags are attached to synthesis supports using a procedure referred to as encoded combinatorial synthesis (56). The identifier tags, a series of secondary amines, are added to the beads in a unique manner during library synthesis. After the encoded beads have been screened, the amines are released by acid hydrolysis, derivatized with dansyl chloride, and determined by LC/MS. Unambiguous decoding of the biologically active components of the library is effectively simplified (i.e., known molecular weight, reproducible chromatography, consistent response) because the analytical method is optimized for detection of a specific set of identifier tags as opposed to the chemical entities they represent.

In the QTOF-based method described by Lane and Pipe (52), a microbore LC/TOF-MS method was used to generate a quick separation of dansylated tags (Table 2) followed by accurate mass measurement of both molecular and product ions formed by MS/MS. The 14-member tag set corresponded to unique molecular weights. The TOF-based approach has a

Table 2 The Affymax Orthogonal 14-Member Tag Set Described by Lane and Pipe, 1999.

Tag	Name	Mol wt	Formula	Mol wt of dansyl derivative
D_3-EP	D_3-N-Ethylpentylamine	118.1549	$C_7H_{14}D_3N$	351.2060
D_3-EH	D_3-N-Ethylheptylamine	132.1706	$C_8H_{16}D_3N$	365.2216
D_3-EH'	D_3-N-Ethylheptylamine	146.1862	$C_9H_{18}D_3N$	379.2373
D_3-EO	D_3-N-Ethyloctylamine	160.2019	$C_{10}H_{20}D_3N$	393.2529
D_3-EN	D_3-N-Ethylnonylamine	174.2175	$C_{11}H_{22}D_3N$	407.2686
D_3-ED	D_3-N-Ethyldecylamine	188.2332	$C_{12}H_{24}D_3N$	421.2842
PO	N-Pentyloctylamine	199.2300	$C_{13}H_{29}N$	432.2801
H'H'	Diheptylamine	213.2457	$C_{14}H_{31}N$	446.2967
H'O	N-Heptyloctylamine	227.2613	$C_{15}H_{33}N$	460.3124
OO	Dioctylamine	241.2770	$C_{16}H_{35}N$	474.3280
PDo	N-Pentyldodecylamine	255.2926	$C_{17}H_{37}N$	488.3437
HDo	N-Hexyldodecylamine	269.3083	$C_{18}H_{39}N$	502.3593
H'Do	N-Heptyldodecylamine	283.3239	$C_{19}H_{41}N$	516.3750
DD	Didecylamine	297.3396	$C_{20}H_{43}N$	530.3906

Source: Ref. 52.

powerful advantage over scanning-type instruments (i.e., quadrupole) for this application and features parallel detection of all masses in the spectrum. The mass chromatograms for the $[M + H]^+$ species of the 14 dansylated tags (Fig. 9) illustrate the chromatographic separation obtained using this methodology. The same microbore LC/TOF-MS method was also used to analyze beads directly.

C. Exact Mass with FT-ICR MS

The use of ESI in combination with FTICR-MS was also described (26,53,54). The feasibility of this approach was demonstrated on combinatorial libraries

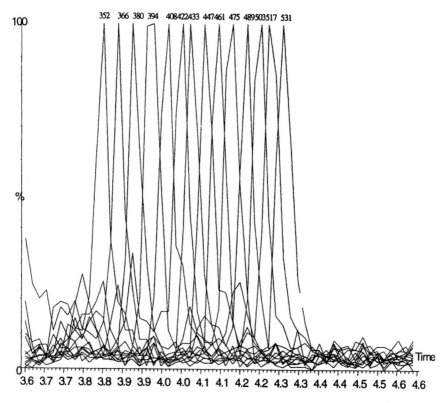

Figure 9 Mass chromatograms for the $[M + H]^+$ species of the 14 dansylated tags described by Lane and Pipe. (From Ref. 52.)

ESI in Combinatorial Chemistry

with flow injection (53). Exact mass measurements were obtained, and 70–80% of the library components were identified. This methodology is particularly useful for combinatorial libraries that contain isobaric components. Similar experiments that use fast gradient HPLC methods on-line with FTICR-MS were recently demonstrated for the accurate mass analysis of mixtures (54). The high resolution exact-mass measurements provide insight into the molecular formulas of unknown compounds and corresponding fragment ions. Using a reverse-phase HPLC method, Speir et al. analyzed a six-component drug mixture in less than 5 min under high mass accuracy conditions. Representative UV and LC/MS chromatograms obtained with the six-component mixture are shown in Fig. 10. Peak widths required the use of an electrostatic ion accumulation system (57) so that acquisition times of 1 s could be achieved. The scan rate was adequate to maintain sufficient chromatographic resolution. Relative mass errors of less than 1 ppm were obtained. Furthermore, on-line LC/FTICR-MS/MS results are possible (with 3 s acquisition time) using this methodology.

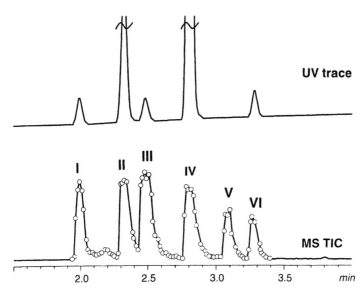

Figure 10 Representative LC/UV and LC/FTICR-MS chromatograms obtained from a six-component mixture as described by Speir et al. The solid circles represent the temporal location of mass spectra with an acquisition time of 1 s. (From Ref. 54.)

VI. ANALYTICAL PERSPECTIVES

The successful application of ESI-LC/MS with combinatorial chemistry is the combined result of four analytical elements: separation sciences, mass spectrometry, information management, and widened scope of application (9). The first, *separation sciences*, or chromatography, whether in preparative or analytical format, provides analytical criteria to compare, refine, develop, and control the critical aspects of high throughput analysis. The combination of a wide variety of HPLC-based technologies and formats with mass spectrometry played a vital role in the acceptance of LC/MS for combinatorial chemistry applications. This achievement is significant because HPLC-based methods are universally recognized within the pharmaceutical industry.

The second element that allowed for the successful application of ESI-LC/MS techniques is *mass spectrometry*. The analytical figures of merit dealing with sensitivity and selectivity provide a powerful platform for analysis. It was not until these analytical attributes could be harnessed into a reliable, reproducible, rugged, and high throughput instrument, however, that mass spectrometric techniques could be taken seriously as an integral tool for drug discovery. Furthermore, the added dimensions of mass analysis using TOF (58) and FTICR (26,54) mass spectrometers provide enhanced limits of detection for the analysis of complex mixtures and unique capabilities for structural identification.

The third element is *information management*. The rate of analysis and subsequent distribution of results has increased tremendously owing to the increased use of combinatorial chemistry–based approaches. From a strictly analytical perspective, LC/MS has demonstrated a unique capability for maintaining high quality performance and a rapid turn around of samples. Yet, it is the accurate and efficient processing of information that has been essential for LC/MS use and acceptance. As a result, LC/MS has developed a unique partnership with software tools responsible for sample tracking, interpretation, and data storage (59,60). Current LC/MS applications are highly dependent on software to integrate key analysis elements that deal with sample preparation, real-time analytical decisions, databasing, distribution and visualization of results (Fig. 11), and prediction of fragmentation.

The fourth element is a *widened scope of application*. The fact that LC/MS is routinely used in combinatorial chemistry activities is a powerful benchmark for acceptance. The increased use of LC/MS in various applications has in turn stimulated new performance levels for sample preparation, high speed separations, automated analysis, information databases, and software

ESI in Combinatorial Chemistry

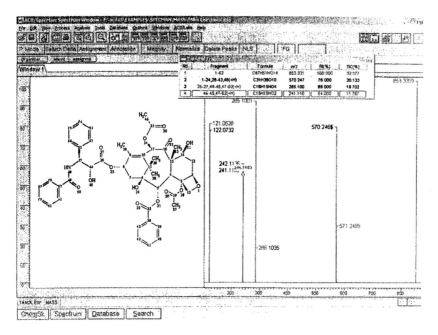

Figure 11 Visualization of molecular fragments using a "lasso tool." The lasso tool is used to identify a particular fragment, and if a signal corresponding to its mass is present in the spectrum, the fragment is highlighted and the corresponding assignment is added to an assignment table. (Courtesy of Advanced Chemistry Development, Toronto, ON, Canada.)

tools, to name a few. Continued advances in integrated approaches that embrace ESI-LC/MS techniques will continue to be an important development for the analysis of combinatorial libraries. For example, the use of an integrated system that includes LC/NMR/MS has been described (61). In this configuration, NMR is used to provide structural and stereochemical information while MS is used to provide molecular weight information. The combination of LC/MS and LC/NMR, on-line within a single analysis, provides a powerful tool for the structural elucidation of components contained in a mixture without isolation (62). Recently, the MS/NMR combination was demonstrated as a screening tool for the assay of a library consisting of approximately 32,000 compounds for molecules that exhibit specific binding to MMP-1 and about 1000 compounds for molecules that bind to RGS4 proteins (63).

VII. CONCLUSIONS

Novel ESI-LC/MS protocols for high throughput drug discovery are developing at an extraordinary pace. Automated approaches capable of generating and screening large libraries of compounds have led to the successful application of ESI-LC/MS to combinatorial chemistry. The continual drive for new applications, motivated by unmet industry needs, has stimulated the tremendous acceptance and growth in LC/MS-based methods. Current trends for further integration of ESI-LC/MS techniques with new analytical formats (i.e., sample preparation, chromatography, mass spectrometry, information management) as well as traditional approaches such as NMR will likely spawn structure-based assays for drug discovery.

REFERENCES

1. MA Gallop, RW Barrett, WJ Dower, SPA Fodor, EM Gordon. Applications of combinatorial technologies to drug discovery. 1. Background and peptide combinatorial libraries. J Med Chem 37:1233–1251, 1994.
2. EM Gordon, RW Barrett, WJ Dower, SPA Fodor, MA Gallop. Applications of combinatorial technologies to drug discovery. 2. Combinatorial organic synthesis, library screening strategies, and future directions. J Med Chem 37:1385–1401, 1994.
3. MC Desai, RN Zuckermann, WH Moos. Recent advances in the generation of chemical diversity libraries. Drug Dev Res 33:174–188, 1994.
4. KS Lam. Application of combinatorial library methods in cancer research and drug discovery. Anticancer Drug Des 12:145–167, 1997.
5. AW Czarnik. Combinatorial chemistry. Anal Chem 70:378A–386A, 1998.
6. G Jung. Combinatorial Chemistry: Synthesis, Analysis, Screening. Weinheim, RFA: Wiley-VCH, 1999.
7. WH Moos, GD Green, MR Pavia. Recent advances in the generation of molecular diversity. Annu Rep Med Chem 28:315–324, 1993.
8. R Storer. Solution-phase synthesis in combinatorial chemistry: Applications in drug discovery. Drug Discovery Today 1:248–254, 1996.
9. MS Lee, EH Kerns. LC/MS applications in drug development. Mass Spectrom Rev 18:187–279, 1999.
10. BN Pramanik, PL Bartner, G Chen. The role of mass spectrometry in drug discovery process. Curr Opin Drug Discovery 2:401–417, 1999.
11. RD Süssmuth, G Jung. Impact of mass spectrometry on combinatorial chemistry. J Chromatogr B Biomed Sci Appl 725:49–65, 1999.
12. C Enjalbal, J Martinez, JL Aubagnac. Mass spectrometry in combinatorial chemistry. Mass Spectrom Rev 19:139–161, 2000.

13. LCE Taylor, RL Johnson, R Raso. Open access atmospheric pressure chemical ionization mass spectrometry for routine sample analysis. J Am Soc Mass Spectrom 6:387–393, 1995.
14. FS Pullen, GL Kerkins, KI Burton, RS Ware, MS Teague, JP Kiplinger. Putting mass spectrometry in the hands of the end user. J Am Soc Mass Spectrom 6:394–399, 1995.
15. DJ Burdick, JT Stults. Analysis of peptide synthesis products by electrospray ionization mass spectrometry. Methods Enzymol 289:499–519, 1997.
16. WL Fitch Analytical methods for quality control of combinatorial libraries. Mol Diversity 4:39–45, 1998–99.
17. V Swali, GJ Langley, M Bradley. Mass spectrometric analysis in combinatorial chemistry. Curr Opin Chem Biol 3:337–341, 1999.
18. Y Dunayevskiy, P Vouros, T Carell, EA Wintner, J Rebek Jr. Characterization of the complexity of small-molecule libraries of electrospray ionization mass spectrometry. Anal Chem 67:2906–2915, 1995.
19. S Cepa, P Searle. Establishing an open access LC/MS lab for pharmaceutical discovery and development. Proc 46th ASMS Conf on Mass Spectrometry and Allied Topics, Orlando, FL 1998, p 283.
20. T Wang, L Zeng, T Strader, L Burton, DB Kassel. A new ultra-high throughput method for characterizing combinatorial libraries incorporating a multiple probe autosampler coupled with flow injection mass spectrometry analysis. Rapid Commun Mass Spectrom 12:1123–1129, 1998.
21. T Wang, J Cohen, DB Kassel, L Zeng. A multiple electrospray interface for parallel mass spectrometric analyses of compound libraries. Comb Chem High Throughput Screen 2:327–334, 1999.
22. V de Biasi, N Haskins, A Organ, R Bateman, K Giles, S Jarvis. High throughput liquid chromatography/mass spectrometric analyses using a novel multiplexed electrospray interface. Rapid Commun Mass Spectrom 13:1165–1168, 1999.
23. JL Whitney, EH Kerns, RA Rourick, ME Hail, KJ Volk, SW Fink, MS Lee. Accelerated structure profiling using automated LC-MS and robotics. Pharm Tech May:6–82, 1998.
24. H Tong, D Bell, T Keiko, MM Siegel. Automated data massaging, interpretation and e-mailing for high throughput open access mass spectrometry. J Am Soc Mass Spectrom 10:1174–1187, 1999.
25. R Richmond, E Gorlach, JM Seifert. High-throughput flow injection analysis-mass spectrometry with networked delivery of colour rendered results: The characterization of liquid chromatography fractions. J Chromatogr A 835:29–39, 1999.
26. N Huang, MM Siegel, GH Kruppa, FH Laukien. Automation of a Fourier transform ion cyclotron resonance mass spectrometer for acquisition, analysis, and e-mailing of high-resolution exact-mass electrospray ionization mass spectral data. J Am Soc Mass Spectrom 10:1166–1173, 1999.

27. JN Kyranos, JC Hogan. High-throughput characterization of combinatorial libraries generated by parallel synthesis. Anal Chem 70:389A–395A, 1998.
28. L Van Hijfte, G Marciniak, N Froloff. Combinatorial chemistry, automation and molecular diversity: New trends in the pharmaceutical industry. J Chromatogr B Biomed Sci Appl 725:3–15, 1999.
29. HN Weller. Purification of combinatorial libraries. Mol Diversity 4:47–52, 1998–99.
30. HN Weller, MG Young, SJ Michalczyk, GH Reitnauer, RS Cooley, PC Rahn, DJ Loyd, D Fiore, SJ Fischman. High throughput analysis and purification in support of automated parallel synthesis. Mol Diversity 3:61–70, 1997.
31. L Zeng, X Wang, T Wang, DB Kassel. New developments in automated prepLC/MS extends the robustness and utility of the method for compound library analysis and purification. Comb Chem High Throughput Screen 1:101–111, 1998.
32. L Zeng, DB Kassel. Developments of a fully automated parallel HPLC/mass spectrometry system for the analytical characterization and preparative purification of combinatorial libraries. Anal Chem 70:4380–4388, 1998.
33. L Zeng, L Burton, K Yung, B Shushan, DB Kassel. Automated analytical/preparative high-performance liquid chromatography-mass spectrometry system for the rapid characterization and purification of compound libraries. J Chromatogr A 794:3–13, 1998.
34. N Shah, M Gao, K Tsutsui, A Lu, J Davis, R Scheuerman, WL Fitch, RL Wilgus. A novel approach to high-throughput quality control of parallel synthesis libraries. J Comb Chem 2:453–460, 2000.
35. L Fang, J Pan, B Yan. LC/UV/ELSD/MS for fast purity and quantity analysis in combinatorial chemistry. Proc 48th ASMS Conf on Mass Spectrometry and Allied Topics, Long Beach, Ca, 2000, pp 1816–1817.
36. MC Ventura, WP Farrell, CM Aurigemma, X Xiong, MO Osonubi, R Lopez, MJ Grieg. Optimizing quality control of combinatorial libraries with SFC/MS. Proc 48th ASMS Conf on Mass Spectrometry and Allied Topics, Long Beach, Ca, 2000, pp 202–203.
37. BD Duléry, J Verne-Mismer, E Wolf, C Kugel, L Van Hijfte. Analyses of compound libraries obtained by high-throughput parallel synthesis: Strategy of quality control by high-performance liquid chromatography, mass spectrometry and nuclear magnetic resonance techniques. J Chromatogr B Biomed Sci Appl 725:39–47, 1999.
38. DC Schriemer, O Hindsgaul. Deconvolution approaches in screening compound mixtures. Comb Chem High Throughput Screen 1:155–170, 1998.
39. S Kaur, L McGuire, D Tang, G Dollinger, V Huebner. Affinity selection and mass spectrometry-based strategies to identify lead compounds in combinatorial libraries. J Protein Chem 16:505–511, 1997.
40. RS Youngquist, GR Fuentes, MP Lacey, T Keough. Generation and screening of combinatorial peptide libraries designed for rapid screening by mass spectrometry. J Am Chem Soc 117:3900–3905, 1995.

41. HM Geysen, CD Wagner, WM Bodnar, CJ Markworth, GJ Parke, FJ Schoenen, DS Wagner, DS Kinder. Isotope or mass encoding of combinatorial libraries. Chem Biol 3:679–688, 1996.
42. I Hughes. Design of self-coded combinatorial libararies to facilitate direct analysis of ligands by mass spectrometry. J Med Chem 41:3804–3811, 1998.
43. DS Wagner, CJ Markworth, CD Wagner, FJ Schoenen, CE Rewerts, BK Kay, HM Geysen. Ration encoding combinatorial libraries with stable isotopes and their utility in pharmaceutical research. Comb Chem High Throughput Screen 1:143–153, 1998.
44. RB van Breeman, C-R Huang, D Nikolic, CP Woodbury, Y-Z Zhao, DL Venton. Pulsed ultrafiltration mass spectrometry: A new method for screening combinatorial libraries. Anal Chem 69:2159–2164, 1997.
45. S Park, L Wanna, ME Johnson, DL Venton. A mass spectrometry screening method for antiaggregatory activity of proteins covalently modified by combinatorial library members: Application to sickle hemoglobin. J Comb Chem 2:314–317, 2000.
46. RG Davis, RJ Anderegg, SG Blanchard. Iterative size-exclusion chromatography coupled with liquid chromatographic mass spectrometry to enrich and identify tight-binding ligands from complex mixtures. Tetrahedron 55:1653–1667, 1999.
47. ML Nedved, S Habibi-Goudarzi, B Ganem, JD Henion. Characterization of benzodiazepine "combinatorial" chemical libraries by on-line immunoaffinity extraction, coupled column HPLC-ion spray mass spectrometry-tandem mass spectrometry. Anal Chem 68:4228–4236, 1996.
48. RB van Breeman, D Nikolic, JL Bolton. Metabolic screening using on-line ultrafiltration mass spectrometry. Drug Metab Dispos 26:85–90, 1998.
49. FE Regnier. Perfusion chromatography. Nature 350:634–635, 1991.
50. NB Afeyan, SP Fulton, FE Regnier. Perfusion chromatography packing materials for proteins and peptides. J Chromatogr 544:267–279, 1991.
51. KF Blom, AP Combs, AL Rockwell, KR Oldenburg, JH Zhang, T Chen. Direct mass spectrometric determination of bead bound compounds in a combinatorial lead discovery application. Rapid Commun Mass Spectrom 12:1192–1198, 1998.
52. SJ Lane, A Pipe. A single generic microbore liquid chromatography/time-of-flight mass spectrometry solution for the simultaneous accurate mass determination of compounds on single beads, the decoding of dansylated orthogonal tags pertaining to compounds and accurate isotopic difference target analysis. Rapid Commun Mass Spectrom 13:798–814, 1999.
53. AS Fang, P Vouros, CC Stacey, GH Kruppa, FH Laukien, EA Wintner, T Carell, J Rebek Jr. Rapid characterization of combinatorial libraries using electrospray ionization Fourier transform ion cyclotron resonance mass spectrometry. Comb Chem High Throughput Screen 1:23–33, 1998.
54. JP Speir, G Perkins, C Berg, F Pullen. Fast, generic gradient high performance liquid chromatography coupled to Fourier transform ion cyclotron resonance

mass spectrometry for the accurate mass analysis of mixtures. Rapid Commun Mass Spectrom 14:1937–1942, 2000.
55. MA Strege. High-performance liquid chromatographic-electrospray ionization mass spectrometric analyses for the integration of natural products with modern high-throughput screening. J Chromatogr B 725:67–78, 1999.
56. S Brenner, RA Lerner. Encoded combinatorial chemistry. Proc Natl Acad Sci USA 89:5181–5183, 1992.
57. P Caravatti. US Patent 4,924,089, 1990.
58. HR Morris, T Paxton, A Dell, J Langhorne, M Berg, RS Bordoli, J Hoyes, RH Bateman. High sensitivity collisionally-activated decomposition tandem mass spectrometry on a novel quadrupole/orthogonal-acceleration time-of-flight mass spectrometer. J Rapid Commun Mass Spectrom 10:889–896, 1996.
59. RK Julian, RE Higgs, JD Gygi, MD Hilton. A method for quantitatively differentiating crude natural extracts using high-performance liquid chromatography-electrospray mass spectrometry. Anal Chem 70:3249–3254, 1998;
60. A Williams, MS Lee, V Lashin. An integrated desktop mass spectrometry processing and molecular structure management system. Spectroscopy 16:38–49, 2001.
61. RM Holt, MJ Newman, FS Pullen, DS Richards, AG Swanson, High-performance liquid chromatography/NMR spectrometry/mass spectrometry: Further advances in hyphenated technology. J Mass Spectrom 32:64–70, 1997.
62. JP Shockcor, SE Unger, P Savina, JK Nicholson, JC Lindon. Application of directly coupled LC-NMR-MS to the structural elucidation of metabolites of the HIV-1 reverse-transcriptase inhibitor BW935U83. J Chromatogr B Biomed Sci Appl 748:269–279, 2000.
63. FJ Moy, K Haraki, D Mobilio, G Walker, R Powers, K Tabei, H Tong, MM Siegel. MS/NMR: A structure-based approach for discovering protein ligands and for drug design by coupling size exclusion chromatography, mass spectrometry, and nuclear magnetic resonance spectroscopy. Anal Chem, 73:571–581, 2001.

5
Electrospray Mass Spectrometry in Contemporary Drug Metabolism and Pharmacokinetics

Mark J. Cole, John S. Janiszewski, and Hassan G. Fouda
Pfizer Inc., Groton, Connecticut

I. INTRODUCTION

The study of drug metabolism comprises the investigation of absorption, distribution, metabolism, and excretion (ADME) of drug molecules in biological systems. It defines all the biological factors that allow a drug to exert its desired therapeutic effects by reaching the site of action in appropriate concentrations for an appropriate period of time. As such, drug metabolism information is essential for drug discovery and development and the commercial success of new therapeutic agents.

The use of mass spectrometry (MS) in drug metabolism and pharmacokinetics predates the development of electrospray ionization. Gas chromatography coupled with mass spectrometry (GC/MS) has played a significant role in quantitative determination of drugs and metabolites in biological fluids (1–3) and in structure identification of drug metabolites (4). The successful coupling of HPLC and mass spectrometry (HPLC/MS) dramatically increased the applicability of MS in drug metabolism and pharmacokinetics. Specifically, the commercial introduction in the late 1980s (5) of reliable HPLC/MS instruments based on atmospheric pressure ionization (API), namely electrospray ionization (ESI) and atmospheric pressure chemical ionization (APCI), resulted in an explosion in the use of HPLC/MS in ADME studies. There are several reasons for this. First, both ESI and APCI feature much higher ionization efficiencies than previous ionization approaches, and this efficiency translates into lower detection limits. Second, ESI and APCI together are applicable to a broad range of structure classes, including high molecular weight drugs and metabolites, whose analysis is not possible by other MS techniques, and to polar and thermally labile compounds, whose analysis by GC/MS requires a time-consuming chemical derivatization. The ionization by ESI is especially soft, making it the method of choice for labile conjugated drug metabolites. Finally, the HPLC/MS coupling via API does not impose significant restrictions on HPLC parameters such as mobile phase solvents, modifiers, and flow rates. Hence, ESI and APCI permit taking full advantage of HPLC for the quantification, mixture analysis, and profiling of drug metabolites.

Early applications of ESI and APCI to drug metabolism and pharmacokinetics emphasized drug development (6–9). After the demonstration of the high analysis speed of HPLC/ESI methods and the realization that such methods are amenable to high throughput automation, HPLC/MS was rapidly extended to discovery applications. In a way, the development of HPLC/MS has significantly contributed to transforming the role of the drug metabolism discipline from its traditional minor contribution to drug discovery to its current major role in drug discovery as well as drug development.

II. MASS SPECTROMETRY PLATFORMS FOR DRUG METABOLISM AND PHARMACOKINETICS

A. Role of Quadrupole Instruments

Electrospray ionization was initially developed on single-quadrupole instruments (10). However, the first commercially successful implementation of ESI was derived from an existing instrument specially suited for atmospheric pressure ionization (Sciex TAGA 6000; Sciex API-III) and as such happened to use the triple-quadrupole (TQMS) implementation (5,11). The early success of the ESI-TQMS combination in pharmaceutical applications established its dominance in the field and led to its adoption throughout the pharmaceutical industry (5). It was widely assumed that all the demonstrated benefits of API for quantitative and structural studies could be realized only with a TQMS platform. As a result, the commercial introduction of API on single-quadrupole (SQMS) platforms lagged behind.

Because quantification is the mainstay application of ESI-MS, and the demand for more quantitative capacity seems unending, it was only natural to consider whether the larger and more costly TQMS platform was truly needed for this application. Not to minimize the importance of the added measure of specificity afforded by MS/MS, it is clear that TQMS often represents more analytical power than necessary for many quantitative applications. By analogy, single-quadrupole GC/MS supported quantitative applications for decades, whereas GC tandem MS (GC/MS/MS) was reserved for the most demanding applications. Recent years witnessed the development and commercialization of several ESI-SQMS instruments.

The main advantage of single-quadrupole instruments is their relatively small size and low cost, providing a sensitive and selective solution for quantification that can be widely distributed. They have been implemented throughout the pharmaceutical industry, especially in the area of drug discovery, where they have supplanted LC/UV in many laboratories. For more demanding quantitative applications, TQMS has greater discrimination against chemical background, providing signal-to-noise enhancement and allowing ultratrace-level quantification. TQMS is also necessary for structure determination studies, because SQMS instruments do not provide a reliable and selective means for collisional dissociation and must rely on low chemical background and absolute LC separation for selectivity.

B. Role of Quadrupole Ion Trap Instruments

Several other types of ESI-MS instrumentation have recently found niches in drug metabolism. The development of ESI-quadrupole ion trap (ITMS) instruments provided the ability to perform multiple stages of tandem mass spectrometry (12–14). Because ion traps measure all ions retained in the trapping step, they do not suffer the sensitivity losses during the full-scan mode that are unavoidable if a TQMS is to acquire a nearly complete product ion spectrum, as illustrated in Fig. 1. In addition, the resonance excitation used for ion dissociation in the trap allows for universal collision conditions and for complete and more efficient conversion of the precursor ions to product ions, as shown in Fig. 2. Because of their retention of sensitivity during full scanning and efficient precursor conversion, these instruments often provide more sensitivity for structure elucidation than TQMS instruments.

Sensitivity in both ITMS and TQMS is dependent on the respective duty cycles. Here, TQMS has an advantage for quantification when operated in selective reaction monitoring (SRM) mode, because the duty cycle approaches 100%. Ion traps do not gain sensitivity in SRM mode, but neither do they lose sensitivity in full-scan mode, because there is no change in the operating duty

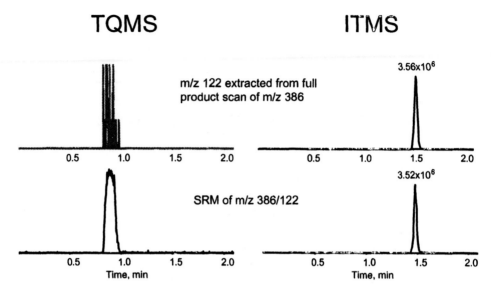

Figure 1 Comparison of sensitivities between the extraction of the major product ion of buspirone from full product scans (top chromatograms) and selected reaction monitoring of the precursor–product ion transitions (bottom chromatograms) for both triple-quadrupole and quadrupole ion trap instruments. It is evident that a significant sensitivity penalty results from using full-scan spectra with linear quadrupoles, whereas sensitivity is retained during full scan with quadrupole ion traps.

cycle between the two modes. Typically, TQMS retains a sensitivity advantage of a factor of 5–10 in the SRM mode over ITMS for quantification (15).

For automated method development and analysis for quantification, resonance excitation and the absence of a sensitivity penalty in full-scan mode is an advantage for ITMS. By leveraging these attributes, data can be automatically collected for all samples, and the experimental and processing methods can be determined automatically after all sample data are collected. The collision conditions are inherently optimized, and the specific ion transitions used for quantification are extracted from the full-scan spectra collected from each sample (the total ion chromatogram from the full scan can be used, but it is often not as selective as a single transitions). All the user need do is supply a text list containing the internal standard and analyte masses of interest for each compound, along with a template containing the sample order (16).

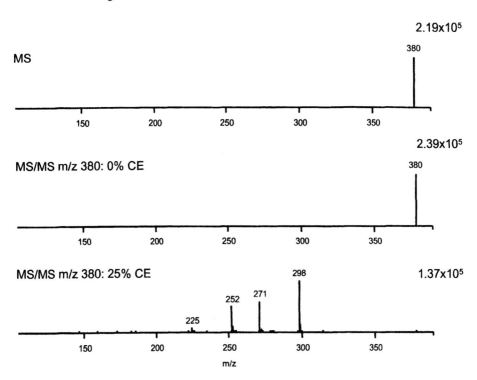

Figure 2 Mass spectrometric (top) and MS/MS (middle, bottom) spectra of the gepirone protonated molecule. The isolation and resonance excitation used for precursor selection and ion dissociation allow for complete and efficient conversion of the precursor ion to product ions. With no collision energy, no change in precursor ion abundance sensitivity is experienced between the MS scan and the MS/MS scan (top, middle spectra). Application of collision energy causes complete conversion of the precursor, with no loss of overall ion abundance (middle, bottom spectra).

C. The Role of Time-of-Flight Instruments

The coupling of ESI with time-of-flight mass spectrometry (TOFMS) expanded the role of these instruments in drug metabolism (17–19). TOFMS instruments are comparable in size and cost to those of SQMS but have several advantages. One of the first characteristics apparent with TOFMS instruments is their ability to operate with relatively high mass resolution and to make accurate mass measurements. Accurate mass measurements can provide structural information not attainable with quadrupole instruments by allowing elemental compositions to be established. An example of this ability is shown in

Fig. 3. Further structure elucidation data can be acquired through in-source collision-induced dissociation, similar to that possible for SQMS. In this manner, a compound of known elemental composition can be fragmented, and each fragment's elemental composition can be determined. However, there is the same reliance as in SQMS on low chemical background and absolute LC separation.

The higher mass resolving power attainable with TOF measurements can also provide a degree of selectivity in quantification that is unattainable with SQMS, and that specificity can be comparable to that of TQMS. This selectivity is a result of the ability to discriminate among mass peaks having similar nominal masses but different exact masses. A comparison of quantitative selectivity is presented in Fig. 4, which compares the discriminating abilities of TOFMS, SQMS, and TQMS measurements of a blank sample containing an endogenous nominal mass interference. In this example, an interfering component with m/z 330.1 contributes signal in the SQMS m/z 329.3 channel of interest, because of the insufficient resolving power of the quadrupole. The selective reaction monitoring (SRM) of m/z 329/132 is specific for the analyte of interest, allowing the TQMS to discriminate against the contribution of the interfering component. The higher resolving power of the TOF instrument also provides additional discrimination against the interference compared to that of SQMS.

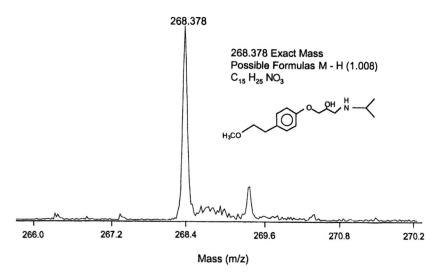

Figure 3 Accurate mass measurement spectrum of metoprolol by TOFMS. From these data, an elemental composition of $C_{15}H_{25}NO_3$ was determined to be most probable.

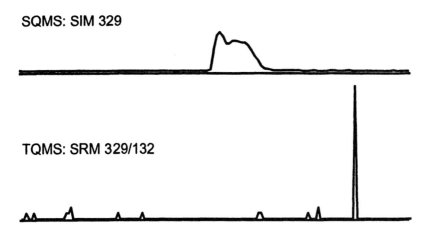
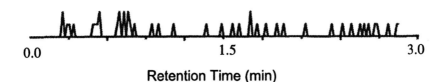

Figure 4 Comparison of the relative selectivities of MS selective ion monitoring (SQMS, top), MS/MS selective reaction monitoring (TQMS, middle), and MS accurate mass measurements (TOFMS, bottom) for the analysis of drug samples with an endogenous interference. Here, accurate mass selectivity is comparable to that obtained with MS/MS.

In TOFMS, a complete mass spectrum is acquired with each pulse of an ion packet. Because pulsing rates can be on the microsecond time scale, relatively large duty cycles can be obtained. These large duty cycles translate directly into increased sensitivity as more of the ions produced in the source contribute to the overall measurement. Like ITMS, TOFMS does not carry a sensitivity penalty for operation in full-scan mode. This feature can be used to automate method development and data acquisition—similar to that described for ITMS, where the appropriate ions for quantification can be determined rapidly postacquisition.

The rapid data collection rates inherent in TOFMS allow many spectra to be acquired over very narrow chromatographic peaks, making TOFMS ideal for high efficiency chromatography and capillary electrophoresis. Figure 5 shows a typical chromatogram from a high speed chromatography run. Quadrupole instruments begin to alias narrow chromatography peaks in full-scan mode due

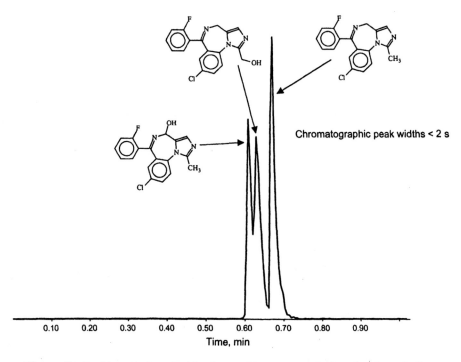

Figure 5 Rapid separation of midazolam and its two monohydroxylated metabolites by HPLC/TOFMS. High speed, efficient chromatography methods produce chromatographic peaks less than 2 s wide. These widths challenge the ability of quadrupole instruments to adequately define them.

to their inefficient duty cycle (mass filter), which results in loss of accuracy, precision, and sensitivity. With TOFMS, spectra can be collected fast enough that a high degree of spectral averaging can be employed to increase the sensitivity of the experiment while still maintaining the good definition of chromatographic peaks required for accurate quantification.

D. The Role of Hybrid Quadrupole/TOF Instruments

Recently developed hybrid quadrupole/TOF instruments (QqTOF) (20,21) provide advantages of both types of instruments, especially for structure elucidation (22). These instruments combine the mass filter and collision cell of a TQMS instrument with a TOF region as the second mass analyzer. A QqTOF spectrometer can thus operate in a true MS/MS mode while providing accurate mass measurements of the resulting product ions. Of course, it can also operate solely as a TOF mass spectrometer to provide accurate mass measurements of molecular species. Unfortunately, unlike a triple-quadrupole mass spectrometer, the QqTOF instrument cannot operate in true precursor or neutral loss modes.

III. DEVELOPMENT APPLICATIONS OF ELECTROSPRAY MASS SPECTROMETRY

In drug metabolism, development ESI is applicable to two types of studies: quantitative analysis to investigate drug pharmacokinetics and structural studies to permit the identification of drug metabolites.

A. Quantitative Analysis

Prior to the introduction of ESI-MS, quantitative analysis of drugs in biological fluids required HPLC with UV detection or, occasionally, electrochemical and fluorescent detection. A smaller number of drugs required gas chromatographic (GC) analysis with various detectors, including mass spectrometers. The latter (GC/MS) was considered the ultimate for trace-level quantitative analysis.

The initial acceptance and rapid expansion of HPLC/MS are due for the most part to the large number of biological samples requiring quantitative analysis. This is quite evident from the number of contract research organizations (CROs) dedicated to HPLC/MS analysis that sprang up in the 1990's. In many of these CROs, and indeed in many laboratories throughout the pharmaceutical industry, most mass spectrometry time is dedicated to quantitative analysis. Applications to clinical samples account for the bulk of published analyses and methodology (4,6,7,9,23–26). Other published methods focused on application to toxicokinetics (27) or preclinical pharmacokinetics and the confirmation of drug residues in food-producing animal tissues (28–30).

One of the main advantages of ESI is its inherently high sensitivity for many classes of compounds. Many of the early published assays demonstrated detection limits much lower than those possible by other common quantitative analysis tools (24) except GC/MS. The increased potency of contemporary therapeutic agents continues to require extremely trace level detection.

Another significant advantage, accounting for the acceptance and wide application of ESI-MS for quantitative analysis, is its specificity, particularly in combination with tandem mass spectrometry. This combination can reduce the requirement for extensive sample preparation and chromatographic resolution and permit shorter analysis times. For many assays, sample preparation is limited to protein precipitation or simple liquid extraction (31,32). Standard solid-phase extraction is used off-line (24), or on-line during the HPLC step (33). Several assays rely on direct plasma injection (34,35), and many sample preparation strategies have been automated (36).

Notwithstanding the assumed specificity of ESI-MS/MS analysis, it is clear that significantly reducing sample cleanup and limiting chromatographic resolution may yield misleading quantitative analytical results. Coeluting drug metabolites, such as conjugates or *N*-oxides, may contribute to the signal from the parent drug because of ion dissociation prior to mass analysis (37). Even more serious, matrix components may suppress or potentiate the analyte signal (38).

Under conditions of minimal HPLC retention, different plasma sources generate substantially different standard curve slopes and inadequate assay accuracy and precision (37). Identical slopes and acceptable accuracy and precision have been obtained by improving chromatographic resolution. Apart from efforts to achieve extensive chromatographic resolution, several other approaches can eliminate matrix effects. These include more extensive sample cleanup (39), the use of a stable isotope internal standard (40), and postcolumn addition of a signal-enhancing HPLC modifier to counteract ion suppression (41). In some instances, switching from ESI to APCI eliminates matrix effects (37,40). In our laboratory, the potential for matrix effects is assessed during the development of quantitative analytical methods by comparing the signal from pure standard solutions with that from plasma extract spiked with the analyte after sample preparation. For rugged method development, biological samples from different individuals and/or multiple sources must be tested for matrix effects.

Because not all compounds are amenable to electrospray ionization, several scientists investigated enhancing the sensitivity and detectability of neutral and nonpolar analytes via chemical derivatization by converting monofunctional organic molecules to "electrospray-active" derivatives (42–44).

B. Definitive Drug Biotransformation

In drug development, the goal of biotransformation studies is the elucidation of metabolite profiles in the preclinical species used for safety assessment and the comparison of these profiles to those in humans. The relationships between preclinical and human metabolic profiles contribute to understanding a drug's safety and efficacy. ESI-MS is currently indispensable for elucidation of the structures of metabolites in in vivo and in vitro matrices.

As opposed to quantitative analysis, which can in most instances be accommodated on simple single-stage mass spectrometers, structure determination greatly benefits from the reliable and selective collisional dissociation provided by tandem mass spectrometers. Some structural information may be obtainable on single-stage instruments by relying on the higher pressure region between the orifice and skimmer cone to effect dissociation (called in-source CID) (45,46). This, however, requires complete LC separation of the chemical species of interest, owing to the inability of single-stage instruments to select an isolated precursor mass. In addition, scattering losses and inefficient dissociation in the high pressure region of an ESI source limit the analytical sensitivity of this technique. By contrast, tandem mass spectrometers are ideal for structure elucidation of metabolites.

The applicability of TQMS to metabolite identification and mixture analysis was well recognized prior to the introduction of ESI (47,48). Several scan modes provide key structural class information. Certain constant neutral loss (CNL) and precursor scans can be employed (49) to determine specifically the presence of certain conjugated drug metabolites, as shown in Table 1. These scan modes help to quickly identify probable conjugate molecular ions from spectra of complex mixtures.

Definitive studies generally use radiolabeled drugs to allow mass balance determination, quantification, and profiling of all drug-related materials in plasma, excreta (urine, feces, and bile), microsomal, hepatocyte, and liver slice incubates. In the course of HPLC profiling of metabolites, radioactivity detection permits pinpointing the retention times of radioactive drug-related metabolites for focused mass spectrometric characterization. A system for

Table 1 Class-Characteristic Fragmentations of Drug Conjugates Under CID Conditions

Conjugate class	Mode (+/−)	Scan
Glucuronides	+/−	CNL 176 u ($-C_6H_8O_6$)
Phenolic sulfates	+	CNL 80 u ($-SO_3$)
Aliphatic sulfates	−	Precursors of m/z 97 (HSO_4)
Sulfonates	−	Precursors of m/z 81 (HSO_3)
Sulfinates	−	CNL 64 u ($-HSO_2$)
Aryl-GSH	+	CNL 275 u ($-C_{10}H_{17}N_3O_6$)
Aliphatic-GSH	+	CNL 129 u ($-C_5H_7NO_3$)
N-Acetylcysteines	−	CNL 129 u ($-C_5H_7NO_3$)
Coenzyme A thioesters	+	Precursors of m/z 428 (ADP^+)
	−	Precursors of m/z 339 and 357
Carnitine butyl esters	+	Precursors of m/z 103 ($C_4H_7O_3$)
Taurines	+	Precursors of m/z 126 ($Tau+H^+$)
Phosphates	−	Precursors of m/z 63 (PO_2^-) and 79 (PO_3^-)

simultaneous HPLC radioactivity and mass spectrometry monitoring, shown in Fig. 6, was first described in 1993 (50) and has since gained wide application (8,51–54). The effluent from the HPLC is split, with the bulk directed to a radioactivity-monitoring detector and the balance used for ESI-MS detection. Both the radioactivity and mass spectrometric data are acquired in the same time domain by a single data system to facilitate matching the two types of information.

As a soft ionization technique, ESI is especially applicable to polar and labile drug metabolites, particularly conjugated metabolites (53,54). Its superior sensitivity permits the characterization of minor metabolites that could not be determined with other approaches. Indeed, the combination of ESI and tandem MS is credited with the identification of several novel and unprecedented metabolic pathways (53,55).

Despite impressive success and significant contribution to drug biotransformation, metabolite identification by combined ESI and TQMS can have several limitations. The mass accuracy of the quadrupole mass analyzer is not sufficient to allow the determination of elemental compositions of product and precursor ions, and, accordingly, structure identification is not always straightforward. Positional isomers and other metabolites can yield isobaric precursor ions of differing structures and convoluted or overlapping product ion spectra. Hybrid quadrupole time-of-flight mass spectrometers were shown to permit accurate mass assignment of precursor and product ions of drug metabolites

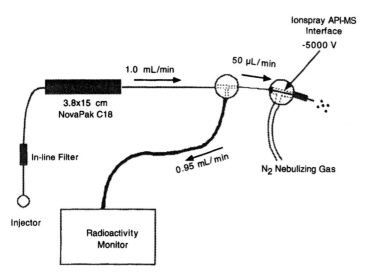

Figure 6 System for simultaneous HPLC radioactivity and mass spectrometry monitoring in which the sample stream is split postcolumn to a radioactivity detector. Radioactivity and mass spectrometric data are collected in the same time domain.

(22,56). It should be noted, however, that using accurate mass assignments of product ions for structural elucidation is possible only if the precursor ion is free of isobaric interferences. Another limitation of TQMS (and also QTOF) is that the generated CID spectrum does not permit distinguishing primary product ions from those formed by secondary or tertiary dissociation steps, further complicating the task of spectral interpretation. Currently the most popular and affordable design for multiple-stage mass analysis (MS^n, where $n > 2$) is the quadrupole ion trap design. This design permits the isolation and trapping of the precursor ion for each stage of mass spectrometry. As a result, ion lineage can be established, and lineage facilitates the interpretation of metabolite collision-induced dissociation (CID) spectra (57).

Pinpointing the exact position of certain metabolic modifications is not always possible by any mass spectrometric approach. During collision-induced dissociation, the energy required to cause the loss of an oxygen atom or a water molecule is often lower than that required for other dissociations. As a result, the position of an N-oxide or a hydroxyl group is not readily evident from the examination of CID spectra. Because of this, unambiguous characterization of certain metabolites requires non-mass spectrometric approaches for structure determination, including classical chemical derivatization (58–60). Chemical derivatization followed by electrospray ionization tandem mass spectrometry also aids the identification of the site of glucuronidation of certain conjugated metabolites (58). The introduction of LC-NMR (61) and enhancements of NMR sensitivity have allowed the application of NMR to low-level drug metabolites. The combination of ESI-MS/MS and parallel or sequential NMR analysis (62) represents an emerging advance for drug biotransformation.

IV. DISCOVERY APPLICATIONS OF ELECTROSPRAY MASS SPECTROMETRY

In the past few years, drug discovery through high speed, high throughput synthesis has received great emphasis in the pharmaceutical industry. Combinatorial chemistry and high speed synthesis dramatically increase the number of new chemical entities (NCEs) produced for discovery efforts. Coupled with licensing of libraries and augmentation of compound production through outsourcing synthesis, the number of NCEs passing into drug discovery practices overwhelms traditional means of selecting optimal lead compounds. The response to this stress on the traditional drug discovery processes has been the development of high throughput screening methods along with the redefinition of which data should be collected in drug discovery and how those data should be used.

Drug metabolism has traditionally played its largest role on the development side of the pharmaceutical industry. A recent industrial trend is to examine the discovery process for insights into how to make better leads and reduce candidate attrition late in development. In this examination, drug metabolism is seen in a new light. Specifically, drug metabolism data are viewed as essential criteria for evaluating each lead compound in discovery. These data help to focus quickly on the most promising leads or key structural features in the vast sea of NCEs by providing information such as metabolic stability (clearance), drug–drug interactions, caco-2 cell absorption (intestinal permeability), and physicochemical parameters. These data can be used to focus on, and select, those compounds or structures in a chemical series that have the best druglike properties and thus the greatest probability of survival as candidates.

Of course, ESI-MS plays a big role in this drug metabolism effort. However, its role in discovery is significantly different from its role in development (63). Although LC/MS is still used for quantification and structure elucidation, the emphasis in discovery is on throughput and efficiency, as traditional drug metabolism experiments are transformed into high throughput screens (64,65). These screens are highly automated, and their LC/MS components are discussed in more detail in the following sections.

A. Discovery Quantification

The primary goal of a discovery program is to select drug candidates with optimum druglike properties and high survivability for further development. Candidate selection based on pharmacokinetic profiles requires analysis of relatively large sets of compounds. In vivo experiments with single compounds can be time-consuming and prohibitive as to the number of animals required. For these reasons, in vivo experiments are usually reserved until late in discovery, when prediction and intuition are used to choose small numbers of compounds from larger series. Given the opportunity, however, drug metabolism scientists prefer to obtain in vivo characterization of as many compounds as reasonable. The timing of these in vivo studies depends on therapeutic approach and on corporate philosophy. However, the need to increase the throughput of in vivo pharmacokinetic studies affects the entire industry.

Cocktail dosing methods, also known as N-in-one dosing and cassette dosing (66–69), have been developed to overcome some of the limitations to in vivo experiments. In these methods, multiple compounds are mixed into a single dose solution and administered to the animal. The cocktails vary in size, but even ones containing only a few compounds provide the advantage of reducing animal handling, sample processing, and sample analysis.

The complicated nature of the resulting samples, however, requires an analytical system with high discriminatory ability. HPLC/MS is central to the success of these methods. The complicated mixtures resulting from these experiments require the selectivity of chromatographic separation, and the low dose concentrations require the sensitivity of mass spectrometric detection.

One drawback to cocktail dosing is the underlying potential for drug–drug interactions and possible saturation of metabolism. To reduce this risk, the amounts of individual compounds in the dose are reduced so that exposure is lessened. Additionally, a reference compound for which the pharmacokinetic profile is known is often included with each cocktail dose. A further drawback is that pharmacodynamic information for each compound is not preserved in cocktail dose experiments. The analysis of mixtures of compounds has a greater potential of suffering from ionization suppression and matrix effects than the analysis of samples containing single analytes. These parameters must be established and understood for each experiment.

Combining multiple samples from singly dosed animals, referred to as cassette analysis, is one approach that avoids drug–drug interactions (70). In this approach, single compounds are administered to animals, and samples from different animals are pooled for analysis. Each sample contains multiple compounds and represents a single timepoint from each animal. Because samples are combined, dilution is inherent, and the requirement for a low limit of detection increases proportionally. Although drug–drug interactions are avoided, compound suppression and matrix effects must still be understood for these types of studies.

Another approach to in vivo studies in drug discovery is plasma pooling (71–73), where appropriate aliquots of the plasma samples obtained from all timepoints are pooled, yielding a single sample. The concentration of this sample is proportional to the pharmacokinetic area under the curve (AUC), which can be used to estimate clearance and bioavailability. This method produces a single sample per animal, preserves pharmacodynamic information, and avoids drug–drug interactions. However, the time profile is lost, hence V_{ss}, V_{beta}, C_{max}, t_{max}, and half-life cannot be determined. Plasma pooling provides a higher analytical throughput but does not increase the throughput of the biological experiment. Increasing both the analytical and biological throughput is accomplished by combining plasma pooling with cocktail dosing (73).

B. Discovery Biotransformation

Collecting metabolic rate data on all compounds naturally raises the issue of metabolic routes, yet metabolite identification is often an area that is underrepresented in drug discovery. Most metabolite identification discovery efforts involve a limited number of compounds and are directed toward advancing a promising candidate rather than to understanding structure–metabolism relationships within and between chemical series. This is due to the difficulty of automating the data collection and the need for expertise in spectral interpretation and biotransformation pathways. The latter is mostly a function of experience and remains the rate-limiting step. Much of the data collection, however, can be automated. Many of the MS experiments are "knowledge-based," lending themselves to automation (for example, product ion scans for [MH + 16] will provide data on possible hydroxylation and oxides), and modern instrumentation provides sufficient instrument control in an appropriate programming or scripting environment. In addition, simple gradient chromatographic methods can be developed that are suitable for a wide range of diverse metabolites. Data collection has been automated in several labs using different approaches (74–78).

The Rapid Automated Biotransformation IDentification (RABID) approach of Cole et al. (74,75) relies on a data-dependent search in which the metabolite search process is dependent upon data collected in a preliminary experiment. The number and types of metabolites searched for vary from sample to sample, depending on the possible metabolism found in the preliminary experiment. In RABID, the preliminary experiment employs a chromatogram constructed of conventional (Q1) mass spectra. A metabolite extractor routine extracts data from this chromatogram using a customizable search list of possible types of metabolism, as shown in Fig. 7. This search list contains all common metabolism possibilities, along with specific, user-added possibilities such as dealkylations. A metabolite finder routine decides which metabolites in the list may be present and determines the retention time(s) for each in the chromatogram. The experiment builder routine divides the chromatogram into collection "zones" and determines which MS/MS structural data to collect in each zone. A second injection of the sample is made, and the appropriate MS/MS data are acquired in each zone during the chromatographic run. A schematic representation of the results of the metabolite finder and experiment builder routines is shown in Fig. 8. In this example, the instrument would collect MS/MS data only for [MH + 16], [MH − 14], [MH + 32], and ["User1"] within zone 1; for [MH + 16], [MH + 32] and ["User2"] within zone 2; etc. In this manner, data are collected most efficiently and scan times can be adjusted

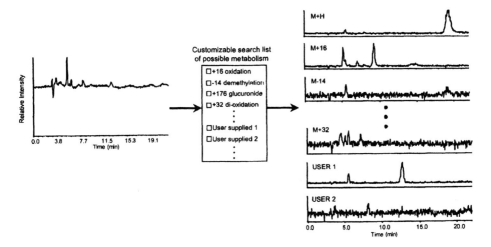

Figure 7 In RABID, the search list (middle) contains common possible types of metabolism and can be customized to include user-defined metabolism or exclude unexpected routes of metabolism. The metabolite extractor routine uses this list to extract the data from a preliminary Q1 ion chromatogram (left), resulting in extracted ion chromatograms for all possible metabolites of interest (right).

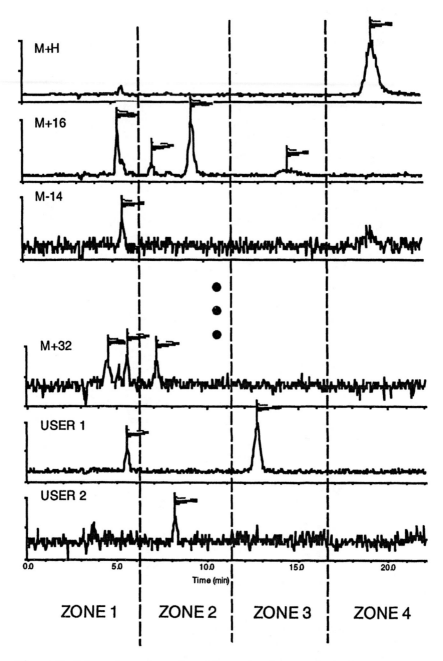

Figure 8 Schematic representation of the results of the metabolite finder and experiment builder routines. During these routines, possible metabolite peaks are found, their retention times are determined, and the chromatogram is divided into collection "zones" for MS/MS data collection during a second chromatographic analysis.

to provide the best ion statistics while properly defining the chromatographic peaks in each zone. RABID requires only two sample injections for collecting metabolic data, and the entire process is automated. Employing rapid gradients with short, narrow-bore columns allows metabolic data to be collected for each sample in less than 15 min.

The Intelligent Automated Mass Spectrometry (INTAMS) system developed by Yu et al. (77) builds on the RABID approach by adding the ability to detect all eluting components. All components are treated as possible metabolites during the collection of structural MS/MS data. In this way INTAMS potentially collects data for any unexpected metabolites present in the sample. Because data may be collected for many species unrelated to the parent compound, this approach underscores the necessity for knowledge of biotransformation processes. This same group also explored data-dependent scanning on the Finnigan LCQ ion trap to automate metabolite identification (78). The LCQ software allows data from a scan to be used in determining the type of experiment to perform in the next scan. By setting appropriate rules, the instrument can collect many of the MS/MS data in a single chromatography run. One shortcoming of dynamically using data from preceding scans is the possibility of missing coeluting metabolite peaks, particularly if they are of low abundance.

Fernandez-Metzler et al. (76) took a more computationally intensive approach to automating the collection of metabolite data, but one that has the potential to be more inclusive and rigorous. Their method relies on correlation analysis of MS spectra, comparing a control sample to an experimental sample to distinguish metabolites from endogenous species. Product ion spectra are collected for the mass species identified from the correlation analysis, and cross-correlation of the spectra is used to identify their relationships with the parent drug spectrum. One advantage of this approach is the ability to accommodate the analysis of metabolites in mixtures of several drugs. The computational requirements make this approach somewhat time-consuming; however, the analysis of drug mixtures can help to offset the computational time.

V. AUTOMATION FOR HIGH THROUGHPUT

The advent of high throughput screening of large numbers of compounds to determine drug metabolism properties early in discovery presents significant challenges to the analyst. HPLC/MS remains the most broadly applicable analytical tool owing to the wide structural diversity and low detection level requirements encompassed by drug metabolism screens. However, synthetic chemistry and biological testing are amenable to plate-based parallel-format experiments, whereas HPLC/MS is a linear technique characterized by serial injection and analysis of individual samples. Parallel-format compound synthesis and biological assays stretch serial analysis techniques like LC/MS to their limits. For LC/MS to impact on and contribute to discovery screening efforts, analyses must be rapid enough to keep pace with parallel screening. Not only must each component of the analytical process, such as chromatography and data review, be as rapid as possible, but also the idle times between system components must be eliminated or minimized. Because many thousands of samples are generated each day, vigilance in eliminating unnecessary idle time is liberally rewarded as each second eliminated per sample analysis amplifies into significant overall throughput gains and run capacities.

The tasks involved in LC/MS analysis are readily divided into four areas: sample preparation and liquid handling, determination of MS conditions, sample analysis, and data processing. Samples need to be prepared, the necessary MS conditions need to be rapidly determined for hundreds of compounds per week, the samples need to be analyzed, and the resulting data need to be reviewed. Several automation and multitasking strategies to accomplish these tasks have been implemented with sufficient speed to keep up with the parallel-format sample generation. Each is described in the following sections.

A. Sample Preparation

Following the wide utilization and demonstrated analytical speed of ESI-HPLC/MS, sample preparation became rate-limiting to higher throughput. To keep pace with the increasing demands for bioanalysis, several parallel and serial approaches for automating sample preparation were proposed. The most productive parallel strategies (batch analysis) use microtiter plate approaches, permitting the preparation of 96 or 384 samples simultaneously. The two main standard approaches for sample preparation are liquid/liquid extraction (LLE) and solid-phase extraction (SPE). Some level of automation of both approaches by microtiter plate format was demonstrated for the HPLC/MS analysis of drugs and metabolites. The availability of 96 microtiter plates containing solid-phase extraction sorbents represents a significant advance in automated sample preparation (79). Both loosely packed beds and particle-loading membranes are currently available in the microtiter plate format. The small bed volume of the membrane allows the elution of the analytes in a very small solvent volume, permitting direct HPLC injection and negating the need for the time-consuming solvent evaporation and reconstitution steps. All SPE steps were automated by the use of a programmable liquid-handling station, resulting in sample processing speed 30 times faster than manual LLE or single-cartridge SPE methods

(36). A generic SPE extraction scheme and a rational, semiautomated approach for sim,ultaneous evaluation and optimization of SPE chemistries were recently described (80). Liquid/liquid extraction has also been automated by the use of microtiter plates (81,82). The main advantages here are the ease of method development and the ready availability of numerous extraction solvents with varying specificity.

Serial on-line sample preparation has also been reported to be in use in a number of laboratories. PROSPEKT is an automated on-line SPE system that processes each sample on an individual disposable precolumn (33,83). Other direct plasma injection systems use a single extraction column that permits retaining the analytes while washing away salts and polar plasma proteins. Following extraction, the analytes are eluted on the analytical column (34). To enhance the throughput of these methods, the extraction and the analytical steps are synchronized such that one sample is extracted while the previous one is analyzed on the analytical column. Despite this enhancement, on-line sample processing with column switching does not currently match the high throughput of parallel methods. Other laboratories have recently demonstrated performing the extraction and analytical steps on a single column, using ultra-high flow rates, permitting the analysis of drugs and metabolites by direct injection of crude plasma (27). Although coined "turbulent flow," this process does not employ true turbulent flow conditions (84).

B. Determination of MS Conditions

Before LC/MS can be applied to quantitative analysis, the most appropriate precursor ion and MS/MS transition must be selected. This selection is an often overlooked, but critical, step that can potentially be time-consuming, because the number of compounds for which MS conditions need to be determined in a high throughput screen easily reaches into hundreds and even thousands on a per-week basis. To perform quantitative analysis on this many different compounds each week, one must be able to rapidly determine MS and MS/MS conditions for each compound and prepare the corresponding injection sequences needed for the actual sample analyses.

An automated MS workstation was developed (85) to allow rapid determination of optimal MS and MS/MS conditions and method building for use in LC/MS quantification. This system uses custom software to determine optimal precursor ions and polarity ($+/-$) as well as product ions and collision energies for 96 compounds in less than 60 min with minimal analyst intervention. The system requires a total of two injections for each compound. The first injection is a Q1 scan that alternately switches between positive and negative ion modes to determine the optimal precursor ion and scan polarity for each compound. The second injection alternately switches between several collision energies (10–50 eV) to determine the optimal product ion and collision energy for each precursor found from the first injection. Finer optimization of the collision energy can easily be accomplished by incorporating additional collision energies; however, the overall scan time will increase. The software chooses the best conditions for each compound and builds an analytical method for all compounds. The software looks for common adducts and losses and compensates for collisionally stable compounds by reverting to a precursor–precursor measurement (basically, selected ion monitoring with additional selectivity gained through using collision gas and energy to lower chemical background). Any compounds falling outside the software's parameters are flagged, and the user is alerted during data review.

A major feature of the custom software is the data review capability. The scanning results are summarized pictorially in a 96-well presentation after completion of data acquisition. Color codes are used to provide information about the collected data. This pictorial data review function provides immediate feedback on data quality and allows the user to select the well location of any compound to view all associated spectra ($+/-$ Q1 scans and high/low energy product ion scans).

C. Sample Analysis

A time dependency on additive steps is inherent in linear serial techniques such as LC/MS. In a multistep process, such as autosampling–LC–MS–data analysis, the overall analysis time is the summation of the times taken for each step. One approach (86) is to recognize the weaknesses inherent in the serial nature of traditional LC/MS analysis and overcome them by focusing on the duty cycle(s) as the ultimate limitation of throughput. Modular analytical systems are created, and the idle times within each module are reduced by making each module as efficient as possible. The key to minimizing idle times between modules is to run modules simultaneously and independently. An important concept is to identify those parts of the duty cycle in which each component is doing *useful* work toward a goal (where the duty cycle is the total time that the component is doing work). We call this the "analytical time constant" and strive to delineate and identify this time constant for each of the components described below.

The useful work done by the autosampler is solely the filling of the injection port. Although part of the overall duty cycle of the instrument, the time taken in needle washing, arm travel, and sample draw does not contribute useful work to the overall process. In fact, these steps add considerable overhead to an analysis and have the greatest impact on throughput speed. The Gilson Multiprobe autosampler is used in a dual-probe mode to make best use of the biological assay format and to deliver sufficient samples to the LC/MS instrumentation so as not to limit throughput. At the start of each autosampler cycle, two samples are aspirated and transferred to the injection ports and delivered sequentially to the dual-column chromatography system (discussed below). The autosampler runs independently of the rest of the system and completes its wash routine and aspirates the next set of samples while the previous set is being processed through the chromatography system. Functionally, the analytical time constant is maximized because the autosampler continuously fills the injection ports as needed by the system. All ancillary steps are performed simultaneously with the analyses. The number of samples that can be processed through a single chromatography system per autosampler cycle time, then, dictates the throughput speed of the system. The number of samples that can be processed is a function of how the chromatography is implemented.

The chromatography requirements for high throughput analysis are universality and speed, because the need to analyze numerous samples for a wide variety of compounds precludes the ability to develop specialized methods. These requirements are met through trade-offs: Speed is achieved at the expense of resolution, and optimum sensitivity is sacrificed for universality. Figure 9 illustrates a concept important for achieving high throughput chromatography: Most of a chromatogram is not used for quantification. This returns to the original concept of the analytical time constant. The chromatographic duty cycle encompasses the entire runtime. However, the part of this duty cycle that is actually performing useful work for quantification is just the elution volume of the analyte, which is approximately 6% of the runtime in the example shown. The remaining 94% of the runtime is important to the chromatography, because it involves the separation and re-equilibration times, but is not important to the overall goal of quantification. Analyses are held hostage to this unused chromatographic space. Important to high throughput, then, is the necessity of taking this unused space off-line from the analysis. The space previously occupied can then be made useful again by allowing additional elutions of analytes to occur, as illustrated in the schematic of Fig. 10.

The chromatography system that allows this type of operation (87–89) is implemented by using dual columns, high pressure mixing, and step gradients. A dual-column switching valve is used, as shown in Fig. 11. In this approach, samples are loaded on the first column in high aqueous mobile phase. The column eluent is directed to waste for a short time, after which the valve switches and the high organic mobile phase is directed through the column, causing analyte elution. In essence, the chromatography system serves to desalt and

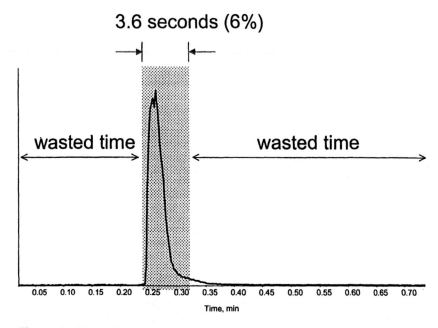

Figure 9 Most of a chromatogram is not used for quantification. The part of the chromatographic duty cycle in which useful work is performed for quantification is just the elution volume of the analyte of interest, which is only 6% of the runtime in this example.

concentrate a sample. Mass spectrometric detection provides analytical selectivity. Chromatographic resolution is minimal, and there is little or no variability in retention times. In fact, retention time shifts are often indicative of system problems. Simultaneously with the elution step, the second sample is flushed from the sample loop onto the second column, and the process repeats such that one column is in the load mode while the other is being eluted. At the column dimensions and flow rates used (1 mm × 15 mm, 1.5 mL/min), re-equilibration time is negligible, allowing nearly simultaneous loading and eluting and nearly parallel processing. In this manner, the deadtimes associated with sample loading and column re-equilibration are minimized, because these steps take place off-line from the actual measurement, and chromatographic peaks from separate samples are stacked closely together. This combination of autosampler and chromatography implementations allows for four injections per minute on a continual basis. Only the speed at which samples can be loaded into the injection ports and the width of the chromatographic peaks limit the throughput.

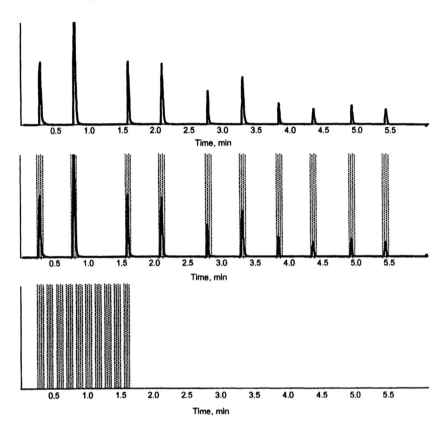

Figure 10 Analyses are hostage to unused chromatographic space. Taking the unused space off-line from the analysis allows occupation by additional elutions of analytes.

The data system provides useful work only in acquiring the data to be used for quantification. Data acquisition software typically consumes 5–15 sec per data file simply opening and closing the file, which does not contribute to throughput. When coupled to a chromatography system capable of presenting a sample injection every 15 s, the data file handling becomes a throughput bottleneck. For this reason, all samples related to a compound are collected into a single data file (90), as shown in Fig. 12. On the surface this seems trivial, but the implications are much deeper and are fundamental to the high throughput process. Although this small amount of time seems unimportant, it subtracts significantly from the throughput of a large sample run.

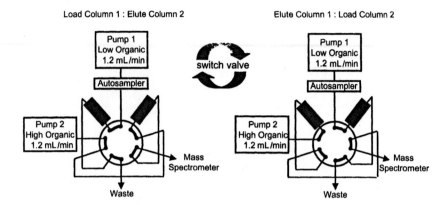

Figure 11 The use of dual columns and a switching valve allows the equilibration, sample loading, and washing of an off-line column while elution occurs on the on-line column. The use of particular mobile phases and valve timing allows stacking of elution volumes to take better advantage of the chromatographic time constants.

Avoiding the deadtimes associated with opening and closing data files can save 8 h over the course of a typical run of 96 compounds (20 samples per compound). Similar time savings are achieved during data processing, where each file must be reopened for review and processing.

D. Data Processing and Review

Coincident with the challenge of high throughput bioanalysis is the challenge of processing and reviewing vast amounts of the resultant raw data. EvaLution (91) is a custom software application uniquely designed for processing data collected from the high throughput quantitative system described. Commercially available software products are not equipped to quantify from data files in which all injections for a given compound are acquired into a single file. EvaLution is customized to process these types of files and extract peak areas into an associated spreadsheet. The application can be used with or without a standard curve, and relevant biological calculations such as P_{app} values (caco-2) and half-life (metabolic lability) are built in. In addition, EvaLution can run in the background during data acquisition, allowing data review as soon as data acquisition is complete. This tight process integration allows higher throughput by providing biologically relevant data very soon after sample analysis.

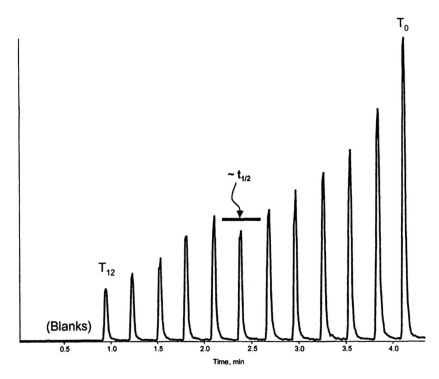

Figure 12 A complete hepatocyte lability experiment collected into a single data file. Avoiding deadtimes associated with opening and closing data files during data collection and processing saves considerable time in the overall process.

Collecting and processing data in this manner has additional features. Because all data for an experiment exist in a single data file, the number of data files produced is greatly reduced. For example, caco-2 experiments produce 20 samples per compound, which, for 96 compounds, would result in 1920 individual files to manage and review. Collecting all injections for a given compound in a single file results in a 20-fold decrease in the number of data files produced (96 files compared to 1920). As stated earlier, the time saved by handling files this way is similar to the time saved in the acquisition step, furthering the throughput of the process. In addition, archiving and retrieval of the data are facilitated by the fact that an entire experiment is contained within a single file and associated with a single name, usually that of the compound analyzed.

The ability to review data is severely taxed in a high throughput environment where the amount of data can overwhelm the capacity to look at, and judge the quality of the results from, every injection. Ironically, data review is more important in a high throughput environment, where there is less time to catch and correct problems further downstream in the process. EvaLution leverages the advantages of human pattern recognition by presenting to the reviewer a visual representation of the entire experiment with all relevant injection data in context. Coupled with the greatly reduced number of files, presenting all data in context allows the reviewer to quickly determine the quality and integrity of both the individual injections and the experiment as a whole. Figure 13 illustrates this point. The top chromatogram is a metabolic lability experiment in which both the biology and analysis were without problem. The quality of the analysis is apparent from the good shape of each chromatographic peak, and the integrity of the biology is quickly deduced from the context of the experimental profile, which shows the expected decrease over time (in the example shown, the samples were run from later to earlier timepoints). The middle chromatogram is an example of an analysis problem, as evident by the poor peak shapes for each injection. A problem with the biological experiment is manifested in the variability from the expected data profile, as shown in the bottom chromatogram. A decision on each experiment can be made in as little as a few seconds because the visual representation of the entire experiment allows a reviewer to take in the experiment as a whole, working in a "pattern recognition" mode rather than looking at individual chromatograms and piecing together peak area numbers. In our experience, all data for 96 compounds (1920 injections) can be reviewed and any needed changes made in as little as 5 min. Fundamentally, data review is the most important part of the high throughput process, because it ensures integrity and relevance between the analytical measurement and the biological representation.

E. Parallel Approaches to Quantification

The recent development and introduction of multisprayer ESI sources (92–95) open the possibility of approaching high throughput quantification through parallel sample analysis using multiple chromatography streams and a single mass spectrometer. This implementation has proved successful for compound characterization and control of multiple fraction collectors during autopurification. Presently, it has yet to be applied to quantification, but several characteristics of this type of system have been explored. A multisprayer source must be indexed to operate with multiple chromatography systems. This means that

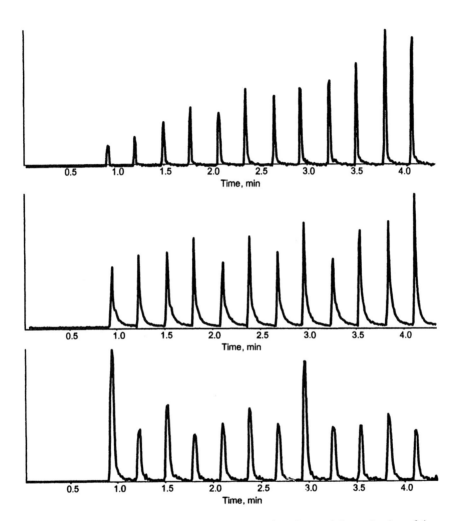

Figure 13 Having all data in context allows quick review and determination of the quality and integrity of both the individual injections and the experiment as a whole. The top data set represents a metabolic stability experiment in which both the biology and analytical data are without problem. An analysis problem is evident in the middle data set from the poor peak shapes, and close attention should be paid to peak integration. A problem with the biological experiment in the bottom data set is manifested in an unexpected stability profile.

the fraction of the time spent measuring each eluting analyte peak is inversely proportional to the number of sprayers. For example, an instrument with an eight-sprayer source would spend only one-eighth of its time measuring a particular sprayer. If the chromatographic peaks are of insufficient width, they could potentially be aliased by the sampling frequency, which would result in quantitative imprecision. Figure 14 represents a multisprayer simulation for a 2–3 s wide chromatographic peak, compared with a single sprayer. For this simulation, data were collected for eight ions in a looped sequence with a cycle time of 1.2 s, in accordance with published data (96). This allowed 150 ms

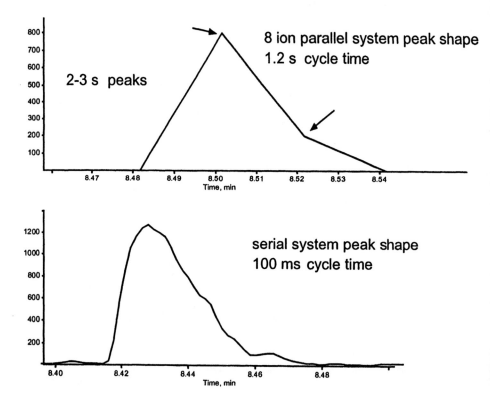

Figure 14 Multisprayer simulation for a 2–3 s wide chromatographic peak compared with that of a single sprayer. The relatively slow cycle time with multisprayers may not allow collection of sufficient data points to adequately define the peak. In this example, only two data points were collected over the peak (arrows), resulting in peak aliasing and poor precision.

sampling times for each sprayer. In practice, 150 ms measurements would not be achieved, because rotation of the sampling device takes up some of this time (50–80 ms). The 1.2 s cycle time used in the example only allowed two to three data points to be collected for each chromatographic peak. Although quality spectra were obtained, the number of data points was insufficient to define the peak with adequate precision. The normal single sprayer has a large duty cycle and allows the peak to be defined with good precision.

To take advantage of multiple sprayer sources for quantification, the chromatography must be adjusted to provide sufficiently wide peaks to avoid their aliasing and discrimination. The multisprayer source gains throughput through multiplexing parallel sample introductions. The system throughput actually decreases, however, with increasing chromatographic speed, owing to the narrower peak widths that limit the number of samples that can be measured in parallel. This is in contrast to the fast serial introduction described earlier, where throughput increases proportionally to the chromatographic speed, because the narrower peaks allow more samples to be analyzed in a shorter time (97). This dichotomy is illustrated in Fig. 15. Figure 16 quantifies the differences in a plot of precision as a function of chromatographic peak width for a single sprayer using serial introduction and a simulated multisprayer using parallel introduction. The data presented here were obtained from a simulation of a multiple-sprayer source based on current multisprayer technology, which was performed as an exploration of the benefits and pitfalls of the approach. The true capabilities and limitations need to be characterized on an actual source, because the simulations do not take

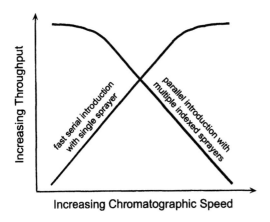

Figure 15 Relative relationship between chromatographic speed and throughput for serial and parallel sample introduction.

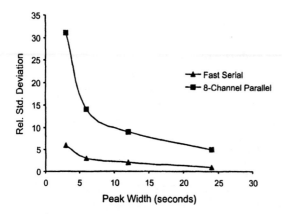

Figure 16 Precision vs. chromatographic speed for a single sprayer using serial introduction and a simulated multisprayer using parallel introduction. The measurements were made at 10 times the limit of detection. Each data point is the average of 30 measurements.

into account performance at limits of detection, liquid loads, carryover, gains in cycle efficiency, or numerous other variables.

REFERENCES

1. WA Garland, ML Powel. Quantitative selected ion monitoring (QSIM) of drugs and metabolites in biological matrices. J Chromatogr Sci 19:392, 1981.
2. FC Falkner, HG Fouda, FG Mullins. A gas chromatographic/mass spectrometric assay for flutroline, a carboline antipsychotic agent with direct derivatization on a moving needle injector. Biomed Mass Spectrom 9:482, 1984.
3. HG Fouda, DC Hobbs, JE Stambaugh. A sensitive assay for the determination of hydroxyzine in plasma; human pharmacokinetics. J Pharm Sci 68:1456, 1979.
4. HG Fouda, J Lukaszewicz, D Clark, B Hulin. Metabolism of the hypoglycemic agent CP-68,722 in the rat. Metabolite identification by gas chromatography mass spectrometry. Xenobiotica 21(7):925, 1991.
5. J Henion, G Schultz, D Mulvana. Integrating atmospheric pressure ionization LC/MS techniques into the pharmaceutical industry. Am Pharm Rev 2(3):43–48, 1999.
6. HG Fouda, MR Nocerini, RP Schneider, CL Gedutis. Quantitative analysis by HPLC-atmospheric pressure chemical ionization. The analysis of the renin inhibitor CP-80, 794 in human serum. J Am Soc Mass Spectrom 2:164, 1991.

7. HG Fouda, RP Schneider. Quantificative determination of the antibiotic azithromycin in human serum by high performance liquid chromatography (HPLC)-atmospheric pressure chemical ionization mass spectrometry: Correlation with a standard HPLC-electrochemical method. Ther Drug Monitoring 17:179, 1995.
8. C Prakash, A Kamel, W Anderson, HG Fouda. Characterization of metabolites of CP-88,059 in rat using HPLC/RAM/ESI/MS/MS. Proc 42nd Annual ASMS Conf on Mass Spectrometry and Allied Topics, Chicago, IL, 1994, p 355.
9. M Avery, D Mitchell, FC Falkner, HG Fouda. Simultaneous determination of tenidap and its stable isotope analogue in plasma by HPLC/API-MS/MS. Biol Mass Spectrom 21:353, 1992.
10. M Yamashita, JB Fenn. Electrospray ion source. Another variation on the free-jet theme. J Phys Chem 88(20):4459–4459, 1984.
11. JE Fulford, T Sakuma, DA Lane. Real-time analysis of exhaust gases using triple quadrupole mass spectrometry. In: M Cooke, AJ Dennis, GL Fisher, eds, Polynuclear Aromatic, Hydrocarbons. Phys Biol Chem, 6th Int Symp, 1982, pp 279–303
12. GJ Van Berkel, GL Glish, SA McLuckey. Electrospray ionization combined with ion trap mass spectrometry. Anal Chem 62(13):1284–1295, 1990.
13. SA McLuckey, GJ Van Berkel, GL Glish, JC Schwartz. Electrospray and the quadrupole ion trap. J Pract Aspects Ion Trap Mass Spectrom 2:89–141, 1995.
14. ME Bier, JC Schwartz. Electrospray-ionization quadrupole ion-trap mass spectrometry. In: RB Cole, ed. Electrospray Ionization Mass Spectrometry, New York: Wiley, 1997, pp 235–289.
15. M Wakefield, L Lopez, L Land. Quantitation of pharmaceutical products using quadrupole and ion trap mass spectrometers. Proc 44th ASMS Conf on Mass Spectrometry and Allied Topics, Portland, OR, 1996, p 614.
16. MJ Cole, J Cunniff, D Drexler, E Hemenway, A Land. Large scale quantitative analysis of pharmaceuticals via LC-MS and LC-MS/MS ion trap mass spectrometry. Proc 46th ASMS Conf on Mass Spectrometry and Allied Topics, Orlando, FL, 1998, p 12.
17. I Lazar, E Lee, A Rockwood, M Lee. General considerations for optimizing a capillary electrophoresis-electrospray ionization time-of-flight mass spectrometry system. J Chromatogr A 829:279–288, 1998.
18. JF Banks, T Dresch. Detection of fast capillary electrophoresis peptide and protein separations using electrospray ionization with a time-of-flight mass spectrometer. Anal Chem 68:480, 1996.
19. G Choudhary, C Horvath, J Banks. Capillary electrochromatography of biomolecules with on-line electrospray ionization and time-of-flight mass spectrometry. J Chromatogr A 828:469–480, 1998.
20. HR Morris, T Paxton, A Dell, J Langhorne, M Berg, RS Bordoli, J Hoyes, RH Bateman. High sensitivity collisionally-activated decomposition tandem mass spectrometry on a novel quadrupole/orthogonal-acceleration time-of-flight mass spectrometer. Rapid Commun Mass Spectrom 10:889–896, 1996.

21. A Shevchenko, I Chernushevich, W Ens, KG Standing, B Thompson, M Wilm, M Mann. Rapid "de novo" peptide sequencing by a combination of nanoelectrospray, isotopic labeling and a quadrupole/time-of-flight mass spectrometer. Rapid Commun Mass Spectrom 11:1015–1024, 1997.
22. G Hopfgartner, IV Chernushevich, T Covey, JB Plomley, R Bonner. Exact mass measurement of product ions for the structural elucidation of drug metabolites with a tandem quadrupole orthogonal-acceleration time-of-flight mass spectrometer. J Am Soc Mass Spectrom 10:1305–1314, 1999.
23. JD Gilbert, TV Olah, DA McLoughlin. High performance liquid chromatography with atmospheric pressure ionization tandem mass spectrometry as a tool in quantitative bioanalytical chemistry. ACS Symp Ser 619:330–350, 1996.
24. D Wang-Iverson, M Arnold, M Jemal, A Cohen. Determination of SQ 33,600, a phosphinic acid containing HMG CoA reductase inhibitor, in human serum by high-performance liquid chromatography combined with ion spray tandem mass spectrometry. Biol Mass Spectrom 21:189–194, 1992.
25. D Garteiz, T Madden, D Beck, W Huie, K McManus, J Abbruzzese, W Chen, R Newman. Quantitation of dolastatin-10 using HPLC/electrospray ionization mass spectrometry. Application in a phase I clinical trial. Cancer Chemother Pharmacol 41(4):299–306, 1998.
26. C Dass. Recent developments and applications of high-performance liquid chromatography-electrospray ionization mass spectrometry. Curr Org Chem 3(2):193–209, 1999.
27. D Zimmer, V Pickard, W Czembor, C Muller. Comparison of turbulent flow chromatography with automated solid-phase extraction in 96 well plates and liquid-liquid extraction used as plasma sample preparation technique for liquid chromatography-tandem mass spectrometry. J Chromatogr A 854:23–35, 1999.
28. MJ Cole, J Lukaszewicz, MA Nowakowski, HG Fouda. An HPLC/ESI-tandem mass spectrometry method for the confirmation of Doramectin residues in cattle liver and fat. 41st Annual ASMS Conf on Mass Spectrometry and Allied Topics, San Francisco, CA, 1993.
29. RP Schneider, MJ Lynch, JF Ericson, HG Fouda. Atmospheric pressure electrospray ionization for mass spectrometry of Semduramicin and other polyether ionophores. Anal Chem 63:1789, 1991.
30. RP Schneider, JF Ericson, MJ Lynch, HG Fouda. Confirmation of danofloxacin residues in chicken and cattle liver by microbore HPLC-electrospray ionization tandem mass spectrometry. Biol Mass Spectrom 22:595, 1993.
31. P Bennett, YT Li, R Edom, J Henion. Quantitative determination of Orlistat (tetrahydrlipostatin, RO 18-0647) in human plasma by high-performance liquid chromatography coupled with ion spray tandem mass spectrometry. J Mass Spectrom 32:739–749, 1997.
32. Y Xia, D Whigan, M Jemal. A simple liquid-liquid extraction with hexane for low-picogram determination of drugs and their metabolites in plasma by

high-performance liquid chromatography with positive ion electrospray tandem mass spectrometry. Rapid Commun Mass Spectrom 13:1611–1621, 1999.
33. A Marchese, C McHugh, J Kehler, H Bi. Determination of Pranlukast and its metabolites in human plasma by LC/MS/MS wit PROSPEKT™ on-line solid phase extraction. J Mass Spectrom 33:1071–1079, 1998.
34. SR Needham, MJ Cole, HG Fouda. Direct plasma injection for high-throughput HPLC/MS quantitative analysis. J Chromatogr B 718:87–94, 1998.
35. M Jemal, M Huang, X Jiang, Y Mao, M Powel. Direct injection versus liquid-liquid extraction for plasma analysis plasma by high performance liquid chromatography with tandem mass spectrometry. Rapid Commun Mass Spectrom 13:2125–2132, 1999.
36. J Janiszewski, R Schneider, K Hoffmaster, M Swyden, D Wells, H Fouda. Automated sample preparation using membrane microtiter extraction for bioanalytical mass spectrometry. Rapid Commun Mass Spectrom 11:1033–1037, 1997.
37. BK Matuszewski, ML Constanzer, CM Chavez-Eng. Matrix effect in quantitative LC/MS/MS analyses of biological fluids: A method for determination of finasteride in human plasma at picogram per milliliter concentrations. Anal Chem 70:882–889, 1998.
38. P Kebarle, L Tang. From ions in solution to ions in the gas phase: The mechanism of electrospray mass spectrometry. Anal Chem 65(22):972A–986A, 1993.
39. D Burhman, PI Price, PJ Rudewicz. Quantitation of SR 27417 in human plasma using electrospray liquid chromatography: A study of ion suppression. J Am Soc Mass Spectrom 7:1099–1105, 1996.
40. SD Clarke, HM Hill, TAG Noctor, D Thomas. Matrix-related modification of ionization in bioanalytical liquid chromatography-atmospheric pressure ionization tandem mass spectrometry. Pharm Sci 2:203–207, 1996.
41. J Yamaguchi, M Ohmichi, S Jingu, N Ogawa, S Higuchi. Utility of postcolumn addition of 2-(2-methoxyethoxy)ethanol, a signal-enhancing modifier, for metabolite screening with liquid chromatography and negative ion electrospray ionization mass spectrometry. Anal Chem 71:5386–5390, 1999.
42. G Van Berkel, J Quirke, R Tigani, A Dilley. Derivatization for electrospray ionization mass spectrometry. 3. Electrochemically ionizable derivatives. Anal Chem 70:1544–1554, 1998.
43. J Quirke, Y Hsu, G Van Berkel. Ferrocene-based electroactive derivatizing reagents for the rapid selective screening of alcohols and phenols in natural product mixtures using electrospray-tandem mass spectrometry. J Nat Prod 63:230–237, 2000.
44. G Zyrek, U Karst. Liquid chromatography-mass spectrometry method for the determination of aldehydes derivatized by the Hantzsch reaction. J Chromatogr 864:191–197, 1999.
45. RF Straub, RD Voyksner. Determination of penicillin G, ampicillin, amoxicillin, cloxacillin and cephapirin by high-performance liquid chromatography-electrospray mass spectrometry. J Chromatogr 647(1):167–181, 1993.

46. JR Perkins, CE Parker, KB Tomer. Nanoscale separations combined with electrospray ionization mass spectrometry: Sulfonamide determination. J Am Soc Mass Spectrom 3(2):139–149, 1992.
47. MS Lee, RA Yost. Tandem mass spectrometry for the identification of drug metabolites. Annu Rep Med Chem 21:313–321, 1986.
48. KM Straub, C Garvie, M Davis. Characterization of polar drug conjugates in biological matrices by FAB-MS/MS. Proc 33rd Annual ASMS Conf on Mass Spectrometry and Allied Topics, San Diego, CA, 1985, pp 472–473.
49. T Baille. The role of LC-ionspray MS/MS in studies of drug metabolism and toxicology. Proc 42nd Annual ASMS Conf on Mass Spectrometry and Allied Topics, Chicago, IL, 1994, p 862.
50. HG Fouda, MJ Avery, K Navetta. Simultaneous HPLC radioactivity and mass spectrometry monitoring. A drug metabolism application. 41st Annual ASMS Conf on Mass Spectrometry and Allied Topics, San Francisco, CA, 1993.
51. MJ Avery, KA Navetta, D Dalvie, FC Falkner, HG Fouda. Identification of Tenidap metabolites in monkey by HPLC/atmospheric pressure ionization mass spectrometry and simultaneous radioactivity monitoring. Proc 42nd Annual ASMS Conf on Mass Spectrometry and Allied Topics, Chicago, IL, 1994, p 61.
52. HG Fouda, MJ Avery, D Dalvie, FC Falkner, LS Melvin, RA Ronfeld. Disposition and metabolism of Tenidap in the rat. Drug Metabol Disp 25(2):140, 1997.
53. RP Schneider, HG Fouda, PB Inskeep. Tissue distribution and biotransformation of Zopolrestat, an aldose reductase inhibitor, in rats. Drug Metab Disp 26:1149, 1998.
54. C Prakash, A Kamel, W Anderson, H Howard. Metabolism and excretion of the antipsychotic drug ziprasidone in rat after administration of a mixture of ^{3}H and ^{14}C labeled ziprasidone. Drug Metab Disp 25:206, 1997.
55. C Prakash, A Kamel, D Cui. Identification of novel benzisothiazole cleaved products of ziprasidone. Drug Metab Disp 25:897, 1997.
56. G Bowers, P Chandrasurin, J Castro-Perez, S Preece. In vitro metabolite identification of delta-opioid agonist GW411484 in human liver S9 using a hybrid quadrupole time of flight mass spectrometer. North American ISSX meeting, Nashville, TN, 1999.
57. S Werness, SR McGown. Metabolite identification using data dependent MSn techniques. 47th Annual ASMS Conf on Mass Spectrometry and Allied Topics, Dallas, TX, 1999, p 3074.
58. T Kondo, K Yoshida, Y Yosimura, M Motohashi, S Tanayama. Characterization of conjugated metabolites of a new angiotensin II receptor antagonist, Candesartan cilexetil, in rats by liquid chromatography/electrospray tandem mass spectrometry following chemical derivatization. J Mass Spectrom 31:873–878, 1996.
59. C Prakash, J O'Donnell, C Khojasteh, T Sutfin, H Fouda. Excretion, pharmacokinetics and metabolism of the substance P receptor antagonist, CJ-11,974 in humans: Identification of polar metabolites by LC/MS/MS and chemical derivatization. 9th North American ISSX Meeting, Nashville, TN, 1999.

60. C Prakash. Strategies for the identification of novel and unusual metabolites, 47th Annual ASMS Conf on Mass Spectrometry and Allied Topics, Dallas, TX, 1999.
61. SD Taylor, B Wright, WE Clayton, ID Wilson. Practical aspects of the use of high performance liquid chromatography combined with simultaneous nuclear magnetic resonance and mass spectrometry. Rapid Commun Mass Spectrom 12:1732–1736, 1998.
62. GB Scarfe, B Wright, E Clayton, S Taylor, ID Wilson, JC Lindon, JK Nicholson. ^{19}F-NMR and directly coupled HPLC-NMR-MS investigation into the metabolism of 2-bromo-4-trifluoromethylanilin in rat: A urinary excretion balance study without the use of radiolabeling. Xenobiotica 28:373–388, 1998.
63. MJ Cole, S Needham, P Jeanville, J Janiszewski. Redefining HPLC/MS approaches to meet the drug discovery challenge. 14th Asilomar Conf on Mass Spectrometry, Pacific Grove, CA, 1998.
64. DJ Tweedie, J Janiszewski, D Johnson. In vitro drug interaction studies in drug discovery: Higher throughput methods. Int Symp Laboratory Automation and Robotics, Boston, MA, 1996.
65. D Tweedie, L Cohen, J Janiszewski, D Johnson, D Mankowski, R Whalen. In vitro inhibition studies in 96-well plates: Higher throughput methods to assess drug interaction potential. Proc Soc Biomol Screening, San Diego, CA, 1997.
66. TV Olah, DA McLoughlin, JD Gilbert. The simultaneous determination of mixtures of drug candidates by liquid chromatography/atmospheric pressure chemical ionization mass spectrometry as an in-vivo drug screening procedure. Rapid Commun Mass Spectrom 11:17–23, 1997.
67. J Berman, K Halm, K Adkinson, J Shaffer. Simultaneous pharmacokinetic screening of a mixture of compounds in the dog using API LC/MS/MS analysis for increased throughput. J Med Chem 40:827–829, 1997.
68. MC Allen, TS Shah, WW Day. Rapid determination of oral pharmacokinetics and plasma free fraction using cocktail approaches: Methods and application. Pharm Res 15(1):93–97, 1998.
69. LW Frick, KK Adkinson, KJ Wells-Knecht, P Woolard, DM Higton. Cassette dosing: Rapid in vivo assessment of pharmacokinetics. Pharm Sci Technol Today 1(1):12–18, 1998.
70. B Kuo, T Van Noord, MR Feng, DS Wright. Sample pooling to expedite bioanalysis and pharmacokinetic research. J Pharm Biomed Anal 16(5):837–846, 1998.
71. RA Hamilton, WR Garnett, BJ Kline. Determination of mean valproic acid serum level by assay of a single pooled sample. Clin Pharmacol Ther 29:408–413, 1981.
72. LE Riad, KK Chan, RJ Sawchuk. Determination of the relative formation and elimination clearance of two major carbamazepine metabolites in humans: A comparison between traditional and pooled plasma analysis. Pharm Res 8:541–543, 1991.
73. CECA Hop, Z Wang, Q Chen, G Kwei. Plasma-pooling methods to increase throughput for in vivo pharmacokinetic screening. J Pharm Sci 87(7):901–903, 1998.

74. MJ Cole, JW Gauthier, EW Luther, HG Fouda. Automated HPLC/tandem mass spectrometric approaches for rapid identification of metabolic pathways. Proc 43rd Annual ASMS Conf on Mass Spectrometry and Allied Topics, Atlanta, GA, 1995, p 569.
75. MJ Cole, JW Gauthier, EW Luther, HG Fouda. Strategies for automated mass spectrometric detection and identification of metabolic pathways. 13th Montreux Symp on Liquid Chromatography Mass Spectrometry, Montreux, Switzerland, 1996.
76. C Fernandez-Metzler, K Owens, T Baille, R King. Rapid liquid chromatography with tandem mass spectrometry-based screening procedures for studies on the biotransformation of drug candidates. Drug Metab Dispos 27:32–40, 1999.
77. X Yu, D Cui, MR Davis. Identification of in vitro metabolites of indinavir by "intelligent automated LC-MS/MS" (INTAMS) utilizing triple quadrupole tandem mass spectrometry. J Am Soc Mass Spectrom 10:175–183, 1999.
78. LL Lopez, X Yu, D Cui, MR Davis. Identification of drug metabolites in biological matrices by intelligent automated liquid chromatography/tandem mass spectrometry. Rapid Commun Mass Spectrom 12:1756–1760, 1998.
79. B Kaye, WJ Herron, PV Mcrae, S Robinson, DA Stopher, RF Venn, WJ Wild. Rapid, solid phase extraction technique for the high throughput assay of darifenacin in human plasma. Anal Chem 68:1658–1660, 1996.
80. J Janiszewski, M Swyden, H Fouda. High throughput method development approaches for bioanalytical mass spectrometry. J Chromatogr Sci 38:2000, pp 255–258.
81. N Zhang, KL Hoffman, W Li, DT Rossi. Semi-automated 96-well liquid-liquid extraction for quantitation of drugs in biological fluids. J Pharm Biomed Anal 22:131–138, 2000.
82. S Steinborner, J Henion. Liquid-liquid extraction in the 96-well plate format with SRM LC/MS quantitative determination of methotrexate and its major metabolite in human plasma. Anal Chem 71:2340–2345, 1999.
83. D Mcloughlin, TV Olah, JD Gilbert. A direct technique for the simultaneous determination of 10 drug candidates in plasma by liquid chromatography-atmospheric pressure chemical ionization mass spectrometry interfaced to a PROSPEKT solid phase extraction system. J Pharm Biomed Anal 15:893–1901, 1997.
84. J Ayrton, RA Clare, GJ Dear, DN Mallett, RS Plumb. Ultra-high flow rate capillary liquid chromatography with mass spectrometric detection for the direct analysis of pharmaceuticals in plasma at sub-nanogram per millilitre concentrations. Rapid Commun Mass Spectrom 13:1657–1662, 1999.
85. KM Whalen, KJ Rogers, MJ Cole, JS Janiszewski. AutoScan: An automated workstation for rapid determination of mass and tandem mass spectrometry conditions for quantitative bioanalytical mass spectrometry. Rapid Commun Mass Spectrom 14:2072–2079, 2000.

86. JS Janiszewski, KJ Rogers, KM Whalen, MJ Cole, TE Liston, H Fouda. A high-capacity LC/MS system for the rapid bioanalysis of samples generated from plate based metabolic screening. Anal Chem 73(7):1459–1501, 2001.
87. J Janiszewski, K Rogers, K Whalen, H Fouda, M Cole. High speed, high capacity analysis of drug metabolism screens. Symp Chem Pharm Struct Anal, Princeton, NJ, 1999.
88. JS Janiszewski, KJ Rogers, KM Whalen, MJ Cole, TE Liston, HG Fouda. A High-capacity LC/MS system for the rapid bioanalysis of samples generated from plate based metabolic screening. Int Symp Lab Automation and Robotics, ISLAR99, Boston, MA, 1999.
89. KJ Rogers, KM Whalen, JS Janiszewski, MJ Cole. Drug discovery aerobics. Getting LC/MS in shape for high speed, high capacity analysis of metabolism screens. 10th Int Symp Pharm Biomed Anal, Washington, DC, 1999.
90. KR Rogers, KM Whalen, JS Janiszewski, MJ Cole. An integrated module approach to enhancing the speed of LC/MS analyses. 24th Int Symp on High Performance Liquid Phase Separations and Related Techniques (HPLC 2000), Seattle, WA, 2000.
91. KM Whalen, KJ Rogers, JS Janiszewski, EW Luther, HG Fouda, MJ Cole, E Duchoslav. EvaLution: A software tool for high-speed processing and review of LC/MS data collected from ADME screening. 48th Annual ASMS Conf on Mass Spectrometry and Allied Topics, Long Beach, CA, 2000.
92. DB Kassel, T Wang, L Zeng. Parallel fluid electrospray mass spectrometer. PCT Int Appl 1999.
93. L Zeng, DB Kassel. Developments of a fully automated parallel HPLC/mass spectrometry system for the analytical characterization and preparative purification of combinatorial libraries. Anal Chem 70(20):4380–4388, 1998.
94. V De Biasil, N Haskins, A Organ, R Bateman, K Giles, S Jarvis. High throughput liquid chromatography/mass spectrometric analyses using a novel multiplexed electrospray interface. Rapid Commun Mass Spectrom 13(12):1165–1168, 1999.
95. O Hindsgaul, DC Schriemer. Electrospray device for mass spectrometer. PCT Int Appl 1999.
96. V De Biasi, N Haskins, K Giles, A Organ. Multiple LC-MS: Parallel and simultaneous analyses of liquid streams by LC-TOF mass spectrometry using a novel eight way interface. 47th Annual ASMS Conf on Mass Spectrometry and Allied Topics, Dallas, TX, 1999.
97. T Covey. Strategies for ultra-high throughput LC/MS/MS. Proc Fed Anal Chem Spectrosc Societies, Vancouver, BC, Canada, 1999.

6
Electrospray Ionization Mass Spectrometry of Peptides and Proteins

Methodologies and Applications

Joseph A. Loo and Greg W. Kilby
Pfizer Global Research and Development, Ann Arbor, Michigan

I. INTRODUCTION

"The secret of life itself [is] how a protein molecule is able to form, from an amorphous substrate, new protein molecules that are made after its own image." This quotation from Linus Pauling in 1937 (1) illustrates the importance of proteins, even during an age when, compared to today's standards, relatively little was known about the structure and function of proteins. Of course, we know now that proteins perform endless functions in biology, and high resolution structures are available for a number of important proteins. Our knowledge base is expanding at an increasing pace. It is for this reason that mass spectrometry has been applied for the study of peptides and proteins for decades. From amino acids, the building blocks of proteins, to larger peptides and eventually to proteins, the progress and development of mass spectrometry can be gauged by the molecular weight range of the method.

It was not too long ago that insulin, a 5 kDa polypeptide hormone (in its monomeric state), was a benchmark for the mass spectrometry community. Ionization methods such as fast atom bombardment (FAB) could be used to determine the molecular weight of insulin. However, the sensitivity for molecules at such large mass was not terribly good. Mass analyzers, such as magnetic sector instruments, were applied because their upper mass limit went typically to a mass-to-charge ratio (m/z) of 10,000. Other desorption or ionization methods, such as plasma desorption (PD), using time-of-flight analyzers, were used to determine molecules as large as insulin and even proteins as large as 35 kDa pepsin (2–4). However, these applications of mass spectrometry were by no means "routine." The field of biological mass spectrometry turned toward the sky in the summer of 1988.

Professor John Fenn presented stunning results using electrospray ionization and a simple low m/z range quadrupole mass analyzer at the American Society for Mass Spectrometry (ASMS) conference in San Francisco in June 1988 (5). One of us (JAL) was among the relatively small audience that listened and watched with amazement as Professor Fenn showed ESI mass spectra of proteins ranging from insulin to 40 kDa alcohol dehydrogenase. After one got used to the plethora of peaks in the mass spectrum for a single protein analyte (i.e., multiple charging), the potential of ESI-MS and the ease with which on-line chromatographic and electrophoretic separations methodologies can be interfaced for protein analysis was obvious (6). Very quickly, ESI mass spectra for 66 kDa bovine serum albumin and its 133 kDa dimer form were published (7), and enhancements in sensitivity, mass range, and instrumentation followed. The number of publications in the scientific literature describing the application of ESI-MS for peptide and protein analysis rapidly increased

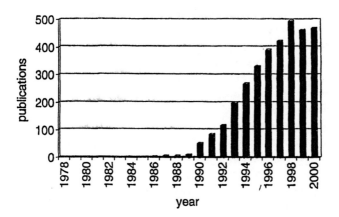

Figure 1 Publications on ESI-MS of peptides and proteins for the years 1978–2000. The literature search used SciFinder (American Chemical Society).

after the 1988 ASMS conference (Fig. 1), as commercial mass spectrometry vendors first supplied ESI interfaces to existing instruments and later developed dedicated ESI-MS systems.

There have been a number of reviews on the applications of ESI-MS for protein analysis (8,9). The intent of this chapter is to highlight some of the newer developments and applications, using examples from our laboratory. Other major new applications, such as proteomics, noncovalent complexes, microfluidics, and combinatorial chemistry, are featured in other chapters in this volume.

II. ESI-MS OF INTACT PROTEINS: METHODS

The measurement of molecular masses of intact peptides and proteins is nearly routine for many bioanalytical laboratories. In fact, the sensitivity of the ESI-MS analysis can be quite high, especially with the advances in low-flow micro- and nanoelectrospray sources (10). Detection limits of low-flow ESI-MS for proteins are typically less than 1 fmol. For example, from a 2 nM solution of 14 kDa lysozyme, a mass spectrum can be obtained from the consumption of 82.5 attomoles (amol) (Fig. 2). However, some classes of proteins have proven to be more difficult to determine by ESI-MS. For example, glycoproteins present special difficulties because there is heterogeneity from the different glycoforms. The successful analysis of extremely heterogeneous samples depends on the resolving power of the mass analyzer, because the individual charge states for each analyte form must be resolved to some degree for a mass measurement.

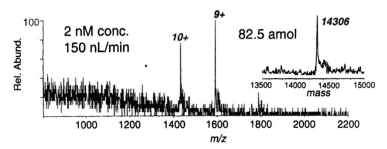

Figure 2 Electrospray ionization mass spectrum of 2 nM solution of hen egg white lysozyme with a microelectrospray source (150 nL/min) and a magnetic sector mass spectrometer. Approximately 82.5 amol of analyte was consumed to generate the spectrum. The inset shows the deconvoluted mass domain spectrum.

The solubility (or rather, the insolubility) of the analyte can present another degree of difficulty. Many hydrophobic peptides and proteins are difficult to handle by ESI-MS because they have limited solubility in the "typical" solvent systems used for ESI (e.g., acetonitrile or methanol with water and an organic acid, such as acetic acid, trifluoroacetic acid, or formic acid) (11–17). Membrane proteins, such as bacteriorhodopsin, fall into this category. However, in some cases, solvent conditions that have been varied to include other organic modifiers such as chloroform can be used successfully.

A. ESI-MS of Protein Mixtures

Mixture analysis also presents a difficult scenario because many mass spectrometers have limited resolving power. The multiple charging feature of electrospray ionization allows large mass molecules to be measured with m/z-limited analyzers. However, as a mixture becomes very complex, the multiplicity of peaks for each analyte can become an issue. Ultrahigh resolving power instruments such as Fourier transform ion cyclotron resonance (FTICR) mass spectrometers can be used successfully to analyze extremely complex mixtures, but this technique has limitations, such as susceptibility to space charge and low ion-storage capacity. Using gas-phase charge exchange reactions, one can reduce the charge distribution for a given analyte to a single or a few low charge states, thereby reducing the complexity of the mass spectra (18–20). However, this requires analyzers with very high m/z limits and/or specialized instruments.

One method to help resolve mixtures composed of differently charged analytes is to use the discriminating feature of ion-counting detectors. For example, the PATRIC array detector on the Finnigan MAT 900 series magnetic sector instruments can discriminate highly charged analytes (large mass proteins) from the low-charge peptides and other contaminants (buffers and salts, polymers, detergents, etc.) (21,22). The mass spectra shown in Fig. 3 illustrate this feature. Lentil lectin is a heterotetrameric $\alpha_2\beta_2$ protein complex, where the α-subunit is a 5.5 kDa polypeptide and the β-subunit is 19 kDa. The lentil lectin sample analyzed was composed of a mixture of five differently processed β-peptides. By varying the array detection voltage, thereby controlling the discrimination power of the detector, either the lower charge α-subunit peptides or the higher charge β-subunit protein can be selectively detected. This charge discrimination feature is also present with quadrupole time-of-flight (QTOF) instruments.

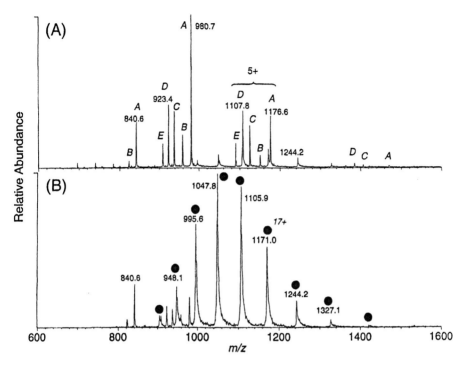

Figure 3 Electrospray mass spectra of lentil lectin proteins, obtained with a magnetic sector mass spectrometer and a PATRIC array detector. (A) With the array detector voltage set to 850 V, the lower charged alpha subunit peptides are preferentially detected ($A = 5877.5$, $B = 5749.4$, $C = 5621.2$, $D = 5534.1$, and $E = 5447.0$ Da). (B) Decreasing the array detector voltage to 675 V allows more selective detection of the more highly charged beta-subunit protein (M_r 19,890.1).

B. Sample Contamination

Other problems for ESI-MS result from samples containing high concentrations of salts and "non-ESI-friendly" buffers. The application of liquid chromatography (LC) with off-line or on-line ESI-MS is a good method for desalting. The use of membrane filtration (23) or on-line microdialysis (24,25) also has been presented with excellent results. However, a very rapid method for desalting solutions for ESI-MS analysis is the application of reverse-phase media, such as the commonly used POROS media.

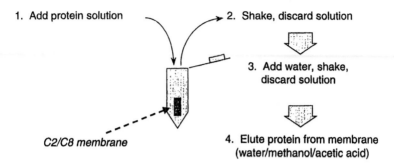

Figure 4 Procedure for using Empore reverse-phase membranes to desalt protein samples suitable for ESI-MS analysis.

We have used other reverse-phase media when acquiring ESI mass spectra from solutions containing high concentrations of salts and buffers. With the use of extraction disks made from reverse-phase separation materials (C2 and C8 membranes, 3M Empore disks), ESI mass spectra can be acquired from a solution containing 3.7 pmol of myoglobin and 0.25 M NaCl, with only 4.6 fmol total protein consumed for the MS measurement (26). The protein is bound to a C2 membrane, washed with water, and eluted off the membrane with a water/methanol/acetic acid solution and submitted to ESI-MS (Fig. 4). The total time for sample cleanup is less than 2 min. Other protein samples in solutions containing high salt concentrations are amenable to this sample cleanup technique. For example, a commercial sample of Benzonase (endonuclease, *Serratia marcescens*) contained 20 mM Tris HCl, 20 mM NaCl, 2 mM $MgCl_2$, and 50% (v/v) glycerol. The simple sample preparation utilizing a C2 membrane yielded the mass spectrum depicted in Fig. 5. The ESI mass spectrum from milk (bovine) is shown in Fig. 6. In an 8 oz (240 mL) serving, milk contains 12 g of carbohydrate, 12 mg sodium, and 8 g of protein. The ESI mass spectrum from milk after cleanup with a C2 membrane shows ions from the predominant 20 kDa casein variant proteins.

III. APPLICATIONS OF ESI-MS

A. Proteomics

Electrospray ionization mass spectrometry has greatly impacted the emerging field of proteomics because it provides a powerful means to identify proteins

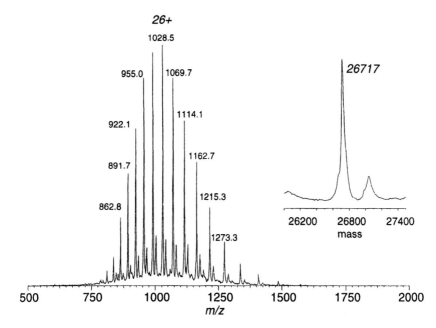

Figure 5 Electrospray mass spectrum of Benonase sample, desalted from a solution containing 20 mM Tris HCl, 20 mM NaCl, 2 mM MgCl$_2$, and 50% (v/v) glycerol with an Empore C2 membrane. The inset shows the deconvoluted mass domain spectrum.

separated by gel electrophoresis. Proteomics is an important and powerful approach that is being integrated into molecular biology research and drug discovery because it allows the examination of the cellular target of drugs, namely proteins. The mapping of proteomes, the protein complements to genomes, from organisms and tissues has been used in these projects for development of high throughput screens, for validating and forwarding new targets for drug discovery, for structure–activity relationship (SAR) development, for exploring mechanisms of action or toxicology of compounds, and for identification of biomarker proteins in disease.

The application of proteomics to study biochemical pathways and to identify potentially important gene products as targets for drug discovery is becoming well established in the literature. The "traditional" approach involves the separation of the highly complex protein mixture, with two-dimensional polyacrylamide gel electrophoresis (PAGE) the most popular method because of its high capacity and separation efficiency. Protein spots are digested in the gel media with trypsin, and the resulting peptides

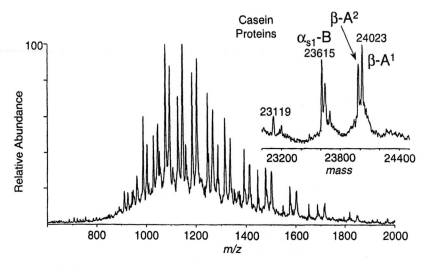

Figure 6 Electrospray ionization mass spectrum of casein proteins obtained from an Empore C2 membrane desalted sample of 2% reduced fat milk. The inset shows the deconvoluted mass domain spectrum.

are extracted from the gel and measured by MALDI- and ESI-based MS methods.

However, the molecular weight of the intact protein can be measured by MS, which provides a more accurate means than is possible with PAGE (typically ± 10%). Successful ESI-MS analysis of proteins separated and extracted from one-dimensional isoelectric focusing (IEF) gels have been published by our laboratory (27). Proteomics is also discussed in Chapters 00.

B. Disulfide Bond Determination

Molecular mass measurement can provide important structural information for proteins. The enhanced precision and accuracy afforded by electrospray ionization and modern MS analyzers allow the differentiation of even a 1 Da difference in mass. With the ultrahigh resolving power afforded by FTICR-MS, the sulfur isotopes can be measured to yield the number of sulfur atoms in a large protein (28). For a more conventional application, this feature of mass measurement can be used to differentiate changes due to amidation or deamidation, single residue mutations, phosphorylation, and other post-translational modifications.

As an illustration, consider proteins with disulfide linkages. Covalent bonds between cysteine residues are important components of a protein's conformation and folding process. The mass difference between a reduced and an oxidized disulfide bond is 2 Da. For small proteins, this mass difference can be readily measured by ESI-MS. Bovine pancreatic trypsin inhibitor (BPTI) is a 55 amino acid residue protein composed of three disulfide linkages in its native state. Figure 7 shows the ESI mass spectra of BPTI from its native disulfide state and the reduced [with the reducing agent dithiothreitol (DTT)] form. An obvious difference in the mass spectra for the two forms is the charge state distribution. One of the first observations that suggested that ESI-MS could be used to probe the conformation of a protein was this difference in charging for disulfide proteins (29). Reducing the Cys–Cys bonds allows the protein to "open up," unfold, and/or become more exposed to solvent, thereby increasing its propensity for charging. This is a common observation for disulfide proteins, and it can be used to give qualitative information on the disulfide

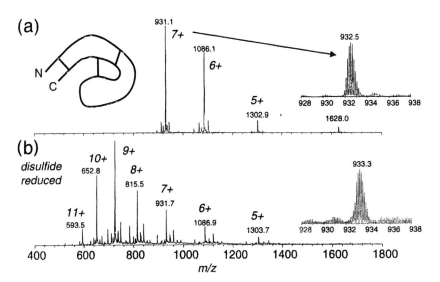

Figure 7 Electrospray ionization mass spectra of bovine pancreatic trypsin inhibitor (a) in its native 3-disulfide state and (b) the disulfide reduced (with DTT) form. The inset shows an expanded view of the 7+-charged ion, showing isotopic resolution and the shift in m/z due to the molecular weight difference between the disulfide reduced and oxidized forms (6 Da). Reducing the disulfide bonds often yields a shift of the multiple charge distribution to higher charging.

bonding state. However, by measuring the mass difference accurately between the two forms, the number of disulfide bonds can be determined (29–31). In the BPTI example, for the 6.5 kDa protein the mass difference upon reduction with DTT is an increase of 6 Da, consistent with three disulfide bonds.

Blocking free sulfhydryl groups with an alkylating agent is a common procedure prior to enzymatic digestion to prevent re-formation of disulfide linkages. Reaction with iodoacetic acid (100 mM concentration, 45°C for 45 min in the dark) results in irreversible carboxymethylation of the reduced cysteine groups. This increases the mass of the protein by 58 Da for every cysteine group present. In such situations, the need for a high resolution mass spectrometer to determine the number of cysteines present in the protein is alleviated. Bovine pancreatic ribonuclease B (RNase B) is a 124 amino acid

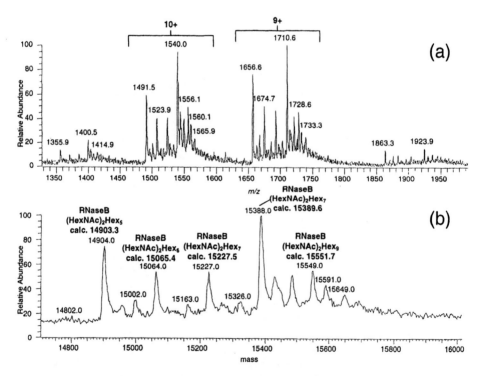

Figure 8 Electrospray ionization mass spectrum (a) and deconvoluted spectrum (b) of bovine pancreatic RNase B obtained with an ion trap mass spectrometer (Finnigan LCQ).

Peptides and Proteins 261

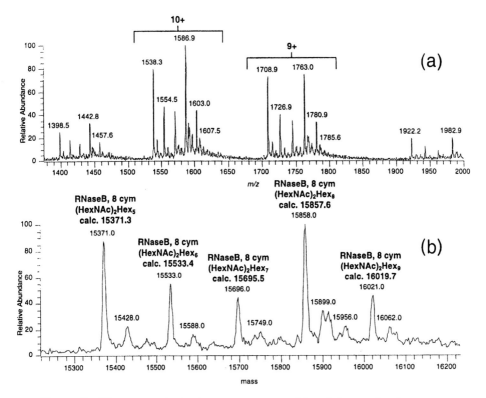

Figure 9 Electrospray ionization mass spectrum (a) and deconvoluted spectrum (b) of disulfide-reduced and carboxymethylated bovine pancreatic RNase B.

protein that has four disulfide linkages and five glycoforms in its native state. Mass spectra of RNase B were collected on an ion trap mass spectrometer while using a concentration of 0.01 mg/mL infused at a flow rate of 1 μL/min using a microelectrospray interface developed in-house. Figure 8 shows the raw and deconvoluted ESI mass spectra of RNase B in its native disulfide bound state. Figure 9 shows the mass spectra of reduced and carboxymethylated RNase B. The most notable difference between Figs. 8 and 9 is the increase in mass of ~469 Da for each of the five glycoforms. This is equivalent to 8 (469/58 = 8.1) carboxymethyl cysteines (Cym). This quick and simple experiment can be leveraged to provide quite powerful information on the total number of cysteine residues present in a protein. Carboxymethylation without DTT reduction can be used to determine the number of free cysteine residues present.

C. Noncovalent Protein Complexes

Electrospray ionization mass spectrometry has shown utility for the study of noncovalently bound complexes, particularly for determination of protein quaternary structure (32,33). The ESI-MS of noncovalent protein complexes can be applied to very large assemblies in excess of 1 MDa, as demonstrated by Heck and coworkers (34). (This applications is also discussed in Chapter 1.) The molecular mass measurement provides a direct determination of the stoichiometry of the binding partners, even for multiligand heterocomplexes. Because the mass measurements are accurate, stoichiometry of the binding partners is readily determined, even for the binding of small molecules (e.g., inhibitors and drug molecules) to targets of larger molecular mass. In most of the cases investigated, the complex stoichiometry is consistent with that known for the condensed phase complex. This feature is the key motivation to permit gas-phase ESI-MS to be used to study solution-phase characteristics.

Multiply charged ions for protein complexes often appear at relatively high mass-to-charge ratios because the noncovalent native complex shows less

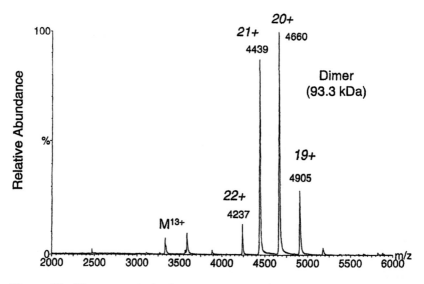

Figure 10 Electrospray ionization mass spectrum of yeast enolase in 10 mM ammonium acetate, pH 6.9, with a QTOF mass spectrometer and a nanoESI source. The predominant ions are consistent for the 93.3 homodimer form of enolase.

charging. This low charging phenomenon has been attributed to the compactness of the native complex structure (35). Enolase is a metalloenzyme involved in the glycolytic pathway, catalyzing the dehydration of 2-phospho-D-glycerate. The yeast enzyme is a 93 kDa homodimeric complex (monomer 46,671 Da); its mass spectrum shows the 93 kDa dimer protein complex as the most abundant species in addition to ions beyond m/z 6000 for a 186 kDa tetrameric complex (Fig. 10).

IV. SEPARATIONS AND SEQUENCE DETERMINATION

A. Capillary LC-MS

A major advantage of electrospray ionization for the analysis of peptides, proteins, and complex mixtures is its ready on-line interface with high performance separation techniques. The ability to register molecular weight with a chromatographic or electrophoretic peak provides important specificity information that is more difficult to obtain with conventional UV detection schemes.

In our laboratory, capillary HPLC/MS is the mainstay for the analysis of complex peptide mixtures. An in-house microelectrospray interface was developed for an ion trap mass spectrometer. It consists of an X-Y-Z translation stage, a stainless steel Valco microvolume connector, and a fused silica 40 μm i.d. × 110 μm o.d. electrospray emitter with the polyimide coating removed at the tip. Figure 11 shows a top-down view of the interface and the microelectrospray interface positioned in the source region of the ion trap mass

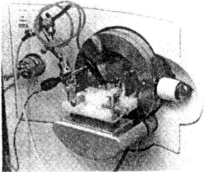

Figure 11 (Left) Top-down view of the micro-electrospray source interface and (right) source positioned in the ion trap (Finnigan LCQ) source region.

spectrometer. Using this interface not only allows greater sensitivity in capillary LC/ESI-MS analysis but also allows very rapid transition to an interface with a capillary electrophoresis (CE)/MS interface. A straight-through Valco microvolume connector is used for LC/MS applications, and a Valco microvolume tee connector is used for CE/MS applications. The microelectrospray source is 10 times as sensitive as the vendor-supplied electrospray source when a constant infusion of 10 μM bradykinin (RPPGFSPFR, M_r 1059.6) at 5 μL/min is used.

It is well known that the presence of trifluoroacetic acid (TFA), even in as small amounts as 0.05% v/v, causes significant signal suppression in LC/ESI-MS. However, reducing the amount of TFA much below 0.05% results in poorer chromatography. Nugent et al. (36) outlined a TFA "doping" technique whereby the peptide sample was injected onto a trap column or cartridge in place of a sample loop. The trap column was then equilibrated with 1% aqueous

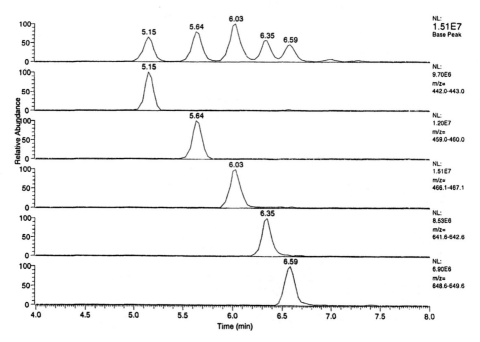

Figure 12 Total ion chromatogram (top) and reconstructed ion chromatograms from capillary LC ion trap MS from a 5 μL injection of 0.1 ng/μL angiotensin peptide standards loaded onto a trapping column (see text).

TFA. The running buffers both contained tenfold less (0.005%) TFA than normally used. However, the higher concentration of TFA used to equilibrate the trap column is enough to retain the good chromatography with short gradients (10 min) but dilute enough to result in much less ion suppression in the ESI-MS process. Figure 12 shows the TIC and RIC from a separation of 5 μL of a mixture of five angiotensin peptide standards (0.1 ng/μL) injected onto a Michrom peptide trap cartridge (0.5 μL volume, 2 ng maximum loading) and then separated on a MagicMS C18 (5μ, 200A, 0.2 mm × 50 mm) column using a 5–95% B gradient over 10 min, where buffer A is 1% acetonitrile, 1% n-propanol, 0.1% acetic acid, and 0.005% aqueous TFA and buffer B is 80% acetonitrile, 10% n-propanol, 0.1% acetic acid, and 0.005% aqueous TFA. The ion intensity with only 0.5 ng of standard injected is approximately four times greater than that observed without a trap column and using solvents with 0.05% TFA with 5 ng of standards injected. Therefore, the total gain in sensitivity is approximately 40-fold without any noticeable loss in chromatographic resolution. An added advantage of using the trap cartridge is that it acts as a guard column for the much more expensive capillary column and therefore extends the useful life span of the latter.

B. Application of Capillary LC/MS for Proteomics

Our laboratory routinely uses the peptide trapping and TFA doping regime combined with the microelectrospray interface for all our capillary LC/ESI-MS and LC/MS/MS work. A major area of application for our laboratory is proteomics. Proteins separated by 2-D gel electrophoresis are identified by means of LC/MS/MS analysis. For example, Fig. 13 shows the TIC obtained from an LC–ion trap–MS/MS analysis of an in-gel tryptic digest of a single protein spot excised from a 2-D gel of an *E. coli* cell lysate (50 μg of total protein loaded onto the gel). The jagged pattern of each of the chromatographic peaks in this figure is characteristic of a data-dependent MS/MS experiment. The spiking peaks in each of the chromatographic peaks is due to a full m/z scan (survey scan), and the concurrent dip in the TIC is an MS/MS event on the most abundant ion determined by the survey scan. In other words, the MS/MS event is determined by the mass spectrometer in a manner dependent upon the results or data of the preceding full m/z scan. Figure 13 shows also the product ion mass spectrum of the precursor ion at m/z 979.1, eluting as shown by the peak at 6.72 min. There are corresponding product ions above the m/z of the precursor ion, indicating that the precursor ion is at least doubly charged. As a general rule, in the case of tryptic digests it is assumed that the most intense ions in the full m/z scan mode on this instrument will be doubly charged.

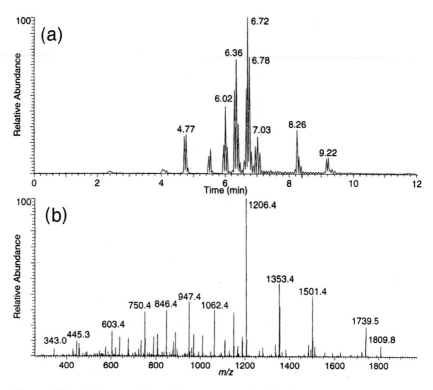

Figure 13 (a) Base peak ion chromatogram of LC/MS/MS (data-dependent scanning) from an in-gel tryptic digest of an *E. coli* protein. (b) The product ion spectrum obtained from a precursor ion at *m/z* 979 with a retention time of 6.72 min.

Figure 14 shows the results (extracted from the Internet Explorer browser) of a database search (owl.fasta) using the Sequest program provided by the vendor on the product ions from the product ion spectrum. These reports are provided in HTML format with active links underlined. As observed in this figure, the top 10 hits from the selected database are listed, with the top two hits shown in greater detail. The top hit in this case reports a relative molecular mass of 1957.2 (assumed 979.1 was doubly charged) and was derived from an *E. coli* protein (flavodoxin 1), whereas the second best hit reports a relative molecular mass of 2935.3 (assuming 979.1 was triply charged) and is a gene product. Clicking on the active link under the "Ions" column for the best hit brings up in the Internet Explorer browser. Figure 15 shows the theoretically generated b-ions (N-terminal fragments) and y-ions

Peptides and Proteins

```
SUMMARY2HTML v.0 (for SEQUEST22 output), Copyright 1996
Molecular Biotechnology, Univ. of Washington, J.Eng/J.Yates
Licensed to Finnigan MAT
01/05/99, 12:59 PM, C:\LCQ\database\owl.fasta, AVG
```

# File*	MassI	MassA	Xcorr*	DelCn*	Ions	Reference*	Sequence**
1 MW98gk120.0165.0167.2	1957.2	1957.1	5.8162	0.624	26/36	FLAV_ECOLI	(-)AITGIFFGSDTGNTENIAK
2 MW98gk120.0165.0167.3	2935.3	2934.5	2.4583	0.149	25/92	A45604	(K)EMVYNIYKNKDTDKKIKAFLETLK

```
1. FLAV_ECOLI    10 (1,0,0,0,0) ( 1 )
2. A45604        10 (1,0,0,0,0) ( 2 )
3. PSEPELA       10 (1,0,0,0,0) ( 4 )
4. CSW_DROME     10 (1,0,0,0,0) ( 3 )
5. RUVB_MYCLE     8 (0,1,0,0,0) ( )
6. APHCPVP1I      8 (0,1,0,0,0) ( )
7. TWST_MOUSE     8 (0,1,0,0,0) ( )
8. SOEMHDRB3D     8 (0,1,0,0,0) ( )
9. TRAC_BACTB     6 (0,0,1,0,0) ( )
10. FPGDCSPB      6 (0,0,1,0,0) ( )
```

Figure 14 SEQUEST (Finnigan MAT) search results from the product ion spectrum shown in Fig. 13.

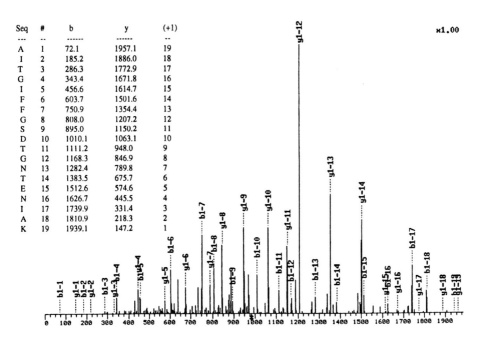

Figure 15 Assignment of product ions observed in the product ion spectrum of Fig. 13 as a result of the SEQUEST search.

(C-terminal fragments) from a peptide having the proposed sequence matched to the product ion spectrum used to generate the search results. These results combined with many of the other product ion spectra from Fig. 14 matching *E. coli* flavodoxin 1 provides confidence that flavodoxin was indeed the protein that was isolated.

C. Capillary Electrophoresis Mass Spectrometry of Peptides

Capillary electrophoresis provides a high resolution separation method for analysis of polypeptide mixtures (as well as a host of other analytes ranging from pharmaceuticals to proteins) and is readily interfaced to ESI-MS by a variety of means. Current CE/MS techniques are not yet as robust as LC/MS, but CE provides a complementary mode of separation (i.e., one based upon electrophoretic mobility rather than hydrophobic interaction). Encouragingly, many CE and mass spectrometry vendors offer CE/MS solutions or mass-spectrometry-friendly interfaces. An uninterrupted electrical contact is essential both for the continued operation of the CE and for the generation of the electrospray droplets when interfacing CE with ESI-MS. The most widely applicable CE/MS interfaces are liquid junction, sheath liquid, and sheathless interfaces.

Although detection limits down to attomole ranges have been reported, CE is generally recognized as having a very low concentration limit of detection (CLOD). To achieve the best resolution and peak shape, it is necessary to inject only very small volumes (low nanoliters) of sample, which forces the use of highly concentrated samples. Several groups developed various preconcentration techniques (37) to attempt to overcome this CLOD. These techniques involve trapping or preconcentrating the samples on some type of C18 stationary phase or hydrophobic membrane in the case of tryptic digest mixtures. We have applied CE/MS to a number of protein characterization projects.

FabI (enoyl acyl carrier protein reductase) from *E. coli* catalyzes the last reductive step of fatty acid biosynthesis. The ESI-MS analysis of intact FabI His-tagged protein shows a molecular mass for the monomer subunit of 30 kDa. However, approximately 40% of the total protein is adducted to a 178 Da moiety. Trypsin digestion followed by CE/MS analysis shows the N-terminus to be the adducted site. Post-translational modification of His-tagged proteins were reported previously to be due to addition of α-N-D-gluconoyl (38,39). Figure 16A shows the TIC of the products of a tryptic digest of FabI; the products were separated by CE using a QTOF mass spectrometer for detection. The components eluting at 20.6 and 23.3 min labeled T1 (GSSHHHHHHSS-GLVPR, M_r 1767.84) and Modified T1 (+178), respectively, are peptides gen-

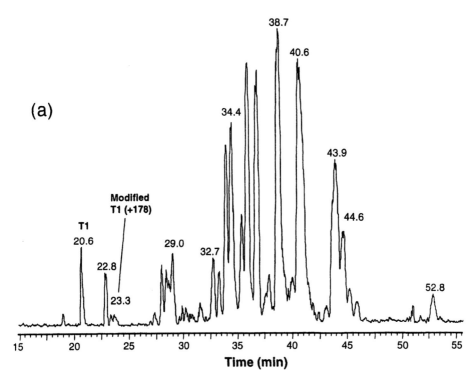

Figure 16a Total ion electropherogram from CE-MS of the tryptic digest of the FabI enzyme. Capillary electrophoresis was performed with a ThermoQuest Crystal 310 unit, and a Micromass QTOF mass spectrometer was used for detection. For CE, a 100 cm long × 50 μm i.d. × 365 μm o.d. fused silica column (conditioned with 0.1 M NaOH) was used (30 kV separation voltage). The CE buffer was 1% formic acid with 10% methanol. A liquid sheath of 75% methanol/0.1% acetic acid at a flow rate of 800 nL/min was used.

erated by the first tryptic cut at the N-terminal side of the protein. These are the peptides that contain the His tag and the putative modified His tag. Figure 16B shows the electrospray mass spectrum of the peak labeled Modified T1, and the inset shows an expansion of the m/z scale highlighting the 4+ charge state of the modified tryptic peptide T1. The mass difference of the tryptic peptide T1 and the modified tryptic peptide T1 is 178.3 Da, which is in very close agreement to the addition of α-N-D-gluconoyl to the His tag.

Another example of the use of CE/MS is a study of the mammalian lens, which is an avascular transparent organ containing approximately 35–50% (by

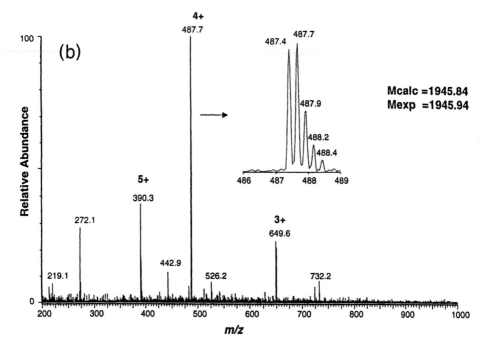

Figure 16b ESI mass spectrum acquired from the online CE/MS analysis of the FabI tryptic digest. The mass spectrum of the peak labeled as the modified T1 tryptic peptide is consistent with post-translational modification of the N-terminus.

weight) soluble protein. These soluble proteins, or lens crystallins, are the major structural proteins of the lens and comprise three major classes: α-, β-, and γ-crystallins. The respective classes of lens crystallins can be separated on the basis of their aggregation behavior by size exclusion chromatography. Figure 17 shows the deconvoluted electrospray mass spectrum of the bovine α/β_H-lens crystallin aggregate. The inset in Fig. 17 is an expansion of a 2-D gel (pH 3–10, Coomassie Blue stained) of the crystallin sample. Previous work (done by this author) had identified the protein with M_r 24,240 as βB3 crystallin, although the published sequence (40) had a calculated M_r of 24,434 (including N-terminal acetylation). This is not altogether unexpected, because the sequence was determined by a combination of Edman sequencing and homology with rodent crystallins. However, a recent cDNA sequence submitted to NCBI (Larry David et al., NCBI AF013259, 1997) gave an M_r of 24,370.2 (including N-terminal acetylation), which is 130 Da higher in mass.

Peptides and Proteins

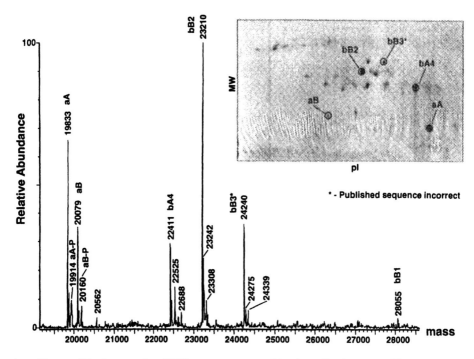

Figure 17 Deconvoluted ESI mass spectrum of bovine α/β_H-lens crystallin aggregate. The inset shows an expansion of a 2-D gel (pH 3–10, Coomassie Blue stained) loaded with 100 μg of the crystallin sample.

It was observed that this cDNA sequence has a methionine residue at the N-terminus (131 Da). Therefore, we hypothesized that this methionine is actually cleaved off before the N-terminus is acetylated. If this were the case, the cDNA sequence would have an M_r of 24,239.0, which is well within experimental error of the mass observed for βB3 crystallin in the electrospray mass spectrum (24,240 Da, Fig. 17).

To confirm this supposition, the protein spot containing βB3 crystallin was excised from the 2-D gel and digested with trypsin, and the subsequent tryptic digest mixture was analyzed by CE/MS. The electrospray tandem mass spectrum from the electrophoretic peak corresponding to the N-terminal tryptic peptide is shown in Fig. 18. This product ion spectrum confirms that the N-terminus is acetylated and that there is no N-terminal methionine residue on this protein. The remaining mass and product ion spectra observed in the entire CE/MS run contain tryptic peptides of βB3

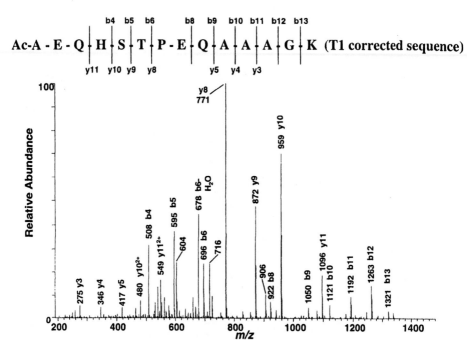

Figure 18 Capillary electrophoresis ion trap MS/MS of the N-terminal tryptic peptide of βB3 (M_r 24,239, N-terminal acetylation), confirming the corrected sequence.

crystallin that make up 76% of the cDNA sequence (less the N-terminal methionine residue). That amount of sequence coverage combined with the relative molecular mass of the corrected cDNA sequence gives confidence that this is indeed the correct sequence for bovine βB3 lens crystallin.

V. MS/MS OF LARGE PEPTIDES AND PROTEINS

The utility of tandem mass spectrometry (MS/MS) for amino acid sequence determination of peptides and proteins has been greatly expanded with the development of electrospray ionization (30,41). The efficiency for collisionally activated dissociation (CAD) is much higher for ESI-produced multiply charged molecules than for their singly charged counterparts. This enhanced efficiency has in turn allowed much larger molecular weight biomolecules to be submitted to CAD. For example, the MS/MS studies of 66 kDa serum albumin proteins yield sequence-specific product ions originating from regions near the N-terminus of the protein sequence (42).

Ion trap mass spectrometry with electrospray ionization has increased the capabilities for near-routine sequencing of polypeptides (43). Its application for protein identification in proteomics-based research has spurred the increased proliferation of ion trap MS. Yet coupling a magnetic sector instrument to an ion trap analyzer provides a powerful and new versatile combination for high resolution mass measurement, high resolution parent ion selection, and sensitive MS/MS (44). This is demonstrated by an ion trap mass analyzer (LCQ) coupled to a forward geometry (E-B) magnetic sector instrument (Finnigan MAT900S-trap). Ions accelerated to 5 kV are deaccelerated prior to entering the ion trap system. The ion trap analyzer has high efficiency for collisionally activated dissociation, particularly for multiply charged peptide ions. Combined with low-flow micro-ESI, the sensitivity for MS/MS and MS^3 with the ion trap analyzer is shown by examples of protein digests (e.g., tryptic peptides from acyl carrier protein) and large peptides (e.g., 2.8 kDa melittin). Less than 75 amol was consumed to measure the product ion spectrum of the $(M + 4H)^{4+}$ molecule of melittin (44). An important advantage of the magnetic sector–ion trap hybrid configuration is the capability for precursor ion selection at high mass resolving power. Resolving power up to 60,000 (FWHM) can be set in MS-1 to differentiate between isobaric precursor ions, thus offering unambiguous product ion spectra in the ion trap stage. For example, a resolving power of 56,000 separates ions from a binary mixture of MW 1031 peptides, Val^5-angiotensin II (monoisotopic mass 1031.5188 Da), and Lys-des Arg^9-bradykinin (monoisotopic mass 1031.5552 Da), or a 36 mDa difference. Distinct product ion spectra are obtained for each peptide from the 2+-charged precursor ions with no contribution in the tandem mass spectra from the other component (45).

The advantage of an ion trap stage of mass analysis is the capability for extended MS/MS (or MS^n) experiments. The MS^n feature is particularly advantageous for deriving sequence information from intact gas-phase protein ions. MS/MS of large polypeptides (3–16 kDa; e.g., melittin, growth hormone releasing factor, ubiquitin, thioredoxin, and hemoglobin alpha and beta chains) shows large multiply charge product ions (45). Additional stages of MS (MS^n, $n > 2$) provide confirmation of the product ion assignments and additional fragments for sequence determination. For example, a small peptide, pancreastatin (5.5 kDa), yields abundant multiply charged products in the product ion spectrum. MS^n to MS^4 provides additional

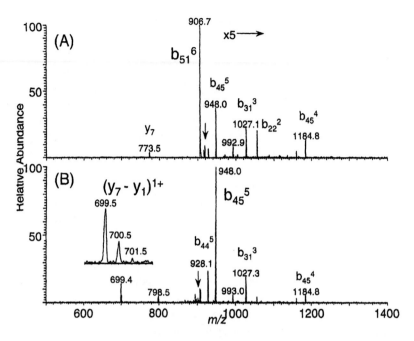

Figure 19 Spectrum from ESI-MS/MS of the $(M + 6H)^{6+}$ molecule of human pancreastatin (1–52, M_r 5508.9) (A) and the MS^3 spectrum of the b_{51}^{6+} product (B) obtained with the magnetic sector–ion trap hybrid mass spectrometer. The arrow denotes the position of the precursor ion in each spectrum.

sequence information not provided in the first-generation product ion spectrum. Using the zoom-scan feature of the ion trap to resolve the product isotope ions makes the assignment of charge-state determination straightforward (Fig. 19).

Dissociation of highly charged precursor ions often yields multiply charged product ions. Interpretation of such spectra is problematic without additional information such as the charge state. However, with ultrahigh resolving power instruments such as FTMS, isotopic resolution easily leads to mass determination of the products (41). Even systems with moderate resolving power (e.g., the quadrupole time-of-flight mass spectrometer) can be successfully applied to MS/MS of intact proteins. Multicharged precursor and product ions to 8–9 kDa can be isotopically resolved to determine charge state. The CAD mass spectra for small proteins such as ubiquitin (8.6 kDa), HIV protease (10 kDa), integrase (17 kDa), and carbonic anhydrase (29 kDa) can be readily interpreted.

Peptides and Proteins

We have used ESI-MS to study proteins and protein complexes relevant to drug discovery efforts against the human immunodeficiency virus (HIV). Several HIV proteins have been identified as possible targets for drug therapies, including protease (PR), integrase (IN), and nucleocapsid (NC) protein. As part of their function, many of these HIV proteins interact with other molecules. In addition to other viral and host cell proteins, HIV infection requires the activity of the three enzymes encoded by the viral *pol* gene: protease, reverse transcriptase (RT), and integrase. Integration of a double-stranded DNA copy of the retroviral genome into the host chromosome is essential for viral replication. To facilitate this process, IN catalyzes a coordinated series of DNA cutting and joining reactions. RT and PR are the targets of several HIV inhibitors currently used in clinical practice. However, HIV integrase protein is potentially a new target for anti-AIDS drugs (46).

The crystal structure of the catalytic core of HIV-1 integrase was solved by Craigie and coworkers (47). The limited solubility of the protein hindered its detailed characterization. However, crystallization was made

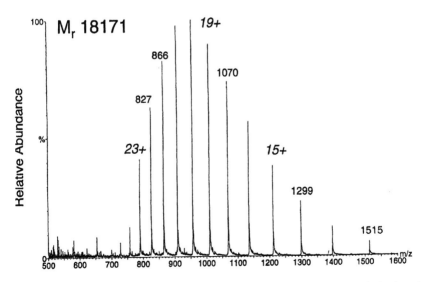

Figure 20 Electrospray ionization mass spectrum of HIV integrase catalytic domain (F185K soluble mutant) from a pH 3 solution (acetonitrile/water) obtained with a QTOF mass spectrometer.

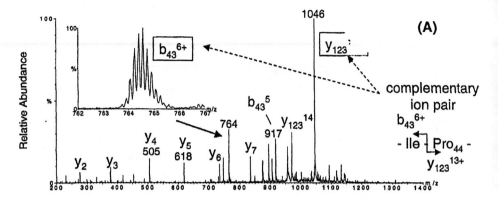

Figure 21A ESI-MS/MS spectrum of the $(M + 19H)^{19+}$ molecule of HIV integrase catalytic domain. The inset shows an expanded view of the b_{43}^{6+} product ion, demonstrating the resolving power of the QTOF system and its capability to determine product ion charge states.

possible by the discovery of a mutation, F185K (Phe, the amino acid at position 185, is replaced by Lys), which enhances solubility but does not affect the biochemical properties of integrase. The ESI mass spectrum of the IN F185K catalytic core (residues 50–212, M_r 18,171) from denaturing solution conditions shows a pattern of multiply charged ions of relatively high charge, consistent for the 18 kDa monomeric species (Fig. 20). ESI-MS/MS of the intact protein yields products from primarily backbone cleavage of bonds N-terminal to proline residues (complementary ion pairs) that provide some confirmation of the primary structure (Fig. 21A). The preference for bond dissociation near proline residues was discussed by Loo et al. (30,48). Further dissociation of these product ions (generated in the ESI interface, or via a "pseudo"-MS³ experiment) generates additional sequence ions (Fig. 21B).

VI. CONCLUSIONS

Over a decade has passed since the initial protein ESI-MS work was carried out in Professor Fenn's laboratory. The field has now progressed to impact biomedically important arenas that were unimaginable just a few years ago. Improvements in ESI source designs, in mass analyzer technology, and in sample handling strategies have allowed researchers to consider both in vitro and in vivo based assays using ESI-MS as the readout mechanism. Certainly, without the ca-

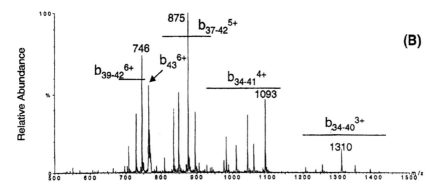

Figure 21B ESI-MS/MS spectrum of the b_{43}^{6+} product ion of HIV integrase catalytic domain. The b_{43}^{6+} product ion was generated by increasing the cone voltage potential in the atmosphere/vacuum interface of the QTOF mass spectrometer to yield a "pseudo"-MS^3 mass spectrum.

pabilities from ESI-MS technology, emerging fields such as proteomics could not have advanced as rapidly as their current pace. Identifying new biological and therapeutic targets for discovering drugs, delineating important biochemical pathways and mechanisms, and providing sensitive assays are only a handful of areas that have benefited from new mass spectrometric techniques.

As to the future of ESI-MS for protein applications, there will be two parallel paths of development, with a great deal of crosstalk between the two streams. One major path will continue to focus on technological advances, improving the sensitivity and resolution of the methodology while expanding its applicability. New developments such as the "lab on a chip" can be interfaced with ESI to miniaturize and reduce sample consumption for MS analysis (see Chapter 0) (49). Another focus will be to expand the range of applications into the biomedical field. However, it will be necessary for the two paths to cross-pollinate, because improvements in technology are more easily facilitated by knowledge of the possible applications.

ACKNOWLEDGMENTS

We thank Kenneth D. Greis for help and advice on the capillary LC/MS and CE/MS projects, and Tracy I. Stevenson for experiments using the QTOF mass spectrometer.

REFERENCES

1. BJ Strasser. Sickle cell anemia, a molecular disease. Science 286:1488–1489, 1999.
2. AG Craig, A Engstrom, H Bennich, I Kamensky. Enhancement of molecular ion yields in plasma desorption mass spectrometry. 35th ASMS Conf on Mass Spectrometry and Allied Topics, Denver, CO, 1987, pp 528–529.
3. MP Lacey, T Keough. Plasma-desorption mass spectrometry of intact enzymes and proenzymes. Rapid Commun Mass Spectrom 3:323–328, 1989.
4. JA Loo, CG Edmonds, RD Smith, MP Lacey, T Keough. Comparison of electrospray ionization and plasma desorption mass spectra of peptides and proteins. Biomed Environ Mass Spectrom 19:286–294, 1990.
5. CK Meng, M Mann, JB Fenn. Electrospray ionization of some polypeptides and small proteins. 36th ASMS Conf on Mass Spectrometry and Allied Topics, San Francisco, CA, 1988, pp 771–772.
6. JB Fenn, M Mann, CK Meng, SF Wong, CM Whitehouse. Electrospray ionization for mass spectrometry of large biomolecules. Science 246:64–71, 1989.
7. JA Loo, HR Udseth, RD Smith. Peptide and protein analysis by electrospray ionization-mass spectrometry and capillary electrophoresis-mass spectrometry. Anal Biochem 179:404–412, 1989.
8. RD Smith, JA Loo, CG Edmonds, CJ Barinaga, HR Udseth. New developments in biochemical mass spectrometry: Electrospray ionization. Anal Chem 62:882–899, 1990.
9. JA Loo, RR Ogorzalek Loo. Electrospray ionization mass spectrometry of peptides and proteins. In: RB Cole, ed. Electrospray Ionization Mass Spectrometry. New York: Wiley, 1997, pp 385–419.
10. M Wilm, M Mann. Analytical properties of the nanoelectrospray ion source. Anal Chem 68:1–8, 1996.
11. DR Barnidge, EA Dratz, AJ Jesaitis, J Sunner. Extraction method for analysis of detergent-solubilized bacteriorhodopsin and hydrophobic peptides by electrospray ionization mass spectrometry. Anal Biochem 269:1–9, 1999.
12. IM Fearnley, JE Walker. Analysis of hydrophobic proteins and peptides by electrospray ionization MS. Biochem Soc Trans 24:912–917, 1996.
13. M le Maire, S Deschamps, JV Moeller, JP Le Caer, J Rossier. Electrospray ionization mass spectrometry on hydrophobic peptides electroeluted from sodium dodecyl sulfate-polyacrylamide gel electrophoresis. Application to the topology of the sarcoplasmic reticulum calcium-ATPase. Anal Biochem 214:50–57, 1993.
14. J Schaller, BC Pellascio, UP Schlunegger. Analysis of hydrophobic proteins and peptides by electrospray ionization mass spectrometry. Rapid Commun Mass Spectrom 11:418–426, 1997.
15. PA Schindler, A Van Dorsselaer, AM Falick. Analysis of hydrophobic proteins and peptides by electrospray ionization mass spectrometry. Anal Biochem 213:256–263, 1993.

16. JP Whitelegge, CB Gundersen, KF Faull. Electrospray-ionization mass spectrometry of intact intrinsic membrane proteins. Protein Sci 7:1423–1430, 1998.
17. JP Whitelegge, J le Coutre, JC Lee, CK Engel, GG Prive, KF Faull, HR Kaback. Toward the bilayer proteome, electrospray ionization-mass spectrometry of large, intact transmembrane proteins. Proc Natl Acad Sci USA 96:10695–10698, 1999.
18. RR Ogorzalek Loo, BE Winger, RD Smith. Proton transfer reaction studies of multiply charged proteins in a high mass-to-charge ratio quadrupole mass spectrometer. J Am Soc Mass Spectrom 5:1064–1071, 1994.
19. JL Stephenson Jr, SA McLuckey. Ion/ion reactions for oligopeptide mixture analysis: Application to mixtures comprised of 0.5–100 kDa components. J Am Soc Mass Spectrom 9:585–596, 1998.
20. M Scalf, MS Westphall, J Krause, SL Kaufman, LM Smith. Controlling charge states of large ions. Science 283:194–197, 1999.
21. JA Loo, R Pesch. Sensitive and selective determination of proteins with electrospray ionization magnetic sector mass spectrometry and array detection. Anal Chem 66:3659–3663, 1994.
22. JA Loo, RR Ogorzalek Loo. Applying charge discrimination with electrospray ionization-mass spectrometry to protein analysis. J Am Soc Mass Spectrom 6:1098–1104, 1995.
23. JA Loo, TP Holler, SK Foltin, P McConnell, CA Banotai, NM Horne, WT Mueller, TI Stevenson, DP Mack. Application of electrospray ionization mass spectrometry for studying human immunodeficiency virus protein complexes. Proteins: Struct Funct Genet Suppl 2:28–37, 1998.
24. JC Severs, RD Smith. Characterization of the microdialysis junction interface for capillary electrophoresis/microelectrospray ionization mass spectrometry. Anal Chem 69:2154–2158, 1997.
25. N Xu, Y Lin, SA Hofstadler, D Matson, CJ Call, RD Smith. A microfabricated dialysis device for sample cleanup in electrospray ionization mass spectrometry. Anal Chem 70:3553–3556, 1998.
26. JA Loo, H Meunster. High sensitivity ESI-MS and MS/MS with a magnetic sector-ion trap hybrid mass spectrometer. 46th ASMS Conf on Mass Spectrometry and Allied Topics, Orlando, FL, 1998, p 1244.
27. JA Loo, J Brown, G Critchley, C Mitchell, PC Andrews, RR Ogorzalek Loo. High sensitivity mass spectrometric methods for obtaining intact molecular weights from gel-separated proteins. Electrophoresis 20:743–748, 1999.
28. SD-H Shi, CL Hendrickson, AG Marshall. Counting individual sulfur atoms in a protein by ultrahigh resolution Fourier transform ion cyclotron resonance mass spectrometry: Experimental resolution of isotopic fine structure in proteins. Proc Natl Acad Sci USA 95:11532–11537, 1998.
29. JA Loo, CG Edmonds, HR Udseth, RD Smith. Effect of reducing disulfide-containing proteins on electrospray ionization mass spectra. Anal Chem 62:693–698, 1990.

30. JA Loo, CG Edmonds, RD Smith. Primary sequence information from intact proteins by electrospray ionization tandem mass spectrometry. Science 248:201–204, 1990.
31. SD Buckel, TI Stevenson, JA Loo. Direct mass spectrometric analyses for protein chemistry studies. In: MZ Atassi, E Appella, eds. Methods for Protein Structure Analysis [Proc Int Conf], 10th. New York: Plenum Press 1995, pp 151–159.
32. JA Loo. Studying noncovalent protein complexes by electrospray ionization mass spectrometry. Mass Spectrom Rev 16:1–23, 1997.
33. JA Loo. Electrospray ionization mass spectrometry: A technology for studying noncovalent macromolecular complexes. Int J Mass Spectrom 200:175–186, 2000.
34. WJH Van Berkel, RHH Van den Heuvel, C Versluis, AJR Heck. Detection of intact megadalton protein assemblies of vanillyl-alcohol oxidase by mass spectrometry. Protein Sci 9:435–439, 2000.
35. KJ Light-Wahl, BL Schwartz, RD Smith. Observation of the noncovalent quaternary associations of proteins by electrospray ionization mass spectrometry. J Am Chem Soc 116:5271–5278, 1994.
36. K Nugent, S Baldwin, K Stoney. Optimizing LC-MS/MS conditions for characterization of peptides and proteins. 45th ASMS Conf on Mass Spectrometry and Allied Topics, Palm Springs, CA, 1997, p 275.
37. Q Yang, AJ Tomlinson, S Naylor. Membrane preconcentration CE. Anal Chem 71:183A–189A, 1999.
38. KF Geoghegan, HBF Dixon, PJ Rosner, LR Hoth, AJ Lanzetti, KA Borzilleri, ES Marr, LH Pezzullo, LB Martin, PK LeMotte, AS McColl, AV Kamath, JG Stroh. Spontaneous α-N-6-phosphogluconoylation of a "His tag" in Escherichia coli: The cause of extra mass of 258 or 178 Da in fusion proteins. Anal Biochem 267:169–184, 1999.
39. Z Yan, GW Caldwell, PA McDonell, WJ Jones, A August, JA Masucci. Mass spectrometric determination of a novel modification of the N-terminus of histidine-tagged proteins expressed in bacteria. Biochem Biophys Res Commun 259:271–282, 1999.
40. GAM Berbers, WA Hoekman, H Bloemendal, WW De Jong, T Kleinschmidt, G Braunitzer. Homology between the primary structures of the major bovine β-crystallin chains. Eur J Biochem 139:467–479, 1984.
41. FW McLafferty, NL Kelleher, TP Begley, EK Fridriksson, RA Zubarev, DM Horn. Two-dimensional mass spectrometry of biomolecules at the subfemtomole level. Curr Opin Chem Biol 2:571–578, 1998.
42. JA Loo, CG Edmonds, RD Smith. Tandem mass spectrometry of very large molecules: Serum albumin sequence information from multiply charged ions formed by electrospray ionization. Anal Chem 63:2488–2499, 1991.
43. KR Jonscher, JR Yates III. The quadrupole ion trap mass spectrometer: A small solution to a big challenge. Anal Biochem 244:1–15, 1997.
44. JA Loo, H Muenster. Magnetic sector-ion trap mass spectrometry with electrospray ionization for high sensitivity peptide sequencing. Rapid Commun Mass Spectrom 13:54–60, 1999.

45. AK Ziberna, H Muenster, JA Loo. High resolution precursor selection and ESI-MS^n for polypeptide sequencing. 47th ASMS Conf on Mass Spectrometry and Allied Topics, Dallas, TX, 1999, pp 2785–2786.
46. M Thomas, L Brady. HIV integrase: A target for AIDS therapeutics. Trends Biotechnol 15:167–172, 1997.
47. F Dyda, AB Hickman, TM Jenkins, A Engelman, R Craigie, DR Davies. Crystal structure of the catalytic domain of HIV-1 integrase: Similarity to other polynucleotidyl transferases. Science 266:1981–1986, 1994.
48. JA Loo, CG Edmonds, RD Smith. Tandem mass spectrometry of very large molecules. 2. Dissociation of multiply charged proline-containing proteins from electrospray ionization. Anal Chem 65:425–438, 1993.
49. D Figeys. Array and lab on a chip technology for protein characterization. Curr Opin Mol Ther 1:685–694, 1999.

7
Structural Analysis of Glycoproteins by Electrospray Ionization Mass Spectrometry

Anthony Tsarbopoulos
GAIA Research Center, The Goulandris Natural History Museum, Kifissia, Greece

Ute Bahr and Michael Karas
J.W.-Goethe University of Frankfurt, Frankfurt, Germany

Birendra N. Pramanik
Schering-Plough Research Institute, Kenilworth, New Jersey

I. INTRODUCTION

The development of electrospray ionization (ESI) (1) and matrix-assisted laser desorption ionization (MALDI) (2) methods has allowed the vaporization and ionization of large biomolecules, thus providing the main analytical tool for determining proteins and other labile biomolecules. That, in turn, has opened the way to elucidate macromolecular structures and study their interactions with small molecules or other macromolecules, which are important aspects in understanding their biological function and developing new therapeutics. Most of these therapeutic macromolecules are produced now by recombinant DNA technology, which has allowed their production in large quantities in cell culture. For these biologically diverse and heterogeneous recombinant protein products, ESI and MALDI mass spectrometry (MS) can provide structural characterization, such as primary structure confirmation through relative molecular mass or molecular weight (M_r) determination, peptide mapping, and sequencing information, as described in many of the previous chapters.

One of the main features of the aforementioned MS methods is their unique ability to determine whether the protein product presents any heterogeneity. This heterogeneity is usually associated with post-translational modifications, such as glycosylation, phosphorylation, proteolytic processing, deamidation, aggregation, and formation of disulfide bonds, with all these contributing to the diversity of the protein product. The extent and the structure of such post-translational processing depend on the nature of the host cell and the conditions of the fermentation and recovery (i.e., isolation and purification) processes. Therefore, characterization of these protein molecules with regard to purity and structure is an essential and challenging task and represents an integral part of the overall records submitted to the regulatory agencies prior to the protein's approval as a drug.

Within the wide variety of post-translational modification of proteins, glycosylation is definitely one of the most difficult to predict. It represents the most common modification for recombinant protein products expressed in mammalian and insect cell lines. These carbohydrates can be important in various cellular processes, such as adhesion, circulation, and immune recognition, because they can serve as binding sites for growth factors, bacteria, viruses, and other messenger molecules. The exact structure of the attached oligosaccharide(s) and the degree of occupancy for each glycosylation site often regulate the mediation of these biological and cellular events (3). For example, the deletion of individual sites for glycosylation and the removal of oligosaccharide end units can result in an enhancement or loss of the protein's biological activity. That was clearly demonstrated for the erythropoietin glycoprotein hormone, where improvement of in vivo activity resulted from the deletion of single glycosylation sites, and loss of in vivo activity resulted from the removal of sialic acid end units (4,5). Therefore, complete structural analysis of a glycoprotein should include the determination of the primary peptide sequence, the identification of the glycosylation sites, and the structural elucidation of the attached oligosaccharide components.

II. MOLECULAR WEIGHT DETERMINATION OF GLYCOPROTEINS

A. Molecular Weight and Heterogeneity

The M_r determination of proteins produced by recombinant DNA techniques is an important first step in their quality control and structure characterization. In ESI mass spectrometric analysis, an aqueous solution of the protein is sprayed at a flow rate of 2–20 µL/min. (ESI-MS analysis using flow rates of 2–20 µL/min will be referred to as microESI.) The generated ions are usually multiply protonated ions with reduced mass-to-charge (m/z) ratios, which makes them easily detected by typical quadrupole systems with a mass range of approximately 2000 Da. The average M_r of the protein can be calculated from two (preferably more) successive multiply charged ions (1). This is demonstrated in the ESI mass spectrum of the deglycosylated CHO interleukin-4 (IL-4) shown in Fig. 1, where all the observed signals from the multiply charged ions give rise to an average M_r value of 14,955.4, as shown in the deconvoluted mass spectrum (Fig. 1, inset).

The use of all the observed m/z values (multiply charged signals) enhances the mass measurement precision and accuracy, which is usually

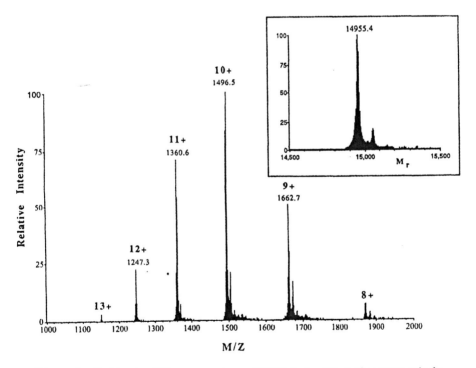

Figure 1 Positive ion ESI mass spectrum of CHO-derived IL-4 after enzymatic deglycosylation with N-Glycanase. The deconvoluted spectrum is shown in the inset, giving a measured M_r value of 14955.4 ± 0.3. (Reprinted with permission from Ref. 8. Copyright (1995) John Wiley & Sons Ltd.)

better than 0.01–0.03% for masses up to 100 kDa (6). The maximum number of charges placed on a protein usually depends on the total number of basic sites (i.e., K, R, H, N-terminus) in the positive ion mode or acidic sites (i.e., D, E, C-terminus) in the negative ion mode. Therefore, for larger proteins, we observe greater charge states, often in the presence of a dilute acid (positive mode) or base (negative mode). Not only is the charge state of each ion envelope increased but also its signal intensity is improved by the addition of low level acid or base solutions (usually of concentration 0.01–0.1%). The concomitant shift of the ion distribution to lower m/z values is also accompanied by a decrease in the m/z spacing between adjacent charge states. This can be critical in identifying the charge state of the ion envelope's components as well as the presence of any protein adducts or variants. The instrument's ability to better resolve adjacent charge states and satellite signals of the same charge becomes more important in the determination of glycoproteins, where carbohydrate heterogeneity is abundant. This heterogeneity arises from the presence of heterogeneous populations of oligosaccharides sharing a basic core structure and/or from the different populations of attached carbohydrate structures at each site ("site heterogeneity"). This carbohydrate heterogeneity generates a great complexity in the ESI mass spectrum, which in turn imposes higher resolving power demands on the employed instrumentation.

This carbohydrate-derived mass spectral complexity is shown in the ESI-MS analysis of the Chinese hamster ovary (CHO) cell–derived IL-4, a mammalian protein that contains two potential glycosylation sites for N-linked carbohydrates (7,8) and shares the same polypeptide backbone with the *E. coli*–derived IL-4. The ESI mass spectrum of CHO IL-4 (Fig. 2A) contains three envelopes of multiply charged ions ranging from 8+ to 10+ charge state, with each envelope containing several peaks corresponding to individual glycoforms and adducts of CHO IL-4. This is better shown in the deconvoluted mass spectrum (Fig. 2B), where the most abundant signals correspond to the mono- and disialylated components. Other higher M_r components indicate the presence of tri- and tetraantennary glycans containing additional lactosamine units (in-chain mass of 365 Da), with mass assignments within 2–3 Da of their theoretical values. The complexity of the mass spectrum of the glycoprotein can be realized by a direct comparison with the mass spectrum of the deglycosylated CHO IL-4 after enzymatic treatment with *N*-Glycanase (Fig. 1). All the spectral signals derived from the heterogeneous glycans have disappeared, and the resulting signals from multiply charged ions provide an accurate M_r determination of the protein part (Fig. 1, inset).

It should be noted that the higher order glycan structures of CHO IL-4, containing carbohydrate components with additional branching, have not been

Structural Analysis of Glycoproteins

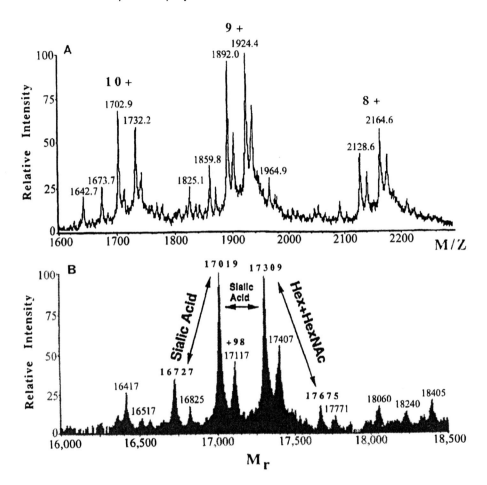

Figure 2 (A) Positive ion ESI mass spectrum of CHO IL-4. (B) The deconvoluted mass spectrum indicates the presence of two sialylated components as well as tri- and tetraantennary glycans containing additional lactosamine units. (Reprinted with permission from Ref. 8. Copyright (1995) John Wiley & Sons Ltd.)

observed in the MALDI mass spectrum, where only signals corresponding to the asialo and mono- and disialylated components were present (9). This rapid assessment of glycosylation at the molecular mass level is invaluable for a preliminary view of the number of different glycoforms present in the mammalian cell–derived protein, and it can be used as a first step for checking the heterogeneity of the manufacturing batches prior to extensive MS mapping (8). This is also useful in glycoprotein screening for certain diseases, such as the differentiation of transferrin in normal and carbohydrate-deficient syndrome patients who usually lack both N-linked disialylated carbohydrate chains (10).

Overall, ESI-MS analysis of glycoproteins exhibits greater speed, higher sensitivity, and much better mass measurement accuracy than the typically employed classical method of SDS-PAGE, which usually gives diffused bands at masses much higher than the actual molecular masses. On the other hand, the success of determining glycoproteins by ESI-MS depends on their relative carbohydrate content; the success decreases as the size or the weight percent of the carbohydrate component increases. For example, ESI-MS analysis of the 44 kDa ovalbumin containing 4% carbohydrate was successful (11), whereas the higher carbohydrate content glycoproteins of CHO IL-4 receptor ($M_r \sim 38$ kDa and 35% carbohydrate) and CHO IL-5 ($M_r \sim 31$ kDa and 15% carbohydrate) did not give any ESI signals on a quadrupole instrument (9,12).

In general, high M_r glycoproteins yield higher charge states and increased formation of adducts, thus often resulting in unresolved weak signals or no signals at all. The same mass dependence of the ESI glycoprotein signals, albeit smaller, is observed when a capillary electrophoresis (CE) separation is interfaced with ESI-MS analysis (13). Nevertheless, the low flow rates of the CE eluent, which are conducive to smaller-size spray droplets and improved desolvation efficiency, lead to better resolved glycoform signals (13,14).

B. Nanoelectrospray and Orthogonal TOF Instruments

Similarly, the recent development of the nanoelectrospray (nanoESI) ion source (15) has increased the sensitivity of ESI-MS analysis by decreasing the size of the generated droplets, thus easing the desolvation of "difficult" molecules such as glycoproteins (16). This is clearly demonstrated in the nanoESI mass spectrum of the 44 kDa ovalbumin (4% carbohydrate content) (Fig. 3), where 11 different glycosylation states are resolved and assigned to high mannose carbohydrate-containing protein. It is evident that the observed mass-resolving power of the glycoform signals is limited only by the chosen experimental resolution or the ultimate resolution of the mass spectrometer, in this case a triple-quadrupole instrument (16). It should be noted that detection and assignment of the complete heterogeneous glycosylation pattern in ovalbumin was very difficult and the information more limited when using microESI on the same type of instrumentation (11). Two reasons are the high degree of residual solvation and the poorly resolved ES ion signals. The nanoESI ion source (Protana) combined with tandem mass spectrometry (triple-quadrupole or ion-trap) has found

Structural Analysis of Glycoproteins

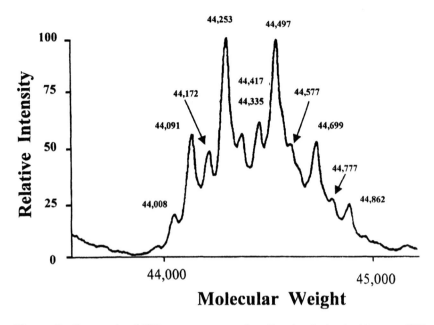

Figure 3 Deconvoluted ESI mass spectrum of ovalbumin obtained with a nanoESI source, showing all the observed glysosylation states. A 1 μL sample of an aqueous 5 pmol/μL ovalbumin solution was injected into the nanoESI capillary. (Reprinted with permission from Ref. 16. Copyright (1996) American Chemical Society.)

immediate applications in the area of identifying and sequencing proteins and glycoproteins isolated from one- or two-dimensional SDS polyacrylamide gels (15,17). Moreover, the interfacing of a nanospray ion source with orthogonal time-of-flight (oTOF) instrumentation (18) led to improved mass measurement accuracy and increased the analytical range, hence providing new momentum to the ESI-MS determination of glycoproteins.

In ESI oTOF instruments (18), ions are analyzed in a TOF mass analyzer instead of the most frequently used quadrupole mass filter or quadrupole ion trap. This overcomes the disadvantage of the limited mass range of most commercial instruments, which is usually restricted to m/z values of 3000–4000. This is especially a limitation in the detection of larger glycoproteins or noncovalent complexes where an upper mass limit of m/z 10,000 may be required. Moreover, TOF instruments provide higher resolving power and sensitivity over a larger mass range than do quadrupole or ion-trap instruments. On the other hand, sensitive structural analysis of peptides and glycopeptides can also be performed with a tandem quadrupole/oTOF (QTOF) mass spectrometer in the low femtomole and attomole range (19).

NanoESI (with flow rates of 10–50 nL/min) was used in the determination of CHO IL-4 on an oTOF mass spectrometer (Fig. 4). The observed ES ion envelopes of multiply charged states range from 9+ to 13+ as opposed to 8+ and 10+ in the mass spectrum obtained with a triple-quadrupole instrument operating with microESI (flow rates of 2–10 μL/min; Fig. 2). The signal-to-noise ratio of the nanoESI mass spectrum is enhanced compared to that of the microESI, as is the mass resolving power, which is 1370 for the most abundant peak (m/z 1732). As in the case of ESI-MS analysis, the mono- and disialylated glycans were the main components, and their measured M_r values of 17,018 and 17,309 are in excellent agreement with the theoretical values (8).

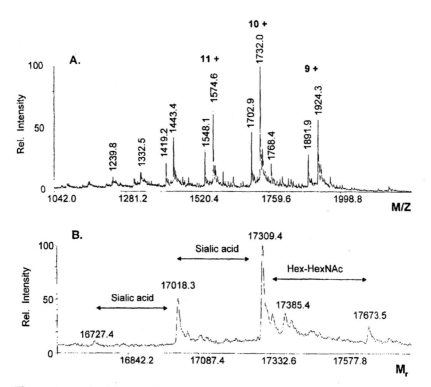

Figure 4 (A) Positive ion ESI mass spectrum of CHO IL-4 obtained in an orthogonal TOF (oTOF) instrument using a nanoESI source. (B) The deconvoluted mass spectrum indicates the presence of the asialo and mono- and disialylated components as well as a triantennary glycan containing an additional lactosamine unit.

Often observed adducts, especially with phosphoric or sulfuric acid, can be reduced by increasing the voltage between nozzle and skimmer, as was done for this experiment. The higher energy collisions of analyte ions with residual gas molecules in this region (pressure ~1 torr) leads to the minimization or even elimination of these adducts. Nonetheless, the voltage has to be carefully selected and monitored to avoid collisional fragmentation of the glycoprotein, especially that of the labile sialic acid end groups. Adduct formation can also be reduced by decreasing the pH of the analyte solution. For example, use of 30% formic acid (pH 1.5) instead of 0.1% formic acid (pH 2.7) leads to a significant reduction in the formation of phosphoric and/or sulfate adducts. Furthermore, the ESI mass spectra from solutions containing 0.1% TFA show no such adduct ions, albeit the ion intensities are lower than when other acids are employed.

The more efficient desolvation of the nanoESI process is particularly beneficial in the determination of larger glycoproteins with a higher carbohydrate content, as shown for the 30 kDa Sf9 IL-4 receptor glycoprotein, which contains about 22% carbohydrate (Fig. 5). Even though microESI analysis yielded no results, the mass spectral signals produced by nanoESI were adequately resolved to achieve direct assignment of the glycoforms attached on the protein backbone (Fig. 5). The enhanced mass-resolving power of the nanoESI experiment leads to reasonable separation of the glycoform signals from those of their sodium adduct ions ($m/\Delta m \sim 1300$) and good mass measurement accuracy. The deconvoluted mass spectrum (Fig. 5 inset) indicates the presence of two sets of glycoforms separated by the residual mass of a fucosylated $Man_3(GlcNAc)_2$ high mannose structure (M_3+Fuc; residual mass of 1039 Da). It should be noted that the nanoESI mass spectrum of the Sf9 IL-4 receptor shows two distributions of multiply charged ions, one centered on the 23+ charge state and another one on the 13+ charge state. Both ion distributions give rise to the same deconvoluted spectrum shown in the inset of Fig. 5. The observation of the two distinct charge distributions may be explained by the presence of two protein conformations, one "tight" (low charge states) and one "open" (high charge states). Increasing the voltage between nozzle and skimmer reduces the most highly charged ions, which are the more labile ones, by collision-induced fragmentation and shifts the "tight" conformation to the "open" one. This can also be seen by the appearance of additional ion signals between the two distributions upon an increase in nozzle voltage.

Similarly, nanoESI MS analysis of the highly sialylated α_1-acid bovine glycoprotein (containing ~40% carbohydrate) gave ion envelopes of narrow charge distribution ranging from 10+ to 14+ between m/z 2000 and m/z 3500 (Fig. 6). The deconvoluted spectrum (Fig. 6 inset) shows an extensive

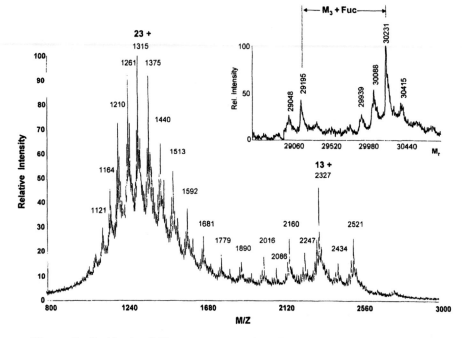

Figure 5 Positive ion ESI mass spectrum of Sf9 IL-4 receptor obtained in an oTOF instrument using a nanoESI source. The deconvoluted mass spectrum (shown in the inset) indicates the presence of two sets of glycoforms separated by the residual mass of a fucosylated Man$_3$(GlcNAc)$_2$ high mannose structure (denoted as M$_3$ + Fuc).

glycosylation pattern, with the most abundant glycoforms having average M_r values of 33,070 and 33,405. The observed heterogeneity is mainly due to the varying content of the *N*-glycolylneuraminic acid residue (in-chain mass of 307 Da).

III. GLYCOPROTEIN MAPPING

A. Glycan Classification

Complete structural analysis of a glycoprotein involves the determination of the protein amino acid sequence and the structures of the attached oligosaccharide moieties and the identification of the glycosylation attachment sites. In most cases, the amino acid sequence of the deglycosylated protein is determined in parallel unless it is known from the corresponding DNA sequence.

Structural Analysis of Glycoproteins

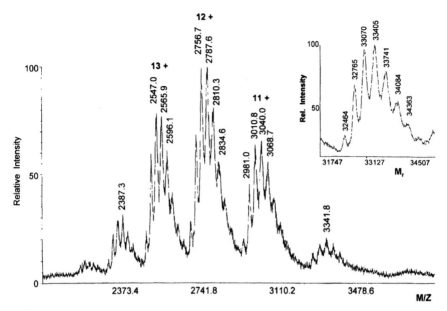

Figure 6 Positive ion ESI mass spectrum of the α_1-acid glycoprotein obtained in an oTOF instrument using a nanoESI source. The deconvoluted mass spectrum (shown in the inset) indicates the protein's high degree of sialylation.

First, mapping of the protein sequence by MS mapping of enzyme-generated peptide mixtures, either directly or after chromatographic separation (20), provides a rapid and accurate confirmation of the expected sequence and identification of any existing modifications. In general, a newly appearing signal in the mass spectrum combined with the absence of an expected signal indicates the modification area of the protein. Identification of the exact location of the protein modification often requires LC separation of the enzyme-derived protein fragments coupled with on-line tandem MS analysis (e.g., LC/ESI-MS/MS), and/or off-line tandem MS analysis and Edman sequencing (21,22). These strategies have been employed in the determination of a variety of posttranslational modifications such as phosphorylation (23), disulfide bonding (24,25), and glycosylation (8,26,27).

The same MS-based mapping approach has been used for the determination of the glycosylation attachment sites and the site-specific glycosylation patterns. The absence of a structural template in the biosynthetic machinery of glycans is responsible for the substantial structural diversity of the glycan chains and the associated microheterogeneity present at each glycosylation site

(28). The glycans of glycoproteins can be classified into two groups, depending on the linkage site. The ones that are linked to the hydroxyl group of serine, threonine, and hydroxyproline residues through an *N*-acetylgalactosamine (GalNAc) residue are denoted as O-linked glycans, whereas those that are linked to the amide group of an asparagine residue through an *N*-acetylglucosamine residue are defined as N-linked glycans. The O-linked glycans contain a GalNAc residue at their reducing termini, and they have structures based on several core structures such as Gal-GalNAc and $(GlcNAc)_n$-GalNAc, where $n = 1$ or 2 (29). In contrast, the N-linked glycans share a common pentasaccharide core structure, the trimannosyl core, consisting of three mannose (Man) and two *N*-acetylglucosamine (GlcNAc) residues (28).

The structures of the N-linked oligosaccharides have been further classified into three main categories: high mannose, complex type, and hybrid, as shown in Fig. 7. High mannose glycoproteins, such as ovalbumin and ribonuclease B (RNase B), usually contain two to nine mannose residues added to the trimannosyl core. Glycoproteins containing complex glycans (e.g., bovine fetuin) show a higher degree of heterogeneity by having a number of GlcNAc, galactose, fucose, and sialic acid residues attached to the core. Thus, they present the greatest structural variation, owing to the possible extension and/or

A High Mannose (M9)

B Complex Type (Triantennary)

C Hybrid Type

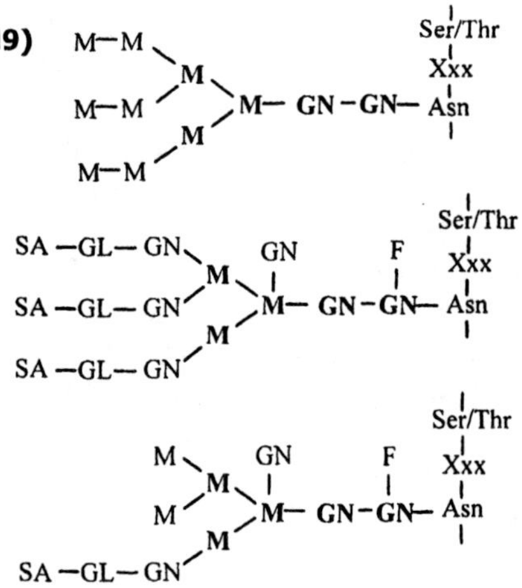

Figure 7 Types of N-linked oligosaccharide structures sharing the common trimannosyl $Man_3(GlcNAc)_2$ (M_3) core structure (Man = M, GlcNAc = GN, Gal = GL, sialic acid = SA, fucose = F). (A) High mannose (M9); (B) complex type (triantennary); (C) hybrid type.

branching of the outer chains, which results in the formation of glycans ranging from mono- to pentaantennary. Finally, the hybrid-type glycoproteins incorporate features of both high mannose and complex-type glycans (28).

B. Mapping of the Protein Sequence and Identification of Glycosylation Sites

Several MS-based mapping strategies have been developed for the identification of the attachment sites of N- and O-linked oligosaccharides and the rapid characterization of the carbohydrate class at specific sites (20–22). The identification of the potential N-glycosylation sites in a protein sequence is aided by the requirement that the sequence positions fulfill the sequence motif (sequon) Asn-X-Ser/Thr, where X is any amino acid except Pro or Asp (28). In contrast to N-glycosylation, there is no sequon for O-glycosylation on Ser or Thr.

The first step toward the location of O- and N-linked glycosylation sites in a glycoprotein is the sequence mapping by combining a specific enzymatic cleavage (proteolytic degradation) with either FAB (20,30), MALDI (31–33), or LC/ESI-MS (8,31,34). That provides a reference MS map that will be used for subsequent identification of the possible glycosylation sites. The next step involves the carbohydrate removal from the glycoprotein by base-catalyzed β-elimination (for O-linked carbohydrates) or N-Glycanese (for N-linked carbohydrates). Removal of the O-linked carbohydrates leads to the concomitant conversion of Ser to Ala ($\Delta m = 16$ Da) and of Thr to α-aminobutyric acid ($\Delta m = 16$ Da), whereas removal of the N-linked carbohydrates converts the glycosylated Asn to Asp ($\Delta m = 1$ Da). The final step is the proteolytic digestion of the deglycosylated protein followed by MS mapping of the resulting peptide mixture. The appearance of new mass spectral signals, which are at lower m/z (for sugars O-linked to Ser or Thr) or higher m/z (for Asn-linked sugars) than those of the respective unglycosylated peptides, reveals those peptides that were originally glycosylated (30,35). In the case of the N-linked carbohydrates, two strategies have been used to differentiate the type of the attached carbohydrate (i.e., complex, high mannose, or hybrid). In the first, enzymatic treatment of the glycoprotein with endoglycosidase H releases the high mannose and hybrid-type oligosaccharides, leaving an N-acetylglucosamine (GlcNAc) residue attached to the peptide's Asn residue, thus resulting in the detection of peptides having a mass that is 203 Da higher than that of the respective unglycosylated peptides. Glycosylation sites containing complex-type oligosaccharide structures are unaffected by the endoglycosidase H treatment. This approach was employed in the FAB carbohydrate mapping of the major envelope glycoprotein gp 120 of HIV-1 (27) and tissue plasminogen activator (36).

The second strategy involves complete hydrolysis of the attached N-linked oligosaccharides with PNGase F (*N*-Glycanase) and comparative MS mapping before and after the treatment, employing FAB, MALDI, and ESI-MS (26,36,37). The resulting new mass spectral signals correspond to the formerly glycosylated peptides, and they are at higher *m/z* than those of the respective unglycosylated peptides by 1 Da because of the attachment site conversion of the Asn residue to Asp. In the structural characterization of interleukin-4 (IL-4) expressed in Chinese hamster ovary (CHO) cells (8), comparative LC/ESI-MS tryptic mapping of CHO IL-4 and its *N*-Glycanase-treated counterpart revealed the glycosylation of the Asn^{38} residue. The main oligosaccharides attached on that site were the mono- and disialylated biantennary complex-type oligosaccharide structures with average M_r values of 5577 and 5868 (Fig. 8; inset), while additional ESI signals depict the heterogeneity of the attached carbohydrates. It should be noted that the separation of adjacent signals at a particular charged state is indicative of the existing carbohy-

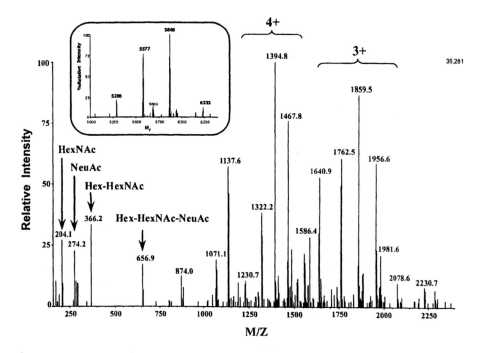

Figure 8 Positive ion ESI mass spectrum of CHO IL-4 tryptic component (peak 11) obtained at a high orifice (OR) potential of 110 V. The deconvoluted glycopeptide M_r values of 5286, 5577, and 5868 are shown in the inset.

drate heterogeneity in the glycopeptide. For example, the 97 and 73 Da mass differences in the triply and quadruply charged states in Fig. 8 readily reveal the existing sialic acid heterogeneity in the glycopeptide.

In addition, carbohydrate heterogeneity can be readily seen in the contour diagram of the trypsin-treated CHO IL-4 (Fig. 9). This two-dimensional diagram displays the distribution of m/z values as a function of retention time (or scan number) with the ion abundance being depicted by the density of the contour profile. In this contour plot, the carbohydrate patterns of glycopeptides appear as diagonal ladder of peaks or cluster of ions, in contrast to peptides that appear as vertical lines (34,38). Once these patterns are located, the mass spectra can be extracted from averaging scans from these glycopeptide-containing regions. In these scans, the average M_r of each glycoform can be calculated using m/z values from all multiply charged ions, whereas the mass separation of the signals at a particular

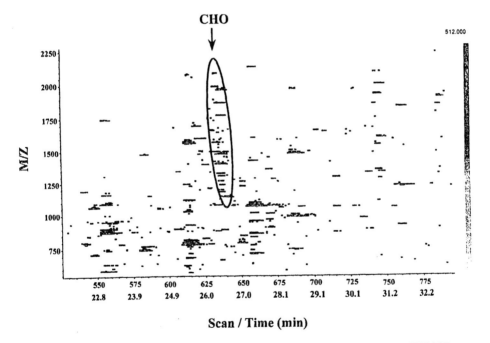

Figure 9 Contour plot of trypsin-digested CHO IL-4 generated from the LC/ESI-MS analysis. The tryptic digest mixture was separated on a 1 mm i.d. reverse-phase (C18) column (~50 pmol injected), and the ESI-MS analysis was carried out at normal OR potential (60 V). The characteristic glycopeptide profile is circled and designated by CHO (region between scan numbers 630 and 645).

charged state can readily reveal the glycoform heterogeneity. The observation of these characteristic patterns is useful in locating sialic acid, high mannose heterogeneity, and/or extended branching, which may serve for the preliminary assignment of the glycosylation pattern in the glycoprotein (see Sect. III. C).

The same strategy was employed in the establishment of the glycosylation sites in CD4, a 55 kDa glycoprotein receptor for the human AIDS virus (26). Tryptic mapping by LC/ESI-MS showed that oligosaccharides were attached on both of the potential glycosylation sites, and their heterogeneity was apparent in the respective mass spectra (26). Similarly, LC/ESI-MS analysis of the trypsin-treated IL-5 receptor α-subunit (IL-5Rα) revealed the four glycosylation sites out of the six potential sites in the protein fulfilling the consensus sequence (39). The mass spectrum of one of the tryptic glycopeptides containing a Man_9 $(GlcNAc)_2$ (M9) high mannose carbohydrate is shown in

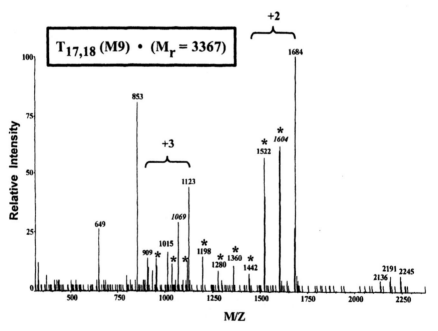

Figure 10 Positive ion ESI mass spectrum of shIL-5Rα tryptic component (peak 10). The ESI-determined M_r value of 3367 (box) is consistent with the attachment of a $Man_9(GlcNAc)_2$ (M9) high mannose carbohydrate to the $T_{17,18}$ tryptic peptide. The mass separation of the asterisk-denoted satellite signals in both doubly and triply charged ES ion envelopes correspond to variations in the mannose (Man) content.

Fig. 10, where an extensive heterogeneity in the mannose content was evident in the ion signals corresponding to doubly and triply charged species.

The identification of the glycopeptide-containing LC fractions among the LC eluting peaks of a complex glycoprotein proteolytic digest can be assisted by the detection of low mass ions that are diagnostic of carbohydrate units. Production of these ions can be accomplished by collision-induced dissociation (CID) in either the declustering region (the region before the first mass filter Q1) or the Q2 collision region of a triple-quadrupole system. In the first method, the fragmentation is induced by increasing the potential difference between the orifice (OR) or nozzle and the skimmer; this difference controls the collision excitation and thus the extent of fragmentation. This "in-source" fragmentation is a highly sensitive method for producing sugar-specific oxonium ions in the ESI mass spectrum (i.e., m/z 162 for Hex$^+$, m/z 204 for HexNAc$^+$, m/z 274 and 292 for NeuAc$^+$, m/z 366 for Hex-HexNAc$^+$, and m/z 657 for NeuAc-Hex-HexNAc$^+$) (8,38). This can be seen in the LC/ESI-MS analysis of the CHO IL-4 tryptic digest at elevated OR potential (110 V), where the reconstructed ion chromatograms for the carbohydrate-specific oxonium ions at m/z 162, 204, 274, and 366 indicate the exact elution time of the respective glycopeptides (Fig. 11).

The observed glycoform distribution and associated average M_r values for this fraction were in agreement with those observed previously in the LC/ESI-MS analysis using a low potential difference and also indicate that the peptide's glycosylation is of the complex type. Nevertheless, glycopeptide signals within the same ion envelope shifted to lower m/z values, probably owing to the partial loss of carbohydrate units induced when a high potential difference was used (8), and a concomitant shift of the peptide ion distribution to lower charge state was observed (8,38). Although this happens to only a small extent, it can be eliminated by either stepping the OR potential with mass or by generating these ions by CID in the Q2 collision region of the triple-quadrupole instrument, as was shown by Carr et al. (40). The former method of stepping the OR potential can be used successfully to generate carbohydrate-specific ions, even though it usually does not provide any useful sequence information on peptide fragments. This sequence information is available when a high potential difference is maintained throughout the scanned mass range of the LC/ESI-MS analysis; this can be useful for confirming peptide assignments, especially when only one charged state of a peptide fragment is present (under normal OR potential) (8,38). In the tandem MS approach, only the precursor ions that produce the carbohydrate marker ions through CID fragmentation in Q2 are detected, accounting for the increased specificity of this approach (40). Nevertheless, the enhancing fragmentation

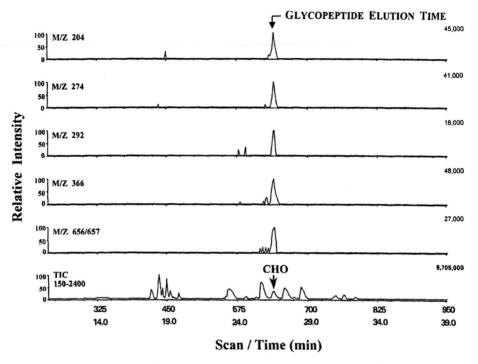

Figure 11 Total ion current (TIC) chromatogram obtained from the LC/ESI-MS analysis of the CHO IL-4 tryptic digest using "in-source" fragmentation (OR = 110 V). The production and detection of carbohydrate-specific oxonium ions at m/z 162, 204, 274, and 366, as shown in the reconstructed ion chromatograms, can be used for locating the glycopeptide-containing tryptic fractions (indicated with CHO).

in the ion declustering region yields the same structural information as the more specific tandem MS method, but with much higher sensitivity and mass-resolving power for the product ions.

In contrast to the N-linked carbohydrates, the O-linked glycans appear to be smaller and less complex in terms of identifying monosaccharide components and branching. Nevertheless, the identification of the O-glycosylation site(s) is more "challenging," and the definition of the O-linked structures is more complicated, because there is an absence of any consensus sequence for O-glycosylation, and the O-linked glycans are located in close proximity to several Ser or Thr residues in a single peptide. These O-glycan moieties can be released by reductive β-elimination, which generates unsaturated amino acid residues at that site, and the glycosylated peptides can be identified by

Structural Analysis of Glycoproteins

comparative MS mapping before and after the carbohydrate release. The glycopeptides with O-linked carbohydrates can be differentiated from the ones containing the N-linked carbohydrates by LC/MS/MS analysis of the digested glycoprotein before and after complete and selective removal of the attached N-linked carbohydrates with *N*-Glycanase (40). Operating in the parent ion scan mode and monitoring the ions at *m/z* 204 and 366, one can find all the precursor glycopeptide $[M + H]^+$ ions that produce these low mass oxonium ions following the *N*-Glycanase treatment. All these precursor ions correspond to intact glycopeptides containing O-linked carbohydrates, because all N-linked oligosaccharides have been previously released by *N*-Glycanase. This strategy has been nicely demonstrated for bovine fetuin, a 42 kDa glycoprotein that contains at least three O-linked carbohydrates along with three that are N-linked (40).

C. Determination of Site Heterogeneity

In the course of the protein mapping and the detection of the protein's glycopeptide fragments by LC/ESI-MS or MS/MS, the glycosylation heterogeneity at each site can be determined through the use of the contour plots, as described previously (34). The mass spectra that correspond to these scans in the contour plots (usually the averaged mass scans of these glycopeptide-containing regions) can readily reveal the glycopeptide heterogeneity. The mass separation of the observed ESI signals within the same-charge ion envelope indicates the presence of certain monosaccharides adjusted for the signal's charge state. For example, complex-type glycopeptides may exhibit a series of signals corresponding to triply charged ions that differ by the residual mass of sialic acid at that charge state (e.g., 97 Da). Similarly, the mass separation of the remaining ESI signals in an envelope corresponding to multiply charged ions should indicate the presence of additional monosaccharide units due to either arm extensions or additional branching.

In the next step, a direct comparison of the mass difference between the observed M_r values (by LC/ESI-MS and even MS/MS) and the known masses of potential candidate peptides should indicate the amino acid portion of the peptide containing the attachment site. By considering the M_r values of the defined glycan structures typically found in glycoproteins derived from a given mammalian cell system (28), the possible structures of the attached carbohydrates can be derived. The screening of the potential peptides, as well as the structural assignment of the attached glycans, is facilitated by the following factors:

1. The good mass accuracy of the measured glycopeptide masses, which is typically within 1–2 Da of the expected mass value and provides a high degree of confidence in the assignment of the glycan structures.
2. The presence of an expected N- and/or C-terminal amino acid residue based on the specific enzymatic cleavage employed.
3. The presence of the carbohydrate acceptor sequence Asn-X-Ser/Thr for N-linked glycosylation (where X is any amino acid except Pro or Asp), which assists in the identification of the potential N-Glycosylation sites.
4. The comparative MS enzymatic mapping of the glycoprotein before and after enzymatic digestion with *N*-Glycanase (see Sect. III. B).

Once the peptide fragment and the possible structure of the attached oligosaccharides are deduced, the glycosylation heterogeneity of the individual attachment site can be established by the mass differences of the observed ions within the envelope of the same charge state. For example, in the mapping of CHO IL-4, monitoring the sugar-diagnostic ions reveals the presence of a glycopeptide eluting at 26.5 min (Fig. 11), which is also confirmed by the contour plots. The mass spectra of these scans in the contour plots can readily reveal not only the termination of the attached glycans with sialic acid units ("capping") but also the variation in the Hex-HexNAc content (Fig. 8). This glycoform information was present in both the triply (centered at *m/z* 1859.5) and quadruply (centered at *m/z* 1394.8) charged ion envelopes. This nicely confirms the presence of complex-type glycans at the Asn^{38} residue.

Once glycopeptides have been identified they can be isolated by reverse-phase HPLC and then sequenced by combining sequential treatment with linkage-specific exoglycosidases and LC/ESI-MS analysis. The complete structural characterization of the attached carbohydrates (linear sequence, linkage, and branching) can be also carried out by isolating the carbohydrate (e.g., after *N*-Glycanase treatment) and using MS/MS and MS^n, as described by Sheeley and Reinhold (41).

IV. FUTURE TRENDS

Whereas MALDI and ESI have developed into routine tools in protein analysis, glycoproteins—especially those with a relatively high carbohydrate content—still form a group of compounds for which the detection of meaningful molecular ion signals by ESI is a major problem. The sheer complexity arising from their heterogeneity and their tendency to form adducts are causes of this problem. The successful analysis by nanoESI-MS reported here points to the strong advantage of the nanoESI process itself and to the decisive prerequisite of forming very small initial droplets. In the cases presented here, the charge state of the glycoprotein ions was still sufficiently high not to pose any problems for quadrupole or quadrupole ion-trap mass analyzers, which have limited mass ranges. However, modern oTOF analyzers will become the spectrometers of choice owing to their considerably higher mass range and their inherently higher sensitivity, useful traits both for the detection of intact glycoproteins and for the investigation of enzymatically cleaved glycopeptides by LC/MS approaches. The advantages of oTOF analyzers will be especially relevant if glycoproteins are to be investigated as constituents of noncovalent complexes, such as in protein–protein adducts or in enzyme–substrate complexes (see Chapter 9). Successful investigation of these complexes will have a great impact on drug discovery research by revealing protein–receptor and/or receptor–drug interactions and also in proteomics, which is vital to functional genomics.

With respect to glycoprotein structural determination by MS/MS, the major limitation of CID is that the preferred pathway is partial or complete loss of the glyco moiety, which prevents its localization in the peptide chain. This may be overcome by electron capture dissociation (ECD) in Fourier transform ion cyclotron resonance MS systems (42). The backbone fragmentation (c- and z-ions) observed in ECD provides peptide sequence information (often complete) and direct evidence for the position of the glycan in an enzyme-derived glycopeptide, as was recently demonstrated (43). The benefits of ECD (i.e., sequence information with a strongly superior sequence coverage) compared to CID and the possibility to apply ECD to larger glycopeptide fragments and even noncovalent complexes (up to 10 kDa) may become of high utility in glycoprotein determinations.

REFERENCES

1. JB Fenn, M Mann, CK Meng, CF Wong, CM Whitehouse. Electrospray ionization for mass spectrometry of large biomolecules. Science 246:64–71, 1989.
2. F Hillenkamp, M Karas, RC Beavis, BT Chait. Matrix-assisted laser desorption/ionization mass spectrometry of biopolymers. Anal Chem 63:1193A–1203A, 1991.
3. DA Cumming. Glycosylation of recombinant protein therapeutics: Control and functional implications. Glycobiology 1:115–130, 1991.

4. S Dube, JW Fisher, JS Powell. Glycosylation at specific sites of erythropoietin is essential for biosynthesis, secretion and biological function. J Biol Chem 263:17516–17521, 1988.
5. E Goldwasser, CK-H Kung, J Eliason. On the mechanism of erythropoietin-induced differentiation. J Biol Chem 249:4202–4206, 1974.
6. RD Smith, JA Loo, CG Edmonds, CJ Barinaga, HR Udseth. New developments in biochemical mass spectrometry: Electrospray ionization. Anal Chem 62:882–899, 1990.
7. T Yokota, T Otsuka, T Mosmann, J Banchereau, T DeFrance, D Blanchard, JE deVries, KI Arai. Isolation and characterization of a human interleukin cDNA clone, homologous to mouse B-cell stimulatory factor 1, that expresses B-cell- and T-cell-stimulating activities. Proc Natl Acad Sci USA 83:5894–5898, 1986.
8. A Tsarbopoulos, BN Pramanik, TL Nagabhushan, TR Covey. Structural analysis of the CHO-derived interleucin-4 by liquid chromatography/electrospray ionization mass spectrometry. J Mass Spectrom 30:1752–1763, 1995.
9. A Tsarbopoulos, M Karas, K Strupat, B Pramanik, L Tattanahalli, TL Nagabhushan, F Hillenkamp. Comparative mapping of recombinant proteins and glycoproteins by plasma desorption and matrix-assisted laser desorption/ionization mass spectrometry. Anal Chem 66:2062–2070, 1994.
10. T Nakanishi, A Shimizu, N Okamoto, A Ingendoh, M Kanai. Electrospray ionization-tandem mass spectrometry analysis of peptides derived by enzymatic digestion of oxidized globin subunits: An improved method to determine amino acid substitution in the hemoglobin "core." J Am Soc Mass Spectrom 6:854–859, 1995.
11. KL Duffin, JK Welply, E Huang, JD Henion. Characterization of N-linked oligosaccharides by electrospray and tandem mass spectrometry. Anal Chem 64:1440–1448, 1992.
12. N Rajan, A Tsarbopoulos, R Kumarasamy, R O'Donnell, SS Taremi, SW Baldwin, GF Seelig, X Fan, B Pramanik, HV Le. Characterization of recombinant human interleucin 4 receptor from CHO-cells: Role of N-linked oligosaccharides. Biochem Biophys Res Commun 206:694–702, 1995.
13. JF Kelly, SJ Locke, L Ramaley, P Thibault. Development of electrophoretic conditions for the characterization of protein glycoforms by capillary electrophoresis-electrospray mass spectrometry. J Chromatogr A 720:409–427, 1996.
14. JF Kelly, L Ramaley, P Thibault. Capillary zone electrophoresis–electrospray mass spectrometry at submicroliter flow rates: Practical considerations and analytical performance. Anal Chem 69:51–60, 1997.
15. M Wilm, A Shevchenko, T Houthaeve, S Breit, L Schweigerer, T Fotsis, M Mann. Femtomole sequencing of proteins from polyacrylamide gels by nano-electrospray mass spectrometry. Nature 379:466–469, 1996.
16. M Wilm, M Mann. Analytical properties of the nanoelectrospray ion source. Anal Chem 68:1–8, 1996.
17. A Shevchenko, M Wilm, O Vorm, M Mann. Mass spectrometric sequencing of proteins from silver-stained polyacrylamide gels. Anal Chem 68:850–858, 1996.

18. AN Verentchikov, W Ens, KG Standing. Reflecting time-of-flight mass spectrometer with an electrospray ion source and orthogonal extraction. Anal Chem 66:99–107, 1994.
19. HR Morris, T Paxton, A Dell, J Langhorne, M Berg, RS Bordoli, J Hoyes, RH Bateman. High sensitivity collisional-activated decomposition tandem mass spectrometry on a novel quadrupole/orthogonal-acceleration time-of-flight mass spectrometer. Rapid Commun Mass Spectrom 10:889–896, 1996.
20. K Biemann. Mass spectrometry of peptides and proteins. Annu Rev Biochem 61:977–1010, 1992.
21. SA Carr, ME Hemling, MF Bean, GD Roberts. Integration of mass spectrometry in analytical biotechnology. Anal Chem 63:2802–2824, 1991.
22. SA Carr, JR Barr, GD Roberts, KR Anumula, PB Taylor. Identification of attachment sites and structural classes of asparagine-linked carbohydrates in glycoproteins. Methods Enzymol 193:501–518, 1990.
23. LN Amankwa, K Harder, F Jirik, R Aebersold. High-sensitivity determination of tyrosine-phosphorylated peptides by on-line enzyme reactor and electrospray ionization mass spectrometry. Protein Sci 4:113–125, 1995.
24. DL Smith, Z Zhou. Strategies for locating disulfide bonds in proteins. Methods Enzymol 193:374–389, 1990.
25. Y Sun, MD Bauer, TW Keough, MP Lacey. Disulfide bond location in proteins. Methods Mol Biol 61:181–210, 1996.
26. SA Carr, ME Hemling, GF Wasserman, RW Sweet, K Anumula, JR Barr, MJ Huddleston, PJ Taylor. Protein and carbohydrate structural analysis of a recombinant soluble CD4 receptor by mass spectrometry. J Biol Chem 264:21286–21295, 1989.
27. CK Leonard, MW Spellman, L Riddle, RJ Harris, JN Thomas, TJ Gregory. Assignment of intrachain disulfide bonds and characterization of potential glycosylation sites of the type 1 recombinant human immunodeficiency virus envelope glycoprotein (gp 120) expressed in chinese hamster ovary cells. J Biol Chem 265:10373–10382, 1990.
28. R Kornfeld, S Kornfeld. Assembly of asparagine-linked oligosaccharides. Annu Rev Biochem 54:631–664, 1985.
29. A Dell, JE Thomas-Oates, ME Rogers, PR Tiller. Novel fast atom bombardment mass spectrometric procedures for glycoprotein analysis. Biochimie 70:1435–1444, 1988.
30. HR Morris, FM Greer. Mass spectrometry of natural and recombinant proteins and glycoproteins. Trends Biotechnol 6:140–147, 1988.
31. AL Burlingame. Characterization of protein glycosylation by mass spectrometry. Curr Opin Biotechnol 7:4–10, 1996.
32. B Stahl, T Klabunde, H Witzel, B Krebs, M Steup, M Karas, F Hillenkamp. The oligosaccharides of the Fe(III)-Zn(II) purple acid phosphatase of the red kidney bean. Determination of the structure by a combination of matrix-assisted laser desorption/ionization mass spectrometry and selective enzymic degradation. Eur J Biochem 220:321–330, 1994.

33. BJ Harmon, X Gu, DIC Wang. Rapid monitoring of site-specific glycosylation microheterogeneity of recombinant human interferon-γ. Anal Chem 68:1465–1473, 1996.
34. V Ling, AW Guzzetta, E Cannova-Davis, JT Stults, WS Hancock, TR Covey, BI Shushan. Characterization of the tryptic map of recombinant DNA derived tissue plasminogen activator by high-performance liquid chromatography-electrospray ionization mass spectrometry. Anal Chem 63:2909–2915, 1991.
35. SA Carr, GD Roberts. Carbohydrate mapping by mass spectrometry: A novel method for identifying attachment sites of Asn-linked sugars in glycoproteins. Anal Biochem 157:396–406, 1986.
36. SA Carr, GD Roberts, A Jurewicz, B Frederick. Structural fingerprinting of Asn-linked carbohydrates from specific attachment sites in glycoproteins by mass spectrometry: Application to tissue plasminogen activator. Biochimie 70:1445–1454, 1988.
37. A Apffel, JA Chakel, WS Hancock, C Souders, TM Timkulu, E Pungor Jr. Application of high-performance liquid chromatography-electrospray ionization mass spectrometry and matrix-assisted laser-desorption ionization time-of-flight mass spectrometry in combination with selective enzymatic modifications in the characterization of glycosylation patterns in single-chain plasminogen activator. J Chromatogr A 732:27–42, 1996.
38. MJ Huddleston, MF Bean, SA Carr. Collisional fragmentation of glycopeptides by electrospray ionization LC/MS and LC/MS/MS: Methods for selective detection of glycopeptides in protein digests. Anal Chem 65:877–884, 1993.
39. A Tsarbopoulos, A Prongay, S Baldwin, R Kumarasamy, J Schwartz, BN Pramanik, HV Le, M Karas, F Hillenkamp. Mass spectrometric analysis of the Sf9 cell-derived interleukin-5 receptor. In: Proc 44th ASMS Conf on Mass Spectrometry and Allied Topics, Portland, OR; American Society for Mass Spectrometry, Santa Fe, NM, 1996, pp 359a–359b.
40. SA Carr, MJ Huddleston, MF Bean. Selective identification and differentiation of N- and O-linked oligosaccharides in glycoproteins by liquid chromatography-mass spectrometry. Protein Sci 2:183–196, 1993.
41. DM Sheeley, VN Reinhold. Structural characterization of carbohydrate sequence, linkage, and branching in a quadrupole ion trap mass spectrometer: Neutral oligosaccharides and N-linked glycans. Anal Chem 70:3053–3059, 1998.
42. RA Zubarev, DM Horn, EK Fridriksson, NL Kelleher, NA Kruger, MA Lewis, BK Carpenter, FW McLafferty. Electron capture dissociation for structural characterization of multiply charged protein cations. Anal Chem 72:563–573, 2000.
43. E Mirgorodskaya, P Roepstorff, RA Zubarev. Localization of O-glycosylation sites in peptides by electron capture dissociation in a Fourier transform mass spectrometer. Anal Chem 71:4431–4436, 1999.

8
Advanced Mass Spectrometric Approaches for Rapid and Quantitative Proteomics

Richard D. Smith, Gordon A. Anderson, Christophe D. Masselon,
Mary S. Lipton, and Ljiljana Paša Tolić
Pacific Northwest National Laboratory, Richland, Washington

Thomas P. Conrads and Timothy D. Veenstra
National Cancer Institute, Frederick, Maryland

I. INTRODUCTION

With the completion of several dozen genome sequences and the first draft of the human genome in 2000, biological research is moving rapidly into the "postgenomic era." Contributing to the movement toward this era are recent advances in robotics, DNA sequencing technology, and computational analysis, all of which are resulting in an increasingly large amount of DNA sequence data, along with an array of experimental and bioinformatic tools increasingly being used for its analysis. In the postgenomic era, studies will be designed to characterize complex cellular "systems" consisting of networks of molecular networks. In the new paradigm, cellular processes are increasingly subject to global study and modeled "from the top down," leading to new understanding of the cellular functions of the individual *system* constituents, their response to environmental perturbations, and the properties arising from the complex nature of their interactions. A major goal is to understand both the molecular and cellular function(s) of the key system components (i.e., proteins). This higher level view of cellular processes will be a basis for understanding the robustness of cellular systems, the possible modular nature of the cellular machinery, the nature of epigenetic and multigenic diseases, and the individual variability in susceptibility to disease as well as for developing predictive capabilities of the effects arising from external perturbations. Additionally, the information may lead to an understanding of the molecular "nodes" that can be targeted for drug development, gene therapy, genetic manipulations, etc.

Whereas a genomic sequence describes those proteins a cell or organism *may express*, proteomic measurements reveal the proteins *actually present* at a given time. The availability of whole genome sequence databases allows one to determine more effectively how genes act in concert and to understand the molecular changes that occur within identical cells under different environmental conditions. Two rapid methods for simultaneously assessing the activities of thousands of genes at the mRNA level have already been developed. These are the microarray assay and the serial analysis of gene expression (SAGE) techniques (1–4). The level of gene expression at the mRNA level, however, is only one of the factors that determine how much of a specific protein is present in a cell, and existing data indicate that protein abundances can show poor correlation with mRNA abundances (5,6). Additionally, characterization of gene expression at the mRNA level does not provide the necessary information to identify post-translational modifications of the protein products, an important factor in determining the activities and properties of numerous proteins.

Proteome level analyses provide more accurate information about biological systems and pathways than gene expression analysis at the mRNA level, because the measurements focus directly on the major class of effector molecules in cellular systems. The ability to make broad proteome measurements ideally requires experimental methods that can reliably identify and quantify changes in the expression of thousands of proteins in a single experiment. Additionally, speed is important because one desires to examine the effects of many different perturbations as well as to follow the time course of proteome change following a single perturbation. Thus, there is the need for a capability that enables many broad proteome measurements. This is not to say, however, that the conventional approach to proteomics based upon gel electrophoresis and one-at-a-time protein analysis would necessarily be displaced by such a capability. Rather, this ideal new capability would enable new types of measurements at the systems level. Conventional methodologies that are well suited for the study of individual or small subsets of proteins obviously will continue to play a major role in more conventional (e.g. hypothesis-driven or "reductionist") approaches to biological research. However, proteome analyses also allow higher level hypotheses to be tested (e.g., as to how networks interact).

Until now, available analytical techniques have fallen far short of the capabilities needed to provide such high throughput measurements of an organism's proteome. The speed at which proteomes are being characterized using present technology lags far behind that at which gene expression at the mRNA level can currently be measured, and even the mRNA technology has significant limitations related to the preparation and use of the arrays. Development of the capabilities to observed rapidly and quantify the array of proteins expressed in cells will provide a powerful means to identify the functions of key proteins. More important, the essential pathways and networks or modules involved in cellular responses to external perturbations will also be identified. The ability to precisely measure changes in the relative abundances of numerous proteins simultaneously enables the identification of the proteins participating in the multiple pathways and gives insights into how cellular networks are linked.

The power and promise of advanced proteomics measurement technologies are implicitly based upon the capability to measure changes in the relative expression of many proteins within a single experiment. This global and quantitative perspective would provide the ability to observe changes within entire sets of cellular pathways and networks, thereby identifying those pathways that are key to a cell's response to a changing environmental condition. Characterization of proteomes should ideally be accomplished with high speed and sensitivity so that comparisons between several different cell types, or the effects of many different types of perturbations, can be made quickly and in a manner that allows proteins expressed at low levels to be studied. This is the aim of the advanced approaches to proteomics we describe in this chapter.

II. PRESENT APPROACHES FOR PROTEOME MEASUREMENTS

Presently there is no rapid and sensitive technique for large-scale analysis of proteins comparable to that for mRNAs. The current proteome analysis technology is predominantly based upon the separation of the protein mixture by two-dimensional polyacrylamide gel electrophoresis (2D-PAGE) (10) combined with mass spectrometric identification of the separated proteins (7–9). Whereas 2D-PAGE has demonstrated the ability to provide a detailed view of up to thousands of proteins expressed by an organism or cell in a single two-dimensional separation (11), it is a relatively slow, labor-intensive, and cumbersome technology. Moreover, 2D-PAGE results from different labs can be difficult to compare, and contaminants and artifacts may be introduced by

PAGE. The sensitivity of this method is limited by the amount of a protein needed for visualization [typically on the order of ~10 femtomoles (1 fmol = 10^{-15} mol) (12,13)] as well as the quantity of protein that can be loaded onto the gel. Indeed, there is growing recognition that global proteome displays based upon 2D-PAGE are largely constrained to the more stable and abundant (e.g., "housekeeping") proteins and that important classes involved in signal transduction and regulation of expression, for example, and other less abundant (and generally less stable) proteins are largely undetected. In addition, 2D-PAGE is also biased against membrane proteins, highly acidic or basic proteins, and very large or small proteins. Because the mass accuracy afforded by the sodium dodecyl sulfate (SDS) gel electrophoresis dimension of 2D-PAGE separations is poor (compared to mass spectrometry), unambiguous identification of a protein based simply on its spot location is often impractical, and intense spots can also often conceal lower level proteins. Recent work has additionally highlighted the fact that single proteins can give rise to multiple "spots" in 2D-PAGE that are sometimes due to different modification states but in other cases arise from unknown causes. Conventionally, protein identification after 2D-PAGE involves the separate extraction and analysis of each protein. This individual protein analysis approach presents a serious hurdle to overcoming the high throughput demands of effective proteomic studies. The need to extract the protein from the gel concomitant along with the need to prepare the sample prior to MS characterization effectively compounds the sensitivity limitations associated with 2D-PAGE separations. The overall 2D-PAGE MS strategy of proteome characterization is not well suited to automation and is unlikely to provide the high throughput needed for future proteomic studies.

As mentioned above, effective proteomic studies would ideally provide the ability to measure changes in the relative abundances of many proteins within a single experiment. Methods to measure relative protein abundances based on 2D-PAGE technology have generally relied on comparison of the protein "spot" intensity of two gels. This method, however, cannot accurately measure small differences in protein abundances or differences in cases where protein spots are poorly resolved and where the means of locating one spot must be reliable and reproducible. Other large-scale quantitative protein expression studies have also been performed using radioactive labeling (6) and stable isotope metabolic labeling followed by PAGE separation and quantitative analysis by scintillation counting or, very recently, by mass spectrometry (6,14). Although these strategies are capable of detecting subtle differences in protein abundances, they also retain the disadvantages associated with 2D-PAGE analysis; that is, there are sensitivity limitations and an inability to provide broad proteome coverage. Considering the limitations associated with the overall 2D-PAGE MS proteome characterization approach, the need clearly exists for alternative strategies to characterize proteomes with greater speed and sensitivity and for more precise methods to measure changes in relative protein expression.

III. PROTEIN IDENTIFICATION USING MASS SPECTROMETRY

Mass spectrometry is already playing a prominent role in proteomics, primarily for the purpose of protein identification. Two generally applicable MS approaches have been developed for protein identification, one based on mass measurements for a set of peptide products derived from a specific protein and the other based on mass spectrometric fragmentation (MS/MS) of one (or more) of these peptides (15). The former, often referred to as peptide mass fingerprinting, involves the use of chemical or enzymatic agents to digest proteins according to the sequence specificity of the agent. A set of peptide fragments unique to each protein is created, and their masses constitute a "fingerprint" used to identify the original protein (16–19). The typical number of polypeptide molecular masses necessary to determine the identity of the protein from the use of low resolution instrumentation has been found to range from three to six (17,20). This number, however, is (as discussed below) highly dependent on the mass measurement accuracy (MMA) and mass resolving power of the measurements.

The second approach to protein identification is based on the information from the mass of one or more polypeptide fragments augmented by the additional (sequence-related) dissociation of the polypeptide using tandem MS (MS/MS). Activation by collisions with an inert gas, by infrared laser irradiation, etc. results in the sequence-related dissociation of polypeptides (21–26). If a suitable protein or genomic database is available, the MS/MS dissociation of a single polypeptide is often sufficient for unambiguous protein identification, even with the use of low resolving power, low MMA quadrupole mass spectrometers (8,15,27–29). The use of electrospray ionization (ESI)-MS is attractive for this application because it is compatible with on-line liquid chromatography (LC) and capillary electrophoresis (CE) (30–35). Although several peptides may be observed within a single mass spectrum, MS/MS analysis is generally limited to a single peptide per experiment (i.e., spectrum).

These MS methods have been incorporated into conventional proteomic strategies starting with 2D-PAGE and now enable protein identification (at the most effective laboratories) at low femtomole levels. Several large laboratories and pharmaceutical companies have implemented robotic methods for the processing of excised 2D-PAGE spots and are increasing throughput to a level capable of identifying hundreds to thousands of the more abundant proteins in a week. Still, this level of performance is far short of what is needed to elevate the speed and coverage of proteomics to the level of mRNA expression measurements. In addition, the sensitivity limitations and the inability of this technique to provide broad proteome coverage for all classes of proteins severely reduce its effectiveness, making it most typically useful to only highly abundant soluble proteins.

IV. THE ATTRIBUTES OF FTICR MASS SPECTROMETRY FOR PROTEOMIC MEASUREMENTS

Since its introduction in 1974 by Marshall and Comisarow (36), Fourier transform ion cyclotron resonance mass spectrometry (FTICR-MS) has undergone extensive advances in instrumentation, computer control, experimental methodology, ion manipulation techniques in the trapped ion cell(s), and postacquisition data processing techniques. FTICR mass spectra can simultaneously provide high sensitivity, ultrahigh resolving power over a broad mass-to-charge ratio (m/z) range, and high MMA (37). The MMA that can be obtained with FTICR instrumentation depends upon magnetic field strength and other instrumental factors. Mass spectra can be acquired with resolving power in excess of 10^5 with low to sub part-per-million (ppm) mass measurement error (38,39). Tandem or multistage mass spectrometry (MS^n) can be performed based on the introduction of a single group of ions that are retained, dissociated by collisions or irradiation, and remeasured, providing sequence information without additional sample consumption.

High sensitivity (and dynamic range) is of crucial practical importance to the study of cellular pathways and networks because many important protein classes will be present only at low concentrations. Not only is high sensitivity crucial for the detection of important regulatory proteins, but it also improves the fidelity of the measurements of changes in the abundances of these low level proteins due to changes in a cell's environment. Increased sensitivity will also allow proteomic measurements to be effectively extended to the study of smaller cell populations or tissue sample sizes (e.g., samples obtained from microdissection). The sensitivity achievable with FTICR can exceed that feasible with other approaches: McLafferty and coworkers (40), Marshall and coworkers (41), and Belov et al. (42) have all demonstrated subattomole level detection of small proteins using ESI-FTICR. These detection levels are orders of magnitude better than those routinely needed to visualize a protein "spot" using 2D-PAGE.

To increase further sensitivity, we incorporated a new "ion funnel" design into the ESI interface that greatly increases the fraction of the ion current effectively transmitted to the FTICR analyzer (42). The ion funnel is based on the use of radio-frequency (RF) ring electrodes of progressively smaller diameter and with opposing RF phases applied to adjacent elements. The result of this arrangement is to confine ions within the funnel. A separate direct current (DC) field is applied to drive ions down the cone of the funnel. Because the electrode size and spacing are approximately constant even though the inner diameter of the ring electrode is variable, the RF voltage needed for confinement of ions does not increase as the inner diameter increases. Thus, the constraint of RF multipoles to small acceptance apertures does not apply for the ion funnel. The mass spectra obtained using the initial funnel design for ion introduction demonstrate a gain in sensitivity of approximately one to two orders of magnitude. The sensitivity achievable using a 3.5 tesla (T) FTICR is illustrated by a detection limit of ~30 zeptomoles (or roughly 18,000 molecules) obtained for small proteins (42). It is important that this high sensitivity can be obtained simultaneously with high resolution and high mass measurement accuracy, which are also of great value in proteomic applications.

Another key attribute of FTICR-MS is the high dynamic range. One of the greatest difficulties in obtaining broad proteome coverage is the wide range of relative protein abundances ($>10^5$). Although some proteins such as actin are stable, highly abundant, and constantly present in most eukaryotic cells, many signaling proteins are unstable and/or are expressed only at a low copy number (43). High dynamic range is important so that these low abundance proteins can be detected in a mixture that also contains more abundant proteins. The FTICR trap has a useful charge capacity that is three to four orders of magnitude larger than that used in quadrupole ion traps (depending upon trap size, magnetic field strength, and other operational details) and that provides the basis for an inherently greater dynamic range for any given spectrum. Studies in our laboratory have shown that species present at lower concentrations can also be effectively accumulated in the FTICR trapped ion cell by applying quadrupolar excitation, while more abundant species are removed by magnetron expansion. The result is that "space" is made in the trap for the selected less abundant species, even during greatly extended ion accumulation periods. The first implementation of this approach used "colored" noise waveforms to apply quadrupole excitation to any desired parts of the mass spectrum, and even did this in a real-time data-dependent mode of operation (44). This approach is illustrated in Fig. 1 for an ESI-FTICR study of a mixture of insulin, ubiquitin, and cytochrome c at various concentrations ranging from 10^{-4} to 10^{-8} M. Under normal "in-trap" ion accumulation conditions, the dynamic range obtained was approximately 100 with the 7 T FTICR used in this study, and the lower abundance species were not evident in the mass spectrum (Fig. 1A). The application of selected-ion accumulation using a colored noise waveform synthesized from the initial spectrum (where notches were applied at the cyclotron frequencies of the major detected species in the initial spectrum) gave greater than two orders of magnitude gain in dynamic range and allowed the detection of cytochrome c present at 10^{-8} M concentration within

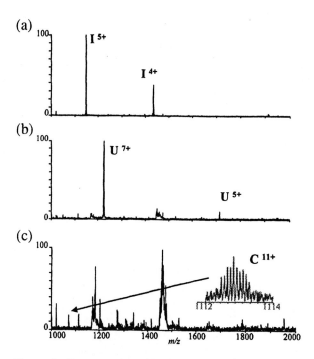

Figure 1 Demonstration of the dynamic range expansion capabilities of FTICR based on the use of data-dependent synthesis of quadrupole colored noise excitation waveforms allowing the detection of cytochrome c in the presence of the higher concentration species. A mixture of cytochrome c (C) (10^{-8} M), ubiquitin (U) (10^{-6} M), and insulin (I) (10^{-4} M) was directly infused into the FTICR spectrometer. The y axis represents relative ion abundance.

the same sample. We are presently implementing a modified version of this approach, with which ions are first selected and accumulated outside the FTICR cell, that is well suited for use with on-line separations of proteins or peptides.

The high MMA obtainable with FTICR also provides an important, and until now largely unrecognized, capability for the high throughput identification of proteins in proteomic studies. As mentioned previously, typical 2D-PAGE approaches to proteomic investigations use the information obtained from the peptide mapping of a single protein or tandem MS of a single peptide to identify proteins. Although the masses of several peptides are generally required, MS/MS data for only a single peptide are often all that is required for unambiguous protein identification.

This fact leads to a concept that opens the door to high throughput approaches for broad proteomic measurements: If the molecular mass of a single peptide can be measured with high enough MMA that its mass is unique among those of all of the possible peptides predicted from a genome, it can then be used as an accurate mass tag (AMT) or an effective biomarker for unambiguous protein identification. The generation of such AMTs would allow the tryptic fragments generated from an entire proteome or complex mixture, for example, to be analyzed and identified with much greater speed, dramatically increasing the throughput for protein identification. The low ppm to sub-ppm level measurements routinely achievable with FTICR (45) (and ~0.1–0.5 ppm measurements may be practical with further refinement and the use of high-field instrumentation) are significantly better than the 5–20 ppm accuracy typical of the best conventional mass spectrometer technologies and the 100–500 ppm accuracy of the most widely used quadrupole and ion trap technologies. Furthermore, the ability to achieve high mass measurement accuracy is largely independent of signal intensity, unlike measurements made with time-of-flight instruments. A subject we discuss later in this chapter is the level of mass measurement accuracy needed so that peptide mass measurements can function as AMTs to give high confidence in protein identification or subsequent peak assignment.

V. PROTEOMIC METHODS BASED ON INTACT PROTEIN ANALYSIS

Approaches to proteomics can be divided into those that are based on intact protein analysis and those that use enzymatic processing. Indeed, the conventional 2D-PAGE approach uses both: 2-D separation of the intact proteins followed by MS analysis of the digestion products from an excised "spot." As discussed above, many of the limitations impeding current proteomic investigations are linked to the separation of proteins by 2D-PAGE. Despite the drawbacks of 2D-PAGE, it presently plays a central role in the study of proteomes and has demonstrated the ability to resolve, in a single separation, up to thousands of proteins expressed by an organism or cell type (11). For the analysis of intact proteins, 2D-PAGE is the standard (it is debatable that it should be referred to as a "gold standard") by which any alternative should be judged.

A. Protein Separation by Capillary Isoelectric Focusing

In this section we discuss alternative methods based on the direct analysis of intact proteins in which we adopt the first of the two dimensions used in 2D-PAGE but do so in a capillary format without the use of gels and with other changes that allow direct compatibility with ESI-MS. The movement from conventional slab-gel-based electrophoretic techniques to a capillary format offers many advantages, such as increased speed, improved separation efficiencies, and improved capability for automation. In the case of capillary isoelectric focusing (CIEF), first demonstrated by Hjertén and Zhu (46), the fused silica capillary is most commonly coated with linear polyacrylamide to eliminate electro-osmotic flow and reduce protein adsorption to the walls (47). CIEF employs a polyampholyte mixture in free solution that, when an electric field is applied that bridges high pH and low pH solutions, sets up a pH gradient in the capillary through which proteins migrate until they have zero net charge and thus focus at their respective isoelectric points (pIs) (48,49). After focusing, protein bands are mobilized to the detector by one of several methods, such as chemical mobilization (anodic or cathodic) or hydraulic mobilization (i.e., pressure-, vacuum-, or gravity-driven flow) (50,51). Separations of analytes having pI differences as small as 0.004 pH unit have been obtained in our laboratory (52).

Whereas CIEF has been used primarily with conventional UV detection, more recently, ESI-MS has been coupled to CIEF for protein analysis (53–55). The use of an information-rich detection method is extremely beneficial when one is dealing with the very complex protein samples used in proteomic analyses. Coupling CIEF separations to MS offers the ability to obtain accurate mass measurements as well as other structurally related information for proteins in complex mixtures. The CIEF-MS combination is thus roughly analogous to a 2D-PAGE separation, in the sense that it provides information on pI and molecular mass, but with greater speed and sensitivity, ease of automation, much more accurate mass measurements, and the possibility of using advanced multidimensional MS methods (56,57).

B. CIEF Combined with FTICR-MS

Current efforts in our laboratory are directed toward developing new approaches based on capillary separations combined with advanced forms of MS for high throughput proteome studies. One of the methods being explored combines CIEF with FTICR mass spectrometry. As discussed previously, FTICR provides mass measurements of much higher resolving power and accuracy than other types of mass spectrometers (e.g., quadrupole, ion trap, time-of-flight instruments), as well as other capabilities for multistage studies (e.g., dissociation studies using MS/MS to extract sequence and other structural information) that are useful for protein identification and studies of protein modifications. Initial results analyzing cellular protein extracts by CIEF-FTICR were presented by Yang et al. (55), and in subsequent work we improved CIEF resolution and used isotope depletion to obtain greater sensitivity and accuracy of molecular mass measurements from FTICR (58). These improvements resulted in the observation of significantly larger numbers of proteins from a proteome in a single CIEF-FTICR experiment than had been observed previously.

Using the CIEF approach avoids many of the problems associated with ingel digestion procedures used with 2D-PAGE, and reduces the time required for subsequent "one-at-a-time" MS analysis. Other advantages of CIEF include the capability to handle extremely small sample sizes, enhanced speed and resolution, ease of automation, and the high sensitivity arising from the natural concentration effect (a factor of 50–100-fold) associated with focusing. In fact, the amount of total protein loaded in the CIEF capillary is typically 100–300 ng, whereas 2D-PAGE can commonly involve from 100 μg to >10 mg of total protein. Very often protein extracts from as many as 20 2D-PAGE separations need to be pooled for analysis by conventional MS methods. In addition, a single CIEF-FTICR analysis from sample preparation to data visualization can be done in only a few hours, whereas 2D-PAGE commonly takes a few days.

The CIEF-FTICR approach allows for proteome-wide analysis of *intact* proteins, which can be crucial for functional proteomic studies. The effectiveness of CIEF for obtaining high resolution separations of complex protein mixtures was initially demonstrated in our laboratory by analyzing cell lysates extracted from *E. coli* (58,59). In our initial efforts, *E. coli* was harvested, lysed, and dual microdialyzed (58) to produce a soluble protein fraction. The resulting protein solutions were prepared by adding Pharmalyte 3–10 to a concentration of 0.5% and were injected to fill the capillary for CIEF. The results, obtained using a 7 T FTICR instrument (60), demonstrate the wealth of mass spectral data that can be obtained because of the ability to detect numerous proteins in a single analysis (Fig. 2). Several proteins are generally detected in each spectrum. Because the FTICR ion trap has a finite total trap charge capacity, the achievable dynamic range can become limited when many proteins elute at the same time. Therefore, the overall dynamic range depends significantly on the CIEF separation quality because the FTICR ion trap can be

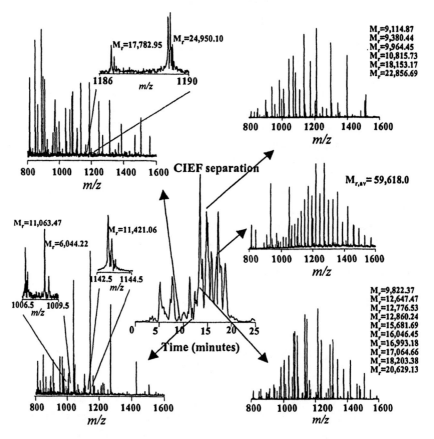

Figure 2 Demonstration of the high density of information obtained in a single CIEF-FTICR analysis of *E. coli* lysate. The reconstructed electropherogram (center) shows several resolved peaks; however, analysis of each scan reveals that several protein masses can be present within a single spectrum. This analysis provided approximately 1000 putative protein masses in a single 30 min experiment.

filled to near capacity with ions from even relatively small CIEF peaks, as discussed earlier. Thus, improved CIEF separations and faster FTICR spectrum acquisition rates will significantly extend the effective dynamic range achievable, and both improvements are presently being pursued in our laboratory along with the use of more advanced methods for dynamic range expansion (61). Even at this point, however, the results illustrate the potential of CIEF-FTICR for the analysis of complex protein mixtures.

Although FTICR provides high MMA, measurements can be significantly improved for high molecular mass proteins by using rare isotope depletion strategies, as shown in Fig. 3 (62). Achievable mass accuracy using natural abundance isotopic distributions (~98.89% ^{12}C, ~99.63% ^{14}N, and 99.985% ^{1}H) can be limited by the resulting broad isotopic envelope of molecular ions (which becomes progressively worse with increasing molecular mass) and the statistical limitations it places on the accurate fitting of experimental and the corresponding "theoretical" isotopic envelopes. Depletion of rare isotopes allows improved assignment of the monoisotopic species (i.e., without ^{13}C, ^{15}N, etc.) and thus avoids the 1–2 Da uncertainties in molecular mass measurements that can result from the fitting of the isotopic envelopes (58). Additionally, the detected signal is effectively "concentrated" into fewer peaks, providing an increase in signal-to-noise ratios, thus improving measurement sensitivity and dynamic range (Fig. 3).

C. Automated Data Processing and Display

Because the CIEF-FTICR experiments result in vast amounts of data that make manual analysis slow and arduous, let alone difficult to visualize, software must be developed to analyze the data effectively as well as to visualize them in a meaningful manner. Analysis of the large and complex data sets arising from CIEF-FTICR experiments and interpretation of the results in the context of available genomic databases were done using software being developed at our laboratory to assist in protein identification, MS/MS data analysis, and database searching (63). This software-based process is fully automated, initially requiring about 4.0 h on a typical PC to process a CIEF-FTICR experiment containing 1000 spectral files (and resulting in the detection of ~1500 putative proteins). Obviously, faster computers or the use of multiple processors can significantly reduce the time required for data processing in the future.

Of critical importance when dealing with large data sets is the presentation of the results in a manner that provides a comprehensible overview, a challenge that obviously depends on the needs of the application. Software developed in our lab to provide automated data analysis also allows visualization of the results

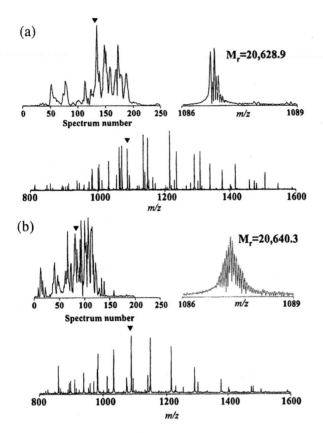

Figure 3 Examples of FTICR mass spectra for *E. coli* proteins obtained during CIEF separations using (A) isotopically depleted media (~98.89% ^{12}C, ~99.63% ^{14}N, and ~99.985% ^{1}H) versus (B) normal growth media. (a) The monoisotopic mass of the protein observed in the FTICR mass spectrum of an extract obtained from *E. coli* grown in isotopically depleted medium can be easily assigned (b) The same protein observed in the FTICR mass spectrum of an extract obtained from *E. coli* grown in normal isotopic abundance medium gives a broad isotopic envelope, making the assignment of the monoisotopic mass more difficult.

in the form of familiar 2D-PAGE displays, as illustrated in Fig. 4. The 2-D displays are produced by plotting molecular mass versus scan number, which can be correlated to the pI (obtained by spiking the CIEF separation with standard proteins of known pI). The spectral intensity in this display is derived from ion abundances and is communicated using variable spot sizes. The 2-D displays

Advanced MS for Rapid Quantitative Proteomics

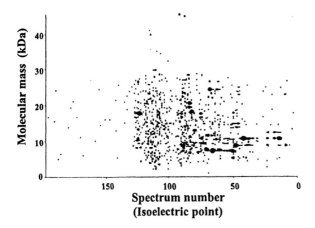

Figure 4 Virtual two-dimensional CIEF-FTICR display from the analysis of lysates from *E. coli* grown in normal medium.

shown of *E. coli* are for the data obtained from the CIEF-FTICR (such as those presented in Fig. 2), and show ~900 spots (i.e., unique putative protein masses) for *E. coli*. Displaying the data in this 2-D format provides a quick and convenient method for visualizing the large amounts of proteomic information and facilitates comparisons of different analyses. These displays, however, do not convey the quality of the mass measurements, and large spots can obscure less abundant species (as is also the case with 2D-PAGE separations). It is also clear that this is not the basis for highly quantitative measurements, because the ionization efficiencies vary somewhat from protein to protein, and run-to-run variations due to sample processing, cleanup, separation, and ESI-FTICR performance can also be significant. Thus, this approach to quantification provides only a rough guide to the magnitudes of protein abundances. Methods we discuss later provide the basis for much greater measurement precision.

VI. IDENTIFICATION OF INTACT PROTEINS BY FTICR

A. Identification Based on Molecular Mass of Intact Proteins

Detection of the proteins is only the first step. There is obviously a need to identify the spots, a task that is greatly assisted by the availability of complete genome sequences. It is important to recognize that the identification process generally needs to be carried out only once for a given organism or sample type. It will be unnecessary to repeat the identification step in most cases owing to the uniqueness of the combination of pI and a highly accurate molecular mass measurement. Identification of a protein using only molecular weight information is problematic, however, because it is generally impractical to predict post-translational modifications and/or proteolytic events, which often play an important role in protein activity (64). Additionally, the prediction of genes remains an imperfect art, and furthermore DNA sequencing errors can lead to incorrect predictions.

There are several different approaches to intact protein identification by mass spectrometry. An example of an identification based solely on a protein's accurate molecular mass and position in the 2-D display is shown in Fig. 5. The mass spectrum for a protein with $M_r = 40,966 \pm 1$ Da (an average

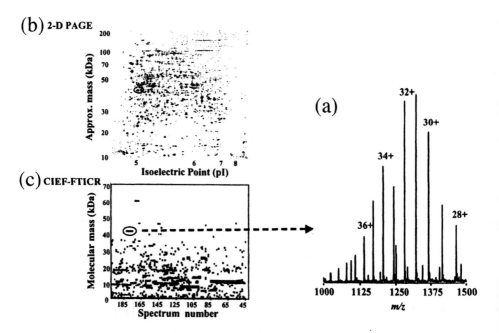

Figure 5 (a) Positive ion mode ESI mass spectrum for phosphoglycerate kinase obtained from CIEF-FTICR analysis of the lysate from *E. coli* grown in isotopically depleted medium. (b) The 2D-PAGE results for *E. coli* are reproduced from SWISS-2DPAGE, Swiss Institute of Bioinformatics, (http://www.expasy.ch/ch2d/). (c) Two-dimensional display of the CIEF-FTICR results for a lysate of *E. coli* grown in isotopically depleted medium generated in our laboratory.

mass calculated from the unresolved charge states of the isotopically depleted protein) was identified as that of phosphoglycerate kinase by searching the experimentally obtained mass against a database of masses for all theoretical *E. coli* proteins. The location of the corresponding spot on a 2-D gel (SWISS-2DPAGE at http://www.expasy.ch/ch2d/, Swiss Institute of Bioinformatics) attributed to phosphoglycerate kinase is shown in Fig. 5B and a 2-D display generated from the CIEF-FTICR in Fig. 5C. Although good correlation between the 2-D CIEF-FTICR and 2D-PAGE displays is evident, because of the differences in isoelectric focusing conditions, the accurate molecular mass is by far the most useful information for identification.

B. Identification Based on MS/MS of Intact Proteins

As stated earlier, pI and accurate molecular mass measurements will be insufficient for identification in many cases. In those cases, one can often apply multistage MS/MS (MS^n) capabilities (65,66), as illustrated in Fig. 6. To demonstrate this capability, a sample of isotopically depleted *E. coli* lysate was mixed 50/50 (v/v) with sheath liquid and directly infused in the FTICR mass spectrometer (Fig. 6A). A peak for a species of 7266.6 Da was trapped using selective ion accumulation (SIA) (Fig. 6B). Although this mass correlated well with the theoretical mass of the cold shock–like protein CspC in the *E. coli* database, MS/MS was performed using sustained off-resonance irradiation–collisionl-induced dissociation (SORI-CID) to verify the identity of the protein by using its fragment "mass tags." The resulting product ion spectrum, shown in Fig. 6C, provides partial sequence information (see assignments), allowing for the unambiguous identification of the protein as CspC.

Ideally, this MS/MS approach would be carried out on-line many times by software-driven data-dependent ion selection during the course of the on-line CIEF mobilization. Initial experiments aimed at demonstrating this technique were applied to the separation of a mixture of two standards, horse heart myoglobin and bovine carbonic anhydrase II (Sigma, St. Louis, MO). The mass spectrum for a single scan taken for the myoglobin peak provides a mass of 16,591 Da. The most abundant ion (or other programmed selection criterion) in each mass spectrum is then used to direct the synthesis of the RF excitation waveform(s) to be used during the SIA, stored waveform inverse Fourier transform (SWIFT) isolation, and SORI-CID events in the subsequent MS/MS experiment. The sequence information derived from the product ion spectrum of a selected ion was then used to verify the identity of one of the peaks as myoglobin. At present, the addition of the MS/MS procedure can extend the total sequence time to as much as 10–12 s, owing to the need for two separate gas

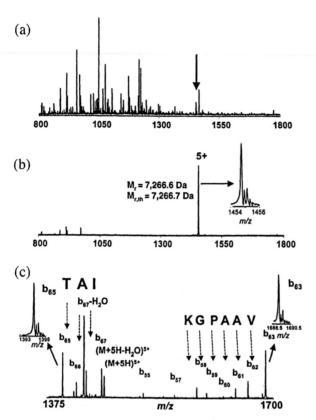

Figure 6 (a) Positive ion mode ESI mass spectrum from direct infusion of a lysate from *E. coli* grown in isotopically depleted media. (b) Selective ion accumulation and SWIFT isolation of the indicated peak in (a). (c) The spectrum obtained after SORI-CID, providing mass fragments for identification of the cold-shock-like protein (CspC).

injection events (ion accumulation and SORI-CID), providing a lower throughput than that obtainable with conventional instrumentation (albeit with higher data quality). Additionally, because the base pressure in the FTICR trap increases with the frequency of sequence repetition, especially during long CIEF-FTICR MS/MS experiments, the mass resolution obtained is not as high. These problems can be greatly alleviated by the use of external ion accumulation methods (decreasing the required pump downtime in the trap) and infrared multiphoton laser-induced dissociation (67) in place of SORI-CID. Implementation of these steps should reduce the time needed for the entire sequence

Advanced MS for Rapid Quantitative Proteomics

for both stages of analysis to less than 3 s. These steps are presently being implemented in our laboratory. However, the use of tandem MS for intact protein identification is most effective for smaller proteins, and as size increases MS/MS becomes less sensitive and the fragmentation less useful. The primary reason for this appears to be the large number of possible dissociation pathways. At the moment, although some success can be achieved for very large proteins, we consider this approach to be generally impractical for routine identification of intact proteins larger than 40–50 kDa.

C. Identification of Intact Proteins by Isotope Labeling

It is obvious that even with the high MMA provided by FTICR, more information in addition to molecular mass is generally necessary to more confidently identify intact proteins. One strategy that we have developed is to supplement the molecular mass with information related to the amino acid composition of the protein. Partial amino acid composition was previously determined by measuring the decay rate of radiolabeled proteins from yeast that had been separated by 2D-PAGE (68). This information, in conjunction with predicted isoelectric point and molecular mass, aids identification of 2D-PAGE-separated proteins. This method, however, still requires 2D-PAGE separation of the proteins and is therefore not amenable to high throughput proteomic analyses.

An alternative approach, developed at our laboratory, uses stable-isotope-labeled amino acids in conjunction with auxotrophic forms of *E. coli* and yeast (69). Using this strategy the auxotrophic organism (i.e., leu^-) is grown in normal medium (containing all the necessary nutrients) as well as medium to which a stable-isotope-labeled amino acid residue has been added (e.g., Leu-d_{10}). The presence of the isotopically labeled Leu in the medium results in the expression of proteins with Leu-d_{10} in place of natural isotopic abundance Leu. The result is two versions of the proteome: one containing natural isotopic abundance and the other containing the labeled Leu-d_{10} residues in place of natural isotopic abundance Leu residues. The cells grown in the different media are then combined to create a mixture contianing two versions of the same proteome. After protein extraction, the mixture is analyzed using CIEF-FTICR-MS. Because the unlabeled and labeled versions of proteins have identical CIEF separation characteristics, the natural isotopic abundance and Leu-d_{10} isotopically labeled versions of the protein are observed within the same FTICR spectrum (Fig. 7). Both protein versions display identical charge state distributions, and the predictive mass differences allow them to be easily assigned as isotopically distinct versions of the same protein. The mass difference between the isotopically distinct proteins is then used to determine

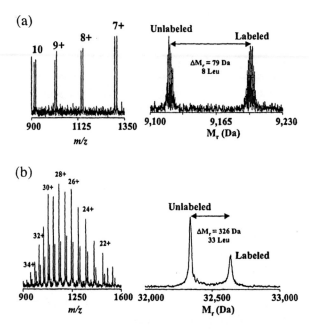

Figure 7 Mass spectra and deconvoluted mass spectra of *E. coli* proteins observed in the CIEF-FTICR analysis of a mixture of lysates from *E. coli* grown in minimal media containing either normal leucine (Leu) or Leu-d_{10}. The intact proteins are identified as (a) PTS system phosphocarrier protein HPr and (b) malate dehydrogenase on the basis of molecular mass measurement and their Leu content.

the number of Leu residues. Knowledge of the measured molecular mass and the number of Leu residues greatly constrains possible protein assignments and generally allows the unambiguous identification of the proteins being observed; extension of this approach should be broadly applicable for high throughput protein identification. An additional benefit of this strategy is that it requires minimal sample handling and processing and therefore is applicable to high throughput proteomic analyses (i.e., all proteins are simultaneously labeled and subjected to analysis). The method has also been extended by the labeling of additional amino acid residues and the determination of proteolytic products, to further increase the confidence in identifications, particularly when there are post-translational modifications. The mass differences between the predicted and experimental accurate mass measurements provide insights into the nature of the mass differences, which in many cases can be attributed to specific types of post-translational modifications.

It should be noted that each of the strategies for identification of intact proteins discussed above has certain disadvantages. Although FTICR provides very accurate mass measurements, our experience has shown that very few proteins can be identified solely on the basis of mass. Although MS/MS of smaller proteins is useful for identification, it is generally not well suited for the identification of large proteins. Specific amino acid isotopic labeling has only been shown using auxotrophic organisms and is not generally applicable to mammalian cell culture or tissue samples. If high throughput identification of intact proteins is to be successful, it will require a combination of several approaches. It is important to emphasize, however, that protein identification will generally need to be conducted only once for any proteome.

VII. APPROACHES TO PROTEOMICS BASED ON GLOBAL PROTEOLYSIS

Although proteome studies analyzing intact proteins by CIEF-FTICR provide a wealth of data, they often do not provide the information necessary for high throughput protein identification for the reasons described above. The alternative, and more classical, approach is to identify proteins by measurements of the peptides obtained from an enzymatic (e.g., tryptic) digest of the proteins. In this approach we use global digestion methods for the proteome mixture and analyze the data in the context of calculated digest products from all the proteins predicted from the subject genome. This strategy is analogous to the approach used in standard proteomic studies employing 2D-PAGE separations followed by protein identification by either MS or MS/MS (15). Whereas the number of peptide masses necessary to make an unambiguous identification of a protein ranges from three to six with the mass-resolving power and accuracy of conventional mass spectrometers (17,20), a major requirement is that the peptides being measured originate from a single protein or a small subset of proteins (i.e., extensive separation is first required). Higher quality MS measurements, however, enable approaches with the potential for very high throughput.

A. Accurate Mass Tags for Protein Identification

The success of MS/MS methods has shown that a single peptide can provide a basis for confident protein identification. If the molecular mass of a single peptide could be measured with high enough MMA, such that its mass was unique among those of all the possible peptides predicted from a genome, it could be used as an *accurate mass tag* (AMT) for protein identification (70).

The generation of such AMTs provides a basis for identifying the proteins with much greater speed and sensitivity. With the very high MMA that can be achieved using FTICR, it is possible to use a single peptide AMT for protein identification. A major issue with this approach is the degree of mass measurement accuracy (MMA) necessary for peptide mass measurements to function as AMTs for unambiguous protein identification. Conventional MS/MS approaches can then be used to fill in ambiguous mass measurements and for the initial validation of AMTs. After AMT validation, subsequent proteome measurements can be based almost exclusively on the use of AMTs, thereby obviating most MS/MS measurements and substantially increasing throughput. This obviates the routine need for additional identification efforts (e.g., using MS/MS), in much the same fashion that a spot on a 2D-PAGE is often attributed to a previously identified protein. The advantage is that this approach should provide much greater reliability in protein identification (due to the MS/MS validation step) as well as a basis for quantitative expression measurements with substantially higher sensitivity than obtained by conventional methods.

Fourier transform ICR provides the most accurate mass measurements currently feasible with mass spectrometry. The results of a tryptic digest of albumin (45), in which all polypeptides were determined to ~0.75 ppm average mass accuracy without the use of an internal calibrant, is shown in Fig. 8. With this high level of MMA, peaks can be confidently assigned to individual tryptic fragments simply on the basis of their mass (45). Thus, a significant number of the peptide masses serve as effective "biomarkers" for the parent protein albumin. This is, in effect, the use of the AMT approach in the context of a single protein rather than an entire proteome. The key question is, How effective can this approach be when it is extended to the complexity and relative abundances of whole proteomes?

It is possible to predict, with some qualifications, the set of proteins possibly expressed by a given organism by translating all hypothetical open reading frames (ORFs) predicted from genomic sequence information. The set of potential polypeptide fragment molecular masses can then be calculated on the basis of predicted amino acid sequences. We have analyzed all of the predicted tryptic polypeptides from yeast and *C. elegans* proteins using all predicted ORFs from the complete genomes of these organisms (71,72). Calculation of the percentage of unique tryptic peptide masses predicted from yeast and *C. elegans* at varying levels of MMA (Fig. 9) shows that the percentage of peptides able to function as AMTs depends on the achievable MMA and the complexity of the particular proteome as well as the peptide mass (larger peptides are more distinctive). For predicted peptides from yeast and *C. elegans*

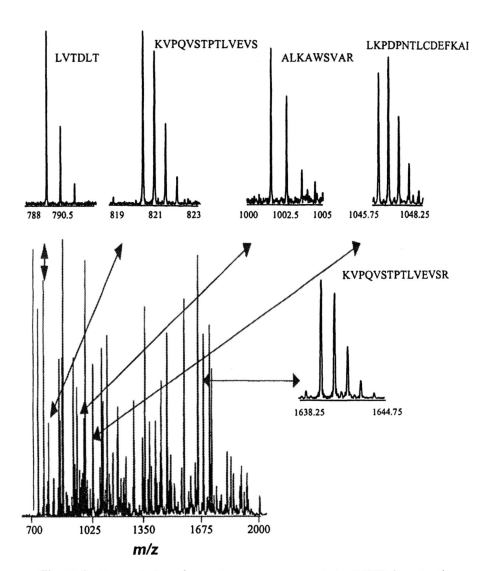

Figure 8 Demonstration of accurate mass measurements by FTICR for a tryptic digest of albumin. The detected fragments encompassed all of the protein. An average mass measurement error of <0.75 ppm was achieved without the use of an internal calibrant.

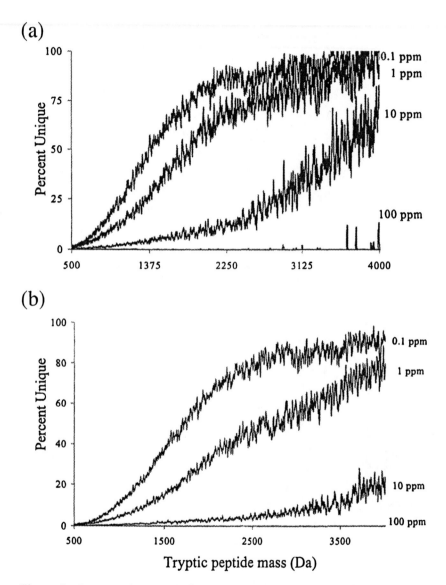

Figure 9 Percent unique tryptic fragments (potential accurate mass tags) as a function of tryptic fragment mass at four different levels of mass measurement accuracy for the predicted proteins of (a) yeast and (b) *C. elegans*.

with a mass of ~2000 Da, approximately 65% and 60%, respectively, have unique masses at a MMA of 0.1 ppm and therefore can act as AMTs to identify a unique protein. Decreasing the MMA to 1 ppm decreases the percent of peptides having unique masses from 65% to 50% in the case of yeast and from 60% to 40% for *C. elegans*. At MMA levels of 10 ppm, less than 10% and 5% of the predicted peptides for yeast and *C. elegans*, respectively, have unique masses, whereas at 100 ppm there are almost no unique masses. The advantages of higher MMA are even greater as the peptide mass increases. Calculations show that for both yeast and *C. elegans*, greater than 80% of the predicted peptides with masses greater than 2500 Da possess a unique mass at

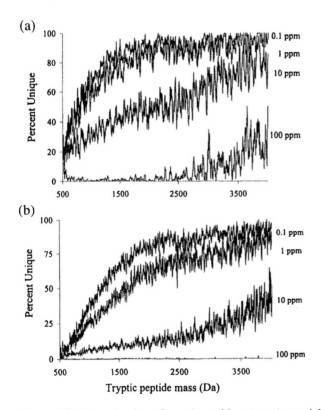

Figure 10 Percent unique Cys-polypeptide masses (potential accurate mass tags) as a function of Cys-polypeptide mass at four levels of mass measurement accuracy for the predicted proteins of (a) yeast and (b) *C. elegans*.

sub-ppm MMA, thus providing a basis for protein identification from more complex mixtures than would otherwise be possible (Fig. 10A). It should be pointed out that in the case of eukaryotic proteins, as the size of the tryptic fragment increases so does the likelihood of post-translational modifications whose presence would necessarily complicate genome database-dependent identification without further MS/MS analysis.

The potential for this approach is supported by the demonstrated capability to make measurements for large numbers of peptides in a single capillary LC-FTICR experiment. Figure 11 shows a recent result for a high pressure high performance reverse-phase capillary LC-FTICR analysis of a yeast protein global tryptic digest. The figure shows only a segment of the 3 h total separation time during which >150,000 isotopic distributions corresponding to >100,000 putative polypeptides were detected.

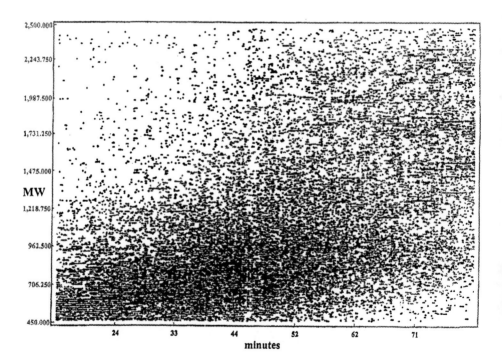

Figure 11 Capillary LC-FTICR partial 2-D display for a 3 h 10,000 psi reverse-phase gradient separation (90 cm column packed with 3 μm particles) in which over 100,000 putative polypeptides were detected from a yeast protein tryptic digest.

B. Accurate Mass Tags for Identification of Post-Translation Modifications

As mentioned previously, a major obstacle faced for eukaryotic protein mass prediction is the often extensive occurrence of post-translational modifications that are not presently predictable from genomic sequence information alone. Of the myriad of post-translational modifications known to occur, perhaps most attention has focused on phosphorylation. Alterations in the phosphorylation state of specific protein tyrosine, serine, threonine, histidine, and lysine residues can significantly alter the biological activity of the target protein, which, if involved in a cell signal pathway, for example, can often lead to large downstream changes in gene expression and hence cellular function (for review see Ref. 73). Largely because it is still unreliable to predict the targets of phosphorylation reactions, and because it is often the case that kinases and phosphatases involved in phosphate transfer reactions have relatively broad enzymatic specificities, progress in such fields as signal transduction has been laborious. A further complication is that the initial signaling events (i.e., alterations in phosphorylation state) often occur at very low levels, which can render them undetectable by conventional means. Mass spectrometry potentially offers large increases in sensitivity and throughput for investigations of post-translationally modified proteins. High MMA measurements of phosphopeptides are particularly attractive owing to the distinctively large mass defect of phosphorus (~30 mDa), which has the net result of offsetting the average mass of phosphopeptides to slightly lower than that of unmodified peptides *of the same nominal molecular mass*, often marking a peptide as phosphorylated simply on the basis of its mass. To address the utility of MMA in the analysis of phosphopeptides in yeast, a database was created in which all tyrosine, serine, threonine, and histidine residues were phosphorylated. With sufficiently high MMA, greater that 80% of the 2 kDa yeast phosphopeptides, for example, can be identified at 0.1 ppm MMA. We project that the extension of the AMT concept to phosphopeptides will be particularly effective if such mass measurement accuracy could be routinely achieved.

C. Accurate Mass Tags for Isotope-Coded Affinity Tags

As we have already discussed in the context of intact proteins, approaches that provide additional information on amino acid composition can greatly aid protein identification. For this purpose Aebersold and coworkers (74) used "isotope coded affinity tags" (to be discussed in further detail later) as

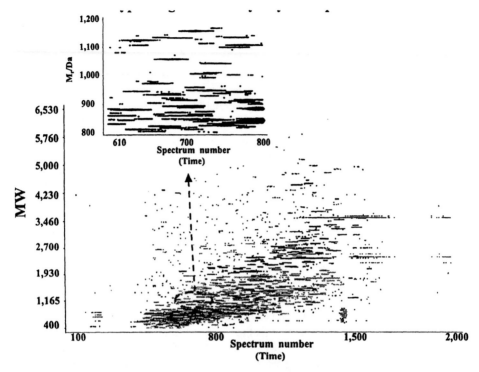

Figure 12 Two-dimensional virtual gel display of the peptide masses detected in a capillary LC-FTICR analysis of a tryptic digest of a cytosolic yeast lysate. More than 10,000 different Cys-polypeptides were observed during this single analysis.

a method to substantially reduce the complexity of the proteome analysis. The important aspect of the approach involves affinity selection of cysteine-containing polypeptides (Cys-polypeptides) by modifying the proteins with a Cys-specific reagent containing a biotin group, and affinity isolation of the modified peptides using immobilized avidin. Not only does this method substantially reduce the complexity of the proteome mixture, it also significantly reduces the MMA required for the generation of AMTs. Of the 918,655 possible tryptic polypeptides within *C. elegans*, only 124,668 contain cysteine residues (a 7.4-fold decrease in mixture complexity), and only 89,697 of these are above 1 kDa. The fractions of unique Cys-polypeptides for both yeast and *C. elegans* as functions of their mass at four different levels of MMA are shown in Figs. 10A and 10B, respectively. It is clear that selecting only Cys-containing tryptic fragments further simplifies the problem of AMT identification and leads to a greater percentage of unique polypeptides. Indeed, at sub-ppm MMA, nearly 95% of the larger Cys-containing tryptic fragments are unique. If the analysis is further constrained by considering only those tryptic fragments in *C. elegans* that contain cysteine residue(s), approximately 80% of the 2000 Da fragments are potentially useful AMTs at 0.1 ppm MMA, compared to only 65% if all tryptic fragments are considered (Fig. 9B).

To obtain the greatest possible numbers of AMTs will not only require high MMA (as discussed above) but will also require the ability to detect as many peptides as possible within the context of a relatively rapid overall separation strategy. As shown in Fig. 12, more than 10,000 peptides (from among over 35,000 peaks!) can presently be observed in a single capillary LC-FTICR analysis of Cys-polypeptides isolated from a yeast lysate. The large number of detectable peptides is due in part to the high resolving power and high dynamic range achievable with FTICR. A single scan from a capillary LC-FTICR analysis of Cys-polypeptides isolated from a yeast lysate is shown in Fig. 13. Although several high intensity peaks are observed in the full spectrum (Fig. 13A), the

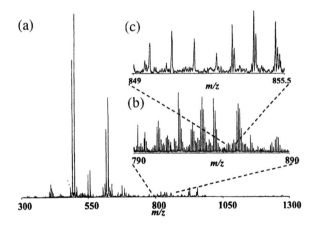

Figure 13 Single spectrum from the LC-FTICR analysis of cysteine-containing polypeptides (Cys-polypeptides) isolated from a tryptic digest of a yeast lysate. Although several highly abundant peptides were observed within the entire *m/z* range (a), expansion of various regions (b and c) shows the presence of less abundant species detectable owing to the high sensitivity, resolution, and dynamic range achievable with FTICR.

advantage of FTICR for peptide detection is clearly demonstrated when regions that appear to contain little information are expanded (Fig. 13B). Even within these regions, large numbers of peptides can readily be observed. Further expansion of one region reveals even more low abundance peptide signals (Fig. 13C). The ability to detect low abundance peptides not only allows the broadest possible proteome coverage, but also, which is more important, allows important regulatory proteins that may be present in low copy number to be studied.

VIII. EXPERIMENTAL GENERATION OF ACCURATE MASS TAGS

A. AMT Validation Using MS/MS

High throughput proteomic studies based on AMTs will generally involve two phases, one in which the proteins are first identified and a second in which many measurements are made and during which changes in expression of proteins are followed using AMTs. Although not distinct, the crucial point is that initially a significant effort will be necessary for broad and confident protein identification. This identification stage will be followed by efforts involving the use of AMTs for high throughput measurements of the response of proteomes to "perturbations." In the initial identification phase, both useful AMTs and ambiguously detected polypeptides will need to be subjected to MS/MS to confirm protein identification, identify modifications, etc. As discussed in Section III, MS/MS studies enable protein identification by acquiring amino acid sequence–related information for distinctive polypeptide fragments. The projected set of polypeptides can be readily calculated for all ORFs for a sequenced genome and constitutes an extremely small number relative to the masses of all possible polypeptides (i.e., masses from all sequence combinations). This fact led to a concept that opens the door to high throughput approaches for broad proteomic measurements: If the molecular mass of a single peptide can be measured with high enough MMA that its mass is unique among those of all possible peptides predicted from a genome, it can then be used as an accurate mass tag or an effective biomarker for unambiguous protein identification (24). The use of such AMTs allows the proteolytic fragments generated from an entire proteome or complex protein mixture to be analyzed and identified with much greater speed.

In this approach we use global digestion methods for a proteome mixture and analyze the data in the context of calculated digest products from all of the proteins predicted from the subject genome (or other appropriate databases). Although sequence-related information obtained by MS/MS and LC elution profiles plays an important role in the generation of AMTs, the attributes of FTICR (especially the high MMA and dynamic range) are important for effective proteome-wide AMT protein identification and quantitation. Technical developments in our laboratory have substantially increased the sensitivity and dynamic range obtainable with FTICR, and this will be important not only in establishing AMTs but also in acquiring MS/MS data for low level peptides. It is important that once an AMT is established and related to a specific protein, it can function to provide a reliable measure of its expression in the context of a comparative display for that protein, without the need for additional identification efforts (e.g., using MS/MS), in much the same fashion that a 2D-PAGE spot is often attributed to a specific previously identified protein.

B. Multiplexed MS/MS

The initial stages of research with any cell type will have a significant need for MS/MS for AMT validation for confident protein identification, as well as for identifying unexpected polypeptides (e.g., those due to mutations, sequencing errors, unknown modifications, unconventional translation). It is well established that even low resolving power MS/MS measurements enable confident protein identification. The throughput with conventional MS instrumentation is actually somewhat greater than with present FTICR capabilities. However, as we recently reported (45) the high MMA provided by FTICR opens the opportunity to analyze a number of peptides simultaneously with multiplexed MS/MS.

The multiplexed MS/MS approach was initially evaluated by extensive computer simulations and experimental studies using infrared multiphoton dissociation (IRMPD) (75). In contrast to the conventional MS/MS approach for protein identification, where an individual polypeptide is sequentially selected and dissociated (21–26), multiplexed MS/MS selects and dissociates several species simultaneously. In the case of polypeptides, a limited set of fragmentation pathways is usually favored, and we have found that high MMA allows most of the fragment ions from·a limited set of parent polypeptides in an MS/MS experiment to be attributed to a specific parent species. Thus, in this approach a set of selected species is dissociated and measured in a single experiment, providing both enhanced speed and sensitivity. Some fragments may remain unassigned to a specific parent, but we have shown that assigned fragments generally provide unambiguous identification of the parent species (75).

In the initial demonstration of multiplexed MS/MS, a set of seven model polypeptides was selected. Polypeptide ions were formed by electrospray ionization from a solution containing the seven peptides and were analyzed using our 11.5 T FTICR mass spectrometer. We used IRMPD for MS/MS, because it does not involve cyclotron excitation of the parent ions or gas introduction (as required by most other activation methods) and thus provides higher throughput and compatibility with on-line separations. The mass spectrum obtained for the mixture is shown in Fig. 14A. For multiplexed analysis, the most abundant species were simultaneously selected using SWIFT (Fig. 14B) and then subjected to IRMPD (Fig. 14C). In the case shown here, 105 species were detected in the multiplexed product ion spectrum (fragments and residual precursor ions). The data set used for the peptide identification consisted of the masses of the seven parent ions and the 105 detected species from the multiplexed MS/MS spectrum.

We evaluated the utility of this approach for protein identification using the set of >19,000 proteins predicted from the full *C. elegans* genome database along with the masses for both the parent species and their fragments. First, a set of all possible tryptic fragments (including partial digestion products) was generated from the *C. elegans* database, and to that we added the sequences of the seven model peptides. The measured mass for each parent species (using a conservative 2 ppm mass accuracy) was then searched against this list, resulting in a set of possible polypeptides (referred to as "candidates") for each parent species. For all candidates, the masses of "y" and "b" fragments and the corresponding ions involving loss of H_2O or NH_3 were computed and compared to the list of masses from the multiplexed product ion spectrum (assuming 2.5 ppm MMA). When three or more fragments in the multiplexed product ion spectrum matched predicted masses for fragments from a specific protein mass, the precursor was considered identified (although in most cases fewer fragments were actually necessary). The five unmodified peptides were uniquely identified using this approach, and there were no "hits" for the two modified peptides (i.e., no false positives). We then expanded the database to include the two post-translational modifications (i.e., all possible tryptic polypeptides in the database were amidated at the C-terminus or/and modified by loss of H_2O at the N-terminus, as occurs in the case of pyroglutamic acid modification). The result was a database four times as larger as the initial one (i.e., original polypeptides plus the polypeptides with the N-terminus, the C-terminus, or both termini modified). In this case, all seven species were uniquely identified (Table 1) using the multiplexed MS/MS information. It should be noted that the *C. elegans* full genome database is roughly equivalent in complexity to that projected for a specific differentiated mammalian cell type.

These initial results demonstrated the high reliability of the multiplexed MS/MS approach and highlight the advantages of FTICR both for enabling multiplexed MS/MS and for providing the high MMA that aids confident protein identification. This approach should enable a substantial increase in throughput

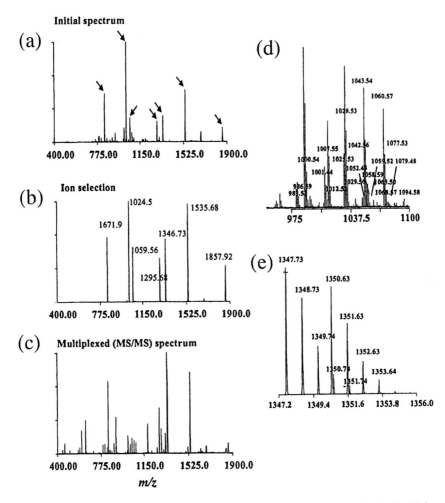

Figure 14 Demonstration of multiplexed MS/MS using FTICR as a basis for high throughput protein identification. (a) The seven most abundant ions from the MS spectrum of a mixture were selected using a SWIFT waveform (b) and subsequently subjected to infrared multiphoton dissociation (c). Expanded views of the m/z range 975–1100 (d) show more than 20 detected isotopic distributions from a total of 105. The m/z range 1347–1356 shows two overlapping but resolved isotopic distributions (e).

Table 1 Search Results in the *C. elegans* Full Genome Database (19,106 Putative Proteins Corresponding to 918,655 Tryptic Segments) Modified to Account for Two Post-Translational Modifications

Parent mass[a]	Number of candidates[b]	Hits with multiplexed MS/MS information[c]	N-Terminus group	C-Terminus group	Identification
1857.918	121	1	NH_2	OH	-Endorphin
1671.910	89	1	pGlu	OH	Neurotensin
1535.685	33	1	NH_2	OH	Fibrinopeptide A
1346.728	166	1	NH_2	NH_2	Substance P
1295.677	81	1	NH_2	OH	Angiotensin I
1059.561	106	1	NH_2	OH	Bradykinin
1024.549	61	1	NH_2	OH	Renin inhibitor

[a] Measured masses of the parent ions.
[b] The number of candidate parent ions found in the database (based on a 2 ppm mass accuracy).
[c] Number of candidates matching three or more fragments in the multiplexed MS/MS spectrum (within a mass error of 2.5 ppm). The seven peptides were identified using the multiplexed MS/MS information with no miss or false positive.

and confidence in protein identification and a decrease in sample consumption (because all species are examined in a single experiment) and should be practical in conjunction with on-line separations. In recent work we implemented multiplexed MS/MS for use with on-line capillary LC-FTICR studies and employed in a "data-dependent" mode of operation. Finally, we note that multiplexed MS/MS of many more species than were used in this initial demonstration should be feasible, further increasing overall throughput for protein identification efforts and other cases where MS/MS is desired. In the long run, the FTICR methods should be able to greatly facilitate obtaining higher throughput for more effective identification, particularly of lower level proteins for which conventional instrumentation lacks sufficient sensitivity.

IX. HIGH THROUGHPUT MEASUREMENTS OF RELATIVE PROTEIN ABUNDANCES

Proteomic approaches for understanding complex biological systems generally involve the quantification of protein abundances. It is also desirable to measure changes in protein abundances in a high throughput manner so that the effect of many "perturbations" on, or changes to, a cell type, tissue, or organ can be determined within a reasonable time period. Obviously, there is also the need to identify the proteins, but it is also important to appreciate that the protein identification step, as for a spot in 2D-PAGE, should generally have to be accomplished only once. Thereafter, the extremely distinctive combination of separation elution time (for example) along with a highly accurate mass measurement can be used to assign identity with very high confidence after appropriate validation. This applies to approaches based on both intact proteins and global enzymatic digestion. However, although the validation of protein identifications will not generally have to be repeated, quantitative information is desired in the many subsequent experiments (as with DNA expression arrays), and this is where the true high throughput demands exist. Although mass spectrometry has not been traditionally used for this purpose, we believe that in the future high throughput expression measurements will constitute one of its major applications.

As discussed earlier, the predominant method of measuring changes in protein expression levels by using current proteomic technology (based, e.g., on 2D-PAGE separations) is to compare protein spot intensities. Not only does this strategy fail to detect subtle changes in protein expression levels and lower abundance proteins, but also the precision of the measurements is much less than is needed to discern many biologically significant changes in protein abundance. Because this technique has not been shown to be amenable to automation for high throughput despite more than a decade of significant effort, it is unlikely to meet the needs of future proteomic studies in which multiple perturbations are compared.

As we have already noted, deriving accurate protein abundance measurements from ESI-MS signal intensities alone is problematic. A major development in the use of mass spectrometry to generate precise measurements of relative protein and peptide abundances is the use of stable isotope labeling methods to provide effective internal standards. Our initial application of this approach was in the context of intact protein measurements, but both our laboratory and others have also demonstrated the approach in several different contexts in conjunction with enzymatic processing (14,59,74). In our original implementation, two isotopically distinct versions of each protein were generated and analyzed simultaneously providing calibrants for all detected proteins, thus enabling precise proteome-wide measurement of changes in protein abundance resulting from cellular perturbations (59). A more general approach is to make measurements for mixtures of the proteome under study along with a stable-isotope-labeled "reference proteome."

A. Metabolic Stable Isotope Labeling

Two general approaches have now been used to create isotopically distinct versions of each protein. In one approach, cells are grown in isotopically defined medium (e.g., ^{15}N-, ^{13}C-enriched or ^{15}N-, ^{13}C-depleted), and the proteins extracted from this source are compared with proteins extracted from cells grown in a natural (or other) isotopic abundance medium (14,59). The second approach incorporates an isotopic label after proteins have been extracted (74). The labeling strategies that use isotopically defined media offer the highest precision (because labeling occurs at the earliest possible point), the broadest proteome coverage, and the least sampling handling but are inapplicable in many instances (e.g., tissue samples). On the other hand, post-isolation isotopic labeling is broadly applicable to proteins extracted from every conceivable source (e.g., tumor samples), can reduce mixture complexity in some labeling schemes, and can aid identification by providing an additional sequence constraint, but does so at the cost of additional sample processing, somewhat decreased protein coverage, and a possible small decrease in precision (due, in greatest part, to the variables that precede the labeling step). Both strategies can be applied to both intact proteins and enzymatically processed proteins.

1. Rare Isotope Depletion Combined with CIEF-FTICR Analysis of Intact Proteins

Our initial demonstration of isotopic labeling strategies for whole proteomes was for the determination of intact proteins using CIEF-FTICR (59). These studies examined the cadmium (Cd^{2+}) stress response in *E. coli* K-12 MG1655, a response previously studied using conventional approaches (74). In these studies, *E. coli* was grown in both normal (i.e., natural isotopic abundance) and rare-isotope-depleted media. An important consideration when using any type of labeling strategy in which the organism is grown in isotopically distinct media is to determine whether the labeling affects the growth or protein expression pattern of the organism. We directly examined this issue and showed that there were no measurable differences in the growth rate of *E. coli* between otherwise identical isotope-depleted and normal media and also that the growth of the organism was perturbed to an identical extent by the addition of Cd^{2+} to both types of media (59). This result is not unexpected owing to the small changes in isotopic composition and to the fact that only rare isotopes were absent (i.e., the only effects expected were due to removal of the normally very small effects on growth due to rare isotope contributions).

The experimental approach involves mixing two cell populations and results in mass spectra that generally show two versions of each protein. This mixing is done prior to sample processing steps, eliminating all experimental variables associated with cell lysis, separation, and mass spectrometric analysis. Therefore, the ratio of the two isotopically different and resolvable versions of each protein accurately reflects any change in relative protein abundance. Differences in relative protein expression levels can be quantitatively determined by using the distinctive isotopic abundances characteristic for each particular media. In the initial demonstration (59), the addition of Cd^{2+} to the cell culture led to a period of growth arrest, presumably due to the production of Cd^{2+}-adaptive proteins, followed by resumption of growth. To examine these proteins and their relative expression, aliquots were removed from the unstressed (normal medium) and stressed (rare-isotope-depleted medium) cultures at different time intervals after Cd^{2+} addition, and the cells were mixed in various combinations before lysis, desalting, and analysis by CIEF-FTICR. Comparative analysis of the resulting CIEF-FTICR data was accomplished using software developed at our laboratory (63), which identified the peaks due to proteins and their isotopically distinct complements (present at predictable mass differences). Using the known isotopic compositions of the growth media, the program calculates the theoretical isotopic peak envelopes for each species and compares them to the experimental data to determine an abundance ratio (AR). Broad effects due to the Cd^{2+} perturbation were observed, with proteins being suppressed (AR <1), induced (AR >1), or showing more complex behaviors. Figure 15 shows representative mass spectra from one experiment, illustrating a few of the types of variation in ARs observed at 45 and 150 min after Cd^{2+} stress. The protein with $M_r = 12,654.5$ Da displayed an AR of 0.59 at $t = 45$ min, which changed at 150 min after Cd^{2+} stress to an AR of 0.86. On the other hand, some proteins seem to have a significant response to Cd^{2+} stress, such as the protein of $M_r = 13,718.1$ Da (Fig. 15), which was not detectable before Cd^{2+} stress (i.e., the depleted version of the protein at $t = 45$ min did not have a normal complement, AR > 30) and was still present but in decreased relative abundance at $t = 150$ min. Proteins exhibiting this behavior seem to have their expression increased (i.e., induced) by the addition of the Cd^{2+}, indicating a possible role of Cd^{2+} in stress response. Other proteins may actually have their production decreased (i.e., suppressed) to divert the cells' resources into producing proteins more essential in responding to the stress. This could be the case for the protein of $M_r = 60,147$ Da, which seems significantly suppressed 45 min after Cd^{2+} stress (AR = 0.25) but bounces back by $t = 150$ min (AR = 0.93). Because our approach circumvents the typical run-to-run variations for measurements of specific proteins, analyses of replicate samples show that variations of <10% in ARs can be obtained. Figure 16 shows the comparative display for the Cd^{2+} stress response experiment at

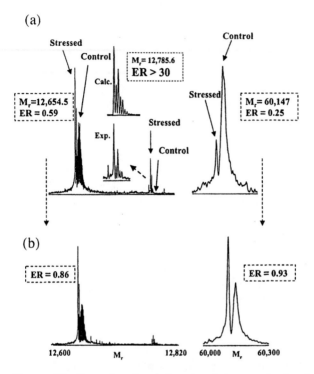

Figure 15 Examples of variations in relative protein abundance ratios observed for CIEF-FTICR analyses 45 min and 150 min after Cd^{2+} stress on *E. coli* grown in normal and isotopically depleted media. (MW is given as the average mass for the natural isotopic abundance.)

$t = 45$ min, with representative spectra showing proteins that were either suppressed (AR = 0.82) or induced (AR > 30) by the addition of Cd^{2+}. The ARs for the 200 most abundant proteins detected in this run ranged from <0.1 to >30. Replicate analyses indicate that the precision of protein expression measurements was ± 6%, far better than that achieved by using 2D-PAGE technology and MS methods without the use of internal standards (Fig. 17).

2. Heavy Isotope Enrichment and FTICR Analysis of Peptides

With the completion of the human genome, increased numbers of proteomic studies will undoubtedly focus on the analysis of human cell lines and tissue samples. The most obvious challenge when analyzing the human proteome will

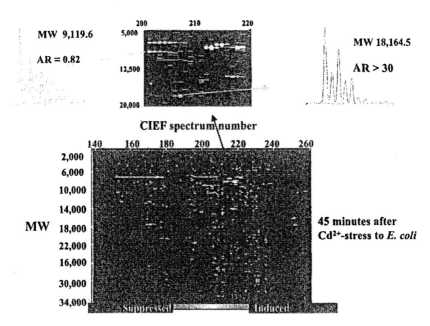

Figure 16 Comparative display of proteins obtained from a CIEF-FTICR analysis of *E. coli* perturbed by Cd^{2+} stress. The analyzed mixture contained equal aliquots of unstressed control culture (grown on standard media) taken at $t = 0$ and Cd^{2+}-stressed culture (grown on isotopically depleted media) taken 45 min after Cd^{2+} stress. In the original changes in relative protein abundance ratios (ARs) are shown using a color scale: Increases in protein abundance relative to nonstressed cells are represented by brighter shades of red and decreases as brighter shades of green. Examples of the ranges in expression ratios obtained are shown in the two spectra presented in the upper left and right corners of this display.

be its complexity. Whereas yeast has approximately 6200 predicted ORFs, the human genome is predicted to code for anywhere from 30,000 to 40,000 genes. To obtain as broad a proteome coverage as possible, it may be necessary to first fractionate the proteome so that samples of reduced complexity can be analyzed individually. The extreme approach for doing this is presently 2D-PAGE, which aims for (but does not achieve) resolution of all proteins. However, 2D-PAGE suffers from substantial limitations, as already noted. What is desired is a faster separation that is more readily automated and that better exploits the power of mass spectrometry.

For this purpose, we are presently developing two-dimensional separation strategies that can be coupled with FTICR analysis. One such strategy in

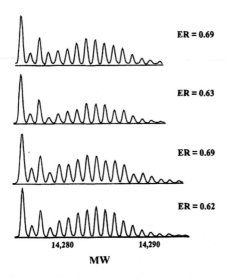

Figure 17 Replicate analysis of a protein observed in four separate experiments analyzing the effects of Cd^{2+} on *E. coli*. The expression ratio between the separate experiments differed by ± 6% for this protein, which demonstrates the high precision measurements obtainable by combining isotopically distinct proteomes so that one serves an effective internal standard.

which proteins are initially fractionated using CIEF was initially demonstrated using the yeast proteome (77). The use of CIEF in the first separation stage serves to both separate and preconcentrate proteins. The CIEF separation is coupled to a robotic fraction collector, enabling <1 uL aliquots to be collected using a 384-well microtiter plate. The fractions are then enzymatically digested with trypsin. Fractions containing the resultant tryptic peptides are then analyzed in a capillary LC column coupled on-line with ESI-FTICR. The number of fractions collected presents a compromise between first-stage separation resolution, the minimum sample size, and the overall speed (because each digested fraction will be subjected to second-stage capillary LC separation combined with high resolution ESI-FTICR analysis). In initial work more than 50 fractions were typically collected from a single CIEF run.

The high resolving power of FTICR provides the capability to measure many peptides in a single scan. In a typical capillary LC-ESI-FTICR experiment analyzing a single CIEF fraction, 95% of the peptide masses obtained matched predicted peptides from the yeast genome database within 5 ppm (using external calibration with a 7 T FTICR instrument). In the initial work, protein identification was also aided by the constraint that multiple peptides observed in a single LC-ESI-FTICR experiment originate from a specific CIEF fraction and a limited subset of proteins. Thus, multiple polypeptides will often be observed from the same protein, improving the confidence of protein identification. Examples of yeast proteins identified on the basis of one or more peptide mass measurements with the corresponding MMA are shown in Table 2. For example, two proteins, ORF YBL004w and ORF YDL239c, were identified from multiple peptides, while the other two proteins, ARS-consensus and ORF YPL272c, were identified using a single peptide mass measurement (714.31 Da and 2577.098 Da, respectively) with a mass measurement error of less than 3 ppm. Many of the peptide mass measurements differed from the calculated peptide mass by less than 1 ppm. Improvements to the mass measurement accuracy and calibration methods and the selective use of data-dependent MS/MS fragmentation to resolve ambiguous assignments would clearly enable more proteins to be identified faster and with greater confidence and sensitivity. In recent work we have routinely achieved <1 ppm MMA for capillary LC-FTICR studies using an 11.5 T magnet.

For the purposes of quantification, we cultured yeast in a medium containing the natural abundance of the isotopes of nitrogen [i.e., ^{14}N (99.6%) and ^{15}N (0.4%)] and a second culture in the same medium enriched in ^{15}N (>98%). Cells grown in normal isotopic abundance and ^{15}N-enriched media

Table 2 Examples of Mass Measurements and Identified Peptide Fragments Originating from Yeast Proteins Collected in a Single CIEF Fraction for On-Line Capillary LC-FTICR Analysis

Protein	Identified peptide	Theoretical peptide mass (Da)	Measured peptide mass (Da)	Error (ppm)
ARS-consensus	HGCGVDK	714.312	714.310	1.6
ORF YPL272c	ENDADWHSDEVTLGTNSSKDDSR	2577.085	2577.09	2.7
ORF YBL004w	IDDLKIEPAR	1168.645	1168.640	4.5
	TFDERNLR	1049.525	1049.525	0.3
	VPELESISK	1000.544	1000.540	2.1
	LVSSFFLK	939.543	939.539	4.2
	QRAIK	614.386	614.388	1.9
ORF YDL239c	LKSELKGKLILSEKIQKNAEDK	2511.463	2511.465	0.6
	CELTLLTK	919.505	919.505	0.2
	NTLKSPNK	900.503	900.505	2.5
	EFHNER	830.367	830.364	2.2
	ALRQK	614.386	614.388	1.9

were mixed, and the soluble proteins were extracted. This combined protein extract was analyzed using the high throughput strategy described earlier in this section. The two isotopic versions of each peptide are easily distinguished owing to the distinctive isotopic distribution of the ^{15}N-labeled peptide. An example of two isotopically distinct versions of a peptide is shown in Fig. 18, with the simulated isotopic distribution expected for each (Fig. 18, inset). The number of nitrogen atoms present in the peptide is obtained from the mass differences between unlabeled and ^{15}N-labeled versions of the peptide (experimental mass difference of 10.970 vs. a calculated difference of 10.968). This knowledge of the number of nitrogen atoms in the peptide provides an additional constraint that is useful for protein identification. For example, a peptide with a mass of 1228.768 Da was detected that was shown, on the basis of the mass difference between its ^{15}N-labeled and unlabeled versions, to contain 22 nitrogen atoms, peptide (21.985). There are 12 peptides in the genome database that are within 5 ppm of the measured mass of 1228.768 Da (Table

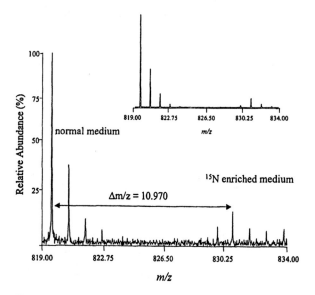

Figure 18 Mass spectrum displaying the natural isotopic abundance and ^{15}N-labeled versions of a peptide observed in the capillary LC separation of a digested CIEF fraction. The presence of two isotopically distinct versions of the same peptide provides the basis for comparative displays of the relative protein expression levels and a measure of the number of nitrogen atoms. The simulated mass spectrum for the two versions of this peptide is shown in the inset.

3). Including the constraint of the number of nitrogen atoms for the peptide, however, allows it to be unambiguously identified as a single peptide originating from the protein translated from ORF YDL184c. Such constraints aid the automated identification efforts that will be important for true high throughput proteomics.

Work in our laboratory has similarly demonstrated the use of ^{15}N metabolic isotope labeling, in conjunction with cysteine affinity tags, to conduct quantitative measurements of mammalian proteomes maintained in culture (79). Again, two separate cultures of mouse B16 melanoma cells were grown in normal isotopic abundance and ^{15}N-enriched media, and equal numbers of cells from both cultures were combined. Proteins were extracted and tagged with a cysteine-specific reagent that contains a biotin moiety. After the proteins were digested with trypsin, cysteine-tagged peptides were isolated by avidin affinity chromatography and analyzed by LC-FTICR using the 11.5 T FTICR instrument developed in our laboratory. As with yeast pairs of differentially labeled Cys-polypeptides were observed whose m/z ratio differed according to the number of nitrogen atoms in the peptide. Portions of several typical spectra are shown in Fig. 19. A key observation, based on the observed isotopic distributions, was that the labeling of the proteins grown in the ^{15}N-enriched medium was effectively complete. Surprisingly, the addition of 10% FBS to the cell

Table 3 Predicted Yeast Peptides Matching a Peptide Molecular Mass of 1228.768 Da Obtained by Tryptic Digestion of a Single CIEF Fraction Analyzed by Capillary LC-FTICR

Protein	Peptide	Nitrogen atoms in peptide	Theoretical peptide mass	Error (ppm)
ORF YDL184c	AKWRKKRTR	22	1228.763	4.2
ORF YKL160w	RKKSTRKPTK	20	1228.773	3.8
ORF YJL148w	KKKEKREKR	20	1228.773	3.8
ORF YCR066w	LIREAKMKIK	16	1228.769	0.7
ORF YOR310c	EKKEKKRKR	20	1228.773	3.8
ORF YOR066w	RILSLLKQMK	16	1228.769	0.7
ORF YBR121c	ILRKSQIPFK	16	1228.766	2.0
LPG5w	VGKNLIFGLLR	16	1228.766	2.0
ORF YPL160w	KSKAAAKKGRGK	20	1228.773	3.8
PS00301	KSKKPPAFAKK	16	1228.766	2.0
YER129w	IKREIAIMKK	16	1228.769	0.7
Putative centromere protein	WKKMLVLAIK	14	1228.773	4.0

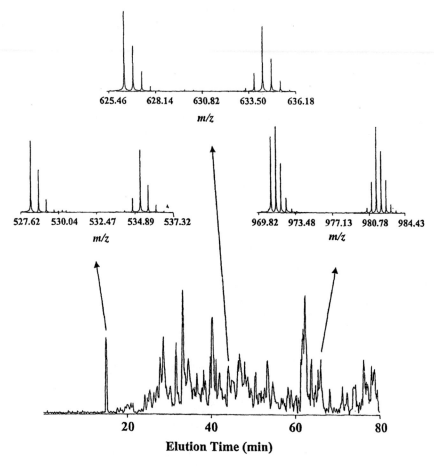

Figure 19 (Top and middle) Examples of Cys-polypeptides observed in the LC-FTICR analysis of peptides isolated from a combined culture of mouse B16 cells grown in normal and ^{15}N-enriched media. Pairs of related peptides can be detected based on the distinctive isotopic distributions of the ^{15}N-labeled peptide and its normal isotopic abundance partner. (Bottom) Total ion chromatogram representing the total ion current detected by the FTICR in the various scans during the capillary separation of the Cys-polypeptides.

culture medium did not affect the metabolic labeling of these cells. This experiment represents the first proteome analysis demonstrating the use of ^{15}N isotopic labeling of mammalian cells in culture to provide a "reference proteome" for quantitative measurements of relative peptide abundances.

Chait and coworkers (14) also recently demonstrated the use ^{15}N isotope labeling techniques to compare relative abundances of specific proteins using MALDI-MS to generate tryptic maps of in-gel digested protein bands. By using known ratios of unlabeled and ^{15}N-labeled Abl-SH2 protein, the experimentally determined intensity ratio was found to be linear ($r = 0.997$) over an abundance ratio of >10:1. Although similarly quantitative, this method is limited to one-at-a-time protein extraction from 2D-PAGE for analysis and thus remains more manually intensive than the one we present here.

B. Post-extraction Stable Isotope Labeling

Other approaches based on affinity capture and peptide analysis, such as the ICAT method more recently reported by Aebersold and coworkers (74), offer some specific advantages (e.g., the capability for sample post-labeling) and disadvantages (e.g., more sample processing, potentially reduced proteome coverage) than the approach proposed here. Whereas labeling strategies using isotopically defined media are limited to cells and tissues compatible with metabolic labeling, the introduction of a post-isolation isotopic label is broadly applicable to proteins extracted from every conceivable source. The ICAT approach uses a thiol-specific reagent that contains a biotin group as well as a linker arm connecting the thiol-reactive group and biotin moiety. The reagent has been produced in a heavy isotope version by replacing eight hydrogen atoms in the linker arm with deuterium atoms.

The labeling and extraction of Cys-polypeptides provides a basis for proteome-wide precise quantification of expression levels and protein identification and significantly reduces the complexity of the polypeptide mixture that needs to be analyzed. To demonstrate this strategy, two mixtures consisting of the same six proteins at known but different concentrations were prepared and analyzed. The protein mixtures were labeled, combined, and analyzed, and the isolated tagged peptides were quantified and sequenced in a single combined LC/MS and LC/MS/MS experiment. All six proteins were unambiguously identified and accurately quantified. Multiple tagged peptides were encountered for each protein. The differences between the observed and expected quantities for the six proteins ranged between 2% and 12%, confirming that good precision is achievable. As a further step, Aebersold and coworkers studied differences in steady-state protein expression in the yeast in two non-glucose-repressed states (74). Cells were harvested from yeast growing in log phase utilizing either galactose or ethanol as the carbon source. Glucose repression causes large numbers of proteins with metabolic functions significant to growth on other carbon sources to be expressed at a low level.

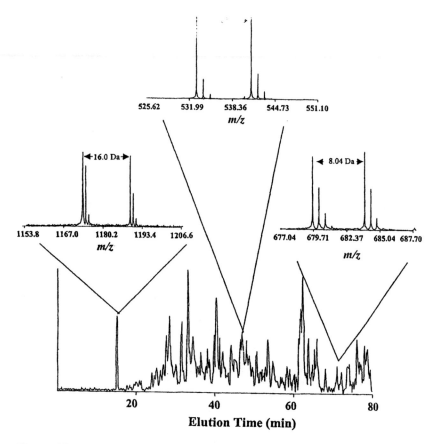

Figure 20 Examples of Cys-polypeptides observed in a capillary LC-FTICR analysis of a mouse B16 melanoma lysate labeled with the heavy- and light-isotope-labeled versions of the ICAT reagent.

A number of genes specific to both growth on galactose and growth on ethanol were detected and quantified.

In collaboration with Aebersold and coworkers, we applied this approach in conjunction with FTICR to develop a high throughput method to obtain simultaneous proteome-wide identification and quantification of proteins isolated from mouse B16 melanoma cells. The studies aimed to evaluate the possible use of the ICAT labeling approach to identify peptides based solely on the accuracy of mass measurements obtainable with FTICR. As discussed previously, the additional Cys constraint generally allows for unambiguous peptide identification at low (not sub-) ppm MMA. To evaluate the labeling efficiency, a soluble protein extract was obtained from mouse B16 cells and divided into two equal fractions. One of the fractions was labeled with the light-isotope version of the ICAT reagent, and the other fraction was labeled with the heavy-isotope version. After digestion and affinity isolation of the derivatized peptides, the sample was analyzed by capillary reverse-phase LC coupled on-line with ESI-FTICR. Almost 2000 pairs of Cys-polypeptides modified with the light- and heavy-isotope-labeled reagents were observed in a single LC-FTICR experiment. To assess this strategy for measurements of relative peptide abundances, the areas of the light and heavy isotopically labeled versions of each peptide were compared. Because the two samples that were derivatized with the light and heavy reagents were presumably identical, the relative abundances of the peptide pairs were expected to be unity. Analysis of all the observed peptide pairs gave a ratio of 1.02 ± 0.11. A few representative spectra showing the presence of differentially labeled Cys-polypeptides, along with their calculated abundance ratios and the peptide identified, are shown in Fig. 20. These results indicate that the ICAT approach, in conjunction with the high MMA of FTICR, has the potential to provide both the high throughput protein identification and quantification needed for future proteomic studies. Clearly, however, the general approach of stable-isotope labeling is broadly attractive, and the best methodology will depend on the purpose and design of the specific application.

X. FUTURE DIRECTIONS

Advances in instrumental and computational technologies, along with the large amounts of genome sequence information continuously being generated for a variety of different cell types and organisms, are leading to revolutionary uses of mass spectrometry for analyzing the protein content of these systems. The possibility now exists for studying hundreds or thousands of proteins within a single experiment instead of focusing on a single species. This new approach to protein analysis potentially allows pathways and networks, as well as the ways in which they affect one another, to be characterized in hours instead of months or years. To reach this exciting goal, proteomic studies must be highly sensitive, high throughput, and quantitative to achieve the broadest possible proteome coverage. Mass spectrometry has an indisputably important role in this enterprise.

The high MMA, resolving power, and sensitivity achievable with ESI-FTICR provide an abundance of new opportunities in proteomics. These include the use of proteolytic fragments as AMTs (i.e., as effective biomarkers) for protein identification. This approach is robust (because many protein fragments will not be modified and can therefore serve as accurate mass tags). Our computer simulations of this approach and experimental results have shown that in cases where the complete genome sequence information is available, an accurate mass measurement for only a single unmodified peptide often allows protein identification. The use of high mass measurement accuracy to produce AMTs, combined with high quality separations accomplished using such techniques as CIEF or capillary LC, will allow the identification of the protein constituents within a proteome in a complex mixture. For protein identification, conventionally carried out on a one-protein-at-a-time basis, the improved quality of mass measurements should enable greater speed and confidence. Given high MMA, it is easy to envision identifying complex mixtures of proteins based solely on the mass of a single peptide. Once proteins are identified the paradigm shifts, and we envision that a major application of mass spectrometry will become high throughput comparative studies of proteomes (e.g., from perturbation studies). Very recent advances in instrumentation and associated methodologies have served to further increase the attributes of the approach and the tractability of mammalian proteomes (80–84). Finally, the ongoing advances in sensitivity should ultimately enable studies at the single cell level.

ACKNOWLEDGMENTS

We acknowledge the contributions of Drs. Mikhail Belov, Hongying Gao, Mikhail Gorshkov, Richard Harkewicz, Suzana Martinovic, Pamela Jensen, Aleksey Tolmachev, and Harold Udseth to the research reviewed in this chapter. We also thank the National Institutes of Health, through NCI grants CA81654 and CA86340, and NINDS grant NS39617, and the U.S. Department of Energy Office of Biological and Environmental Research for support of portions of the research described in this chapter. Pacific Northwest National Laboratory is operated by Battelle Memorial Institute for the U.S. Department of Energy under contract DE-AC06-76RLO 1830.

REFERENCES

1. R Himmelreich, H Plagens, H Hilbert, B Reiner, R Herrmann. Comparative analysis of the genomes of the bacteria *Mycoplasma pneumoniae* and *Mycoplasma genitalium*. Nucleic Acids Res 25:701–712, 1997.

2. M Schena, D Shalon, RW Davis, PO Brown. Quantitative monitoring of gene expression patterns with a complementary DNA microarray. Science 270:467–470, 1995.
3. D Shalon, SJ Smith, PO Brown. A DNA microarray system for analyzing complex DNA samples using two-color fluorescent probe hybridization. Genome Res 6:639–645, 1996.
4. MD Adams. Serial analysis of gene expression: ESTs get smaller. Bioessays 18:261–262, 1996.
5. L Anderson, J Seilhammer. A comparison of selected mRNA and protein abundances in human liver. Electrophoresis 18:533–537, 1997.
6. SP Gygi, Y Rochon, BR Franza, R Aebersold. Correlation between protein and mRNA abundance in yeast. Mol Cell Biol 19:1720–1730, 1999.
7. JI Garrels, CS McLaughlin, JR Warner, B Futcher, GI Latter, R Kobayashi, B Schwender, T Volpe, DS Anderson, R Mesquita-Fuentes, WE Payne. Proteome studies of *Saccharomyces cerevisiae*: Identification and characterization of abundant proteins. Electrophoresis 18:1347–1360, 1997.
8. AJ Link, LG Hays, EB Carmack, JR Yates. Identifying the major proteome components of *Haemphilus influenzae* type-strain NCTC 8143. Electrophoresis 18:1314–1334, 1997.
9. A Shevchenko, ON Jensen, AV Podtelejnikov, F Sagliocco, M Wilm, O Vorm, P Mortensen, A Shevchenko, H Boucherie, M Mann. Linking genome and proteome by mass spectrometry: Large-scale identification of yeast proteins from two dimensional gels. Proc Natl Acad Sci USA 93:14440–14445, 1996.
10. PH O'Farrell. High resolution two-dimensional electrophoresis of proteins. J Biol Chem 250:4007–4021, 1975.
11. J Klose, U Kobalz. Two-dimensional electrophoresis of proteins: An updated protocol and implications for a functional analysis of the genome. Electrophoresis 16:1034–1059, 1995.
12. A Shevchenko, M Wilm, O Vorm, M Mann. Mass spectrometric sequencing of proteins from silver stained polyacrylamide gels. Anal Chem 68:850–858, 1996.
13. M Wilm, A Shevchenko, T Houthaeve, S Breit, L Schweigerer, T Fotsis, M Mann. Femtomole sequencing of proteins from polyacrylamide gels by nano-electrospray mass spectrometry. Nature 379:466–469, 1996.
14. Y Oda, K Huang, FR Cross, D Cowburn, BT Chait. Accurate quantitation of protein expression and site-specific phosphorylation. Proc Natl Acad Sci USA 96:6591–6596, 1999.
15. JR Yates, AL McCormack, J Eng. Mining genomes with MS. Anal Chem 68:A534–A540, 1996.
16. WJ Henzel, TM Billeci, JT Stults, SC Wong, C Grimley, C Watanabe. Identifying proteins from two-dimensional gels by molecular mass searching of peptide fragments in protein sequence databases. Proc Natl Acad Sci USA 90:5011–5015, 1993.
17. M Mann, P Hojrup, P Roepstorff. Use of mass spectrometric molecular weight information to identify proteins in sequence databases. Biol Mass Spectrom 22:338–345, 1993.

18. P James, M Quadroni, E Carafoli, G Gonnet. Protein identification by mass profile fingerprinting. Biochem Biophys Res Commun 195:58–64, 1993.
19. JR Yates III, S Speicher, PR Griffin, T Hunkapiller. Peptide mass maps: A highly informative approach to protein identification. Anal Biochem 214:397–408, 1993.
20. DJ Pappin, P Hojrup, AJ Bleasby. Rapid identification of proteins by peptide-mass fingerprinting. Curr Biol 3:327–332, 1993.
21. KA Cox, JD Williams, RG Cooks, RE Kaiser. Quadrupole ion trap mass-spectrometry: Current applications and future directions for peptide analysis. Biol Mass Spectrom 21:226–241, 1992.
22. MJ Huddleston, MF Bean, SA Carr. Collisional fragmentation of glycopeptides by electrospray ionization LC/MS and LC/MS/MS: Methods for selective detection of glycopeptides in protein digests. Anal Chem 65:877–884, 1993.
23. K Jonscher, G Currie, AL McCormack, JR Yates. Matrix-assisted laser desorption of peptides and proteins on a quadrupole ion trap mass spectrometer. Rapid Commun Mass Spectrom 7:20–26, 1993.
24. JA Loo, CG Edmonds, RD Smith. Primary sequence information from intact proteins by electrospray ionization tandem mass spectrometry. Science 248:201–204, 1990.
25. RD Smith, JA Loo, CG Edmonds, CJ Barinaga, HR Udseth. New developments in biochemical mass spectrometry: Electrospray ionization. Anal Chem 62(9):882–899, 1990.
26. AJ Tomlinson, S Naylor. A strategy for sequencing peptides from dilute mixtures at the low femtomole level using membrane preconcentration-capillary electrophoresis-tandem mass spectrometry (mPC-CE-MS/MS). J Liquid Chromatogr 18:3591–3615, 1995.
27. A Ducret, I Van Oostveen, JK Eng, JR Yates, R Aebersold. High electrospray tandem mass spectrometry throughput protein characterization by automated reverse-phase chromatography. Prot Sci 7:706–719, 1998.
28. JR Yates. Mass spectrometry and the age of the proteome. J Mass Spectrom 33:1–19, 1998.
29. AL McCormack, DM Schieltz, B Goode, S Yang, G Barnes, D Drubin, JR Yates. Direct analysis and identification of proteins in mixtures by LC/MS/MS and database searching at the low-femtomole level. Anal Chem 69:767–776, 1997.
30. RD Smith, JA Loo, CJ Barinaga, CG Edmonds, HR Udseth. Capillary zone electrophoresis and isotachophoresis-mass spectrometry of polypeptides and proteins based upon an electrospray ionization interface. J Chromatogr 480:211–232, 1989.
31. JH Wahl, DC Gale, RD Smith. Sheathless capillary electrophoresis electrospray ionization mass spectrometry using 10 μm id capillaries: Analyses of tryptic digests of cytochrome c. J Chromatogr A 659:217–222, 1994.
32. JF Banks, JP Quinn, CM Whitehouse. LC/ESI-MS determination of proteins using conventional liquid chromatography and ultrasonically assisted electrospray. Anal Chem 66:3688–3695, 1994.

33. A Apffel, JA Chakel, WS Hancock, C Souders, T M'Timkulu, E Pungor Jr. Application of high-performance liquid chromatography electrospray ionization mass spectrometry and matrix-assisted laser-desorption ionization time-of-flight mass spectrometry in combination with selective enzymatic modifications in the characterization of glycosylation patterns in single-chain plasminogen activator. J Chromatogr A 732:27–42, 1996.
34. JC Severs, AC Harms, RD Smith. A new high-performance interface for capillary electrophoresis electrospray ionization mass spectrometry. Rapid Commun Mass Spectrom 10:1175–1178, 1996.
35. D Figeys, I van Oostveen, A Ducret, R Aebersold. Protein identification by capillary zone electrophoresis/microelectrospray ionization-tandem mass spectrometry at the subfemtomole level. Anal Chem 68:1822–1828, 1996.
36. MB Comisarow, AG Marshall. Fourier transform ion cyclotron resonance mass spectrometry. Chem Phys Lett 25:282–283, 1974.
37. JE Bruce, SA Hofstadler, BE Winger, RD Smith. Characterization of ribonuclease-B heterogeneity and the identification and removal of phosphate adducts by high resolution electrospray ionization Fourier transform ion cyclotron resonance mass spectrometry. Int J Mass Spectrom Ion Process 132:97–107, 1994.
38. JP Speir, MW Senko, DP Little, JA Loo, FW McLafferty. High-resolution tandem mass spectra of 37–67 kDa proteins. J Mass Spectrom 30:39–42, 1995.
39. MV Gorshkov, L Paša Tolić, HR Udseth, GA Anderson, BM Huang, JE Bruce, DC Prior, SA Hofstadler, L Tang, LZ Chen, JA Willett, AL Rockwood, MS Sherman, RD Smith. Electrospray ionization-Fourier transform ion cyclotron resonance mass spectrometry at 11.5 tesla: Instrument design and initial results. J Am Soc Mass Spectrom 9:692–700, 1998.
40. GA Valaskovic, NL Kelleher, FW McLafferty. Attomole sensitivity analysis of recombinant proteins by picospray CE/FTICR-MS. 44th ASMS Conf on Mass Spectrometry and Allied Topics, Portland, OR, 1996.
41. MR Emmett, E Jäverfalk, GS Jackson, FM White, CL Hendrickson, AG Marshall. Development of a high sensitivity, low flow rate LC/ESI FT-ICR mass spectrometer. 46th ASMS Conf on Mass Spectrometry and Allied Topics, Orlando, FL, 1998.
42. ME Belov, MV Gorshkov, HR Udseth, GA Anderson, AV Tolmachev, DC Prior, R Harkewicz, RD Smith. Initial implementation of an electrodynamic ion funnel with FTICR mass spectrometry. J Am Soc Mass Spectrom 11:19–23, 2000.
43. TD Pollard. Cytoplasmic contractile proteins. J Cell Biol 91:156S–165S, 1981.
44. JE Bruce, GA Anderson, RD Smith. "Colored" noise waveforms and quadrupole excitation for the dynamic range expansion of Fourier transform ion cyclotron resonance mass spectrometry. Anal Chem 68:534–541, 1996.
45. JE Bruce, GA Anderson, J Wen, R Harkewicz, RD Smith. High-mass-measurement accuracy and 100% sequence coverage of enzymatically digested bovine serum albumin from an ESI-FTICR mass spectrum. Anal Chem 71:2595–2599, 1999.
46. S Hjertén, M-D Zhu. Adaption of the equipment for high-performance electrophoresis to isoelectric focusing. J Chromatogr 346:265–270, 1985.

47. F Kilar, S Hjerten. Fast and high resolution analysis of human serum transferrin by high performance isoelectric focusing in capillaries. Electrophoresis 10:23–29, 1989.
48. JR Mazzeo, IS Krull. Coated capillaries and additives for the separation of proteins by capillary zone electrophoresis and capillary isoelectric focusing. Biotechniques 10:638–645, 1991.
49. F Foret, O Muller, J Thorne, W Gotzinger, BL Karger. Analysis of protein fractions by micropreparative capillary isoelectric focusing and matrix-assisted laser desorption time-of-flight mass spectrometry. J Chromatogr A 716:157–166, 1995.
50. R Rodriguez Diaz, T Wehr, M Zhu. Capillary isoelectric focusing. Electrophoresis 18:2134–2144, 1997.
51. T Manabe, H Miyamoto, A Iwasaki. Effects of catholytes on the mobilization of proteins after capillary isoelectric focusing. Electrophoresis 18:92–97, 1997.
52. Y Shen, F Xiang, TD Veenstra, EN Fung, RD Smith. High-resolution capillary isoelectric focusing of complex protein mixtures from lysates of microorganisms. Anal Chem 71:5348–5353, 1999.
53. W Tang, AK Harrata, CS Lee. Two-dimensional analysis of recombinant *E. coli* proteins using capillary isoelectric focusing electrospray ionization mass spectrometry. Anal Chem 69:3177–3182, 1997.
54. L Yang, Q Tang, AK Harrata, CS Lee. Capillary isoelectric focusing-electrospray ionization mass spectrometry for transferrin glycoforms analysis. Anal Biochem 243:140–149, 1996.
55. L Yang, CS Lee, SA Hofstadler, L Paša-Tolić, RD Smith. Characterization of microdialysis acidification for capillary isoelectric focusing-microelectrospray ionization mass spectrometry. Anal Chem 70:3235–3241, 1998.
56. S Guan, AG Marshall, MC Wahl. MS/MS with high detection efficiency and mass resolving power for product ions in Fourier transform ion cyclotron resonance mass spectrometry. Anal Chem 66:1363–1367, 1994.
57. T Solouki, L Paša-Tolić, GS Jackson, SG Guan, AG Marshall. High-resolution multistage MS, MS2, and MS3 matrix-assisted laser desorption/ionization FT-ICR mass spectra of peptides from a single laser shot. Anal Chem 68:3718–3725, 1996.
58. PK Jensen, L Paša-Tolić, GA Anderson, JA Horner, MS Lipton, JE Bruce, RD Smith. Probing proteomes using capillary isoelectric focusing-electrospray ionization Fourier transform ion cyclotron resonance mass spectrometry. Anal Chem 71:2076–2084, 1999.
59. L Paša-Tolić, PK Jensen, GA Anderson, MS Lipton, KK Peden, S Martinović, N Tolic, JE Bruce, RD Smith. High throughput proteome-wide precision measurements of protein expression using mass spectrometry. J Am Chem Soc 121:7949–7950, 1999.
60. BE Winger, SA Hofstadler, JE Bruce, HR Udseth, RD Smith. High-resolution accurate mass measurements of biomolecules using a new electrospray ionization ion cyclotron resonance mass spectrometer. J Am Soc Mass Spectrom 4:566–577, 1993.

61. JE Bruce, GA Anderson, RD Smith. "Colored" noise waveforms and quadrupole excitation for the dynamic range expansion of Fourier transform ion cyclotron resonance mass spectrometry. Anal Chem 68:534–541, 1996.
62. AG Marshall, MW Senko, WQ Li, M Li, S Dillon, SH Guan, TM Logan. Protein molecular mass to 1 Da by C-13, N-15 double-depletion and FT-ICR mass spectrometry. J Am Chem Soc 119:433–434, 1997.
63. GA Anderson, JE Bruce. Pacific Northwest National Laboratory, Richland, WA, 1995, unpublished work.
64. AJ Link, K Robinson, GM Church. Comparing the predicted and observed properties of proteins encoded in the genome of Escherichia coli K-12. Electrophoresis 18:1259–1313, 1997.
65. SA Hofstadler, JH Wahl, R Bakhtiar, GA Anderson, JE Bruce, RD Smith. J Am Soc Mass Spectrom 5:894–899, 1994.
66. MW Senko, JP Speir, FW McLafferty. Collisional activation of large multiply charged ions using Fourier transform mass spectrometry. Anal Chem 66:2801–2808, 1994.
67. DP Little, JP Speir, MW Senko, PB O'Connor, FW McLafferty. Infrared multiphoton dissociation of large multiply charged ions for biomolecule sequencing. Anal Chem 66:2809–2815, 1994.
68. JI Garrels, CS McLaughlin, JR Warner, B Futcher, GI Latter, R Kobayashi, B Schwender, T Volpe, DS Anderson, R Mesquita-Fuentes, WE Payne. Proteome studies of *Saccharomyces cerevisiae*: Identification and characterization of abundant proteins. Electrophoresis 18:1347–1360, 1997.
69. TD Veenstra, S Martinović, GA Anderson, L Paša-Tolić, RD Smith. Proteome analysis using selective incorporation of isotopically labeled amino acids. J Am Soc Mass Spectrom 11:78–82, 2000.
70. TP Conrads, GA Anderson, TD Veenstra, L Paša-Tolić, RD Smith. Utility of accurate mass tags for proteome-wide protein identification. Anal Chem 72:3349–3354, 2000.
71. The Yeast Genome Directory. Nature 387(6632 suppl):5, 1997.
72. The *C. elegans* Sequencing Consortium. Genome sequence of the nematode *C. elegans*: A platform for investigating biology. Science 282:2012–2018, 1998.
73. T Pawson, JD Scott. Signaling through scaffold, anchoring, and adaptor proteins. Science 278:2075–2080, 1997.
74. SP Gygi, S Rist, SA Gerber, F Turecek, MH Gelb, R Aebersold. Quantitative analysis of complex protein mixtures using isotope-coded affinity tags. Nature Biotechnol 17:994–999, 1999.
75. C Masselon, GA Anderson, R Harkewicz, JE Bruce, L Paša-Tolić, RD Smith. Accurate mass multiplexed tandem mass spectrometry for high throughput polypeptide identification from mixtures. Anal Chem 72:1918–1924, 2000.
76. P Ferianc, A Farewell, T Nystrom. The cadmium-stress stimulon of *Escherichia coli* K-12. Microbiology 144:1045–1050, 1998.

77. H Gao, Y Shen, TD Veenstra, R Harkewicz, GA Anderson, JE Bruce, L Paša Tolić, RD Smith. Two dimensional electrophoretic/chromatographic separations combined with electrospray ionization-FTICR mass spectrometry for high throughput proteome analysis. Microcolumn Sep 12:383–390, 2000.
78. RD Smith, L Pa˘sa Toli´c, MS Lipton, PK Jensen, GA Anderson, Y Shen, TP Conrads, HR Udseth, R Harkewicz, ME Belov, C Masselon, TD Veenstra. Rapid quantitative measurements of proteomes. Electrophoresis 22:1652–1668, 2001.
79. DR Goodlett, JE Bruce, GA Anderson, B Rist, L Paša Tolić, O Fiehn, RD Smith, R Aebersold. Protein identification with a single accurate mass of a cysteine-containing peptide and constrained database searching. Anal Chem 72:1112–1118, 2000.
80. ME Belov, MV Gorshkov, HR Udseth, GA Anderson, RD Smith. Zeptomole-sensitivity electrospray ionization-Fourier transform ion cyclotron resonance mass spectrometry of proteins. Anal Chem 72:2271–2279, 2000.
81. T Kim, AV Tolmachev, R Harkewicz, DC Prior, GA Anderson, HR Udseth, RD Smith, TH Bailey, S Rakov, JH Futrell. Design and implementation of a new electrodynamic ion funnel. Anal Chem 72:2247–2255, 2000.
82. ME Belov, EN Nikolaev, GA Anderson, KJ Auberry, R Harkewicz, RD Smith. Electrospray ionization-Fourier transform ion cyclotron mass spectrometry using ion pre-selection and external accumulation for ultra-high sensitivity. J Am Soc Mass Spectrom 12:38–48, 2001.
83. T Kim, HR Udseth, RD Smith. Improved ion transmission from atmospheric pressure to high vacuum using a multi-capillary inlet and electrodynamic ion funnel interface. Anal Chem 72:5014–5019, 2000.
84. ME Belov, EN Nikolaev, GA Anderson, HR Udseth, TP Conrads, TD Veenstra, CD Masselon, MV Gorshkov, RD Smith. Design and performance of an ESI interface for selective external ion accumulation coupled to a Fourier transform ion cyclotron resonance mass spectrometer. Anal Chem 73:253–261, 2001.

9
Detection of Noncovalent Complexes by Electrospray Ionization Mass Spectrometry

A. K. Ganguly
Stevens Institute of Technology, Hoboken, New Jersey

Birendra N. Pramanik and Guodong Chen
Schering-Plough Research Institute, Kenilworth, New Jersey

Anthony Tsarbopoulos
GAIA Research Center, The Goulandris Natural History Museum, Kifissia, Greece

I. INTRODUCTION

Noncovalent interactions involving enzyme and substrate, protein and ligand, protein and protein, and antigen and antibody elicit many biological activities including cellular functions. Examples of important noncovalent complexes include *ras*-guanosine diphosphate (GDP) (protein–ligand), FKBP-rapamycin (protein–inhibitor complex), hepatitis C virus (HCV) (protein–metal–peptide complex), gamma-interferon (γ-IFN) (protein–protein homodimer), and human immunodeficiency virus (HIV)-1 protease (homodimer of two subunits). Diseases are often caused by the disruption of normal cellular processes that may involve noncovalent complexes. Thus, an important first step toward understanding some of the disease processes is to fully comprehend the structures and mechanism of these noncovalent complexes. Specific therapeutic compounds can then be designed to restore normal cellular functions. This strategy has become an important aspect of modern drug discovery research.

An example of the implication of noncovalent interactions in diseases is the involvement of *ras* protein in transduction of cell surface signals in cancer research. The *ras* oncogenic proteins exist in two interconvertible states: the *ras*-GDP (inactive state) and *ras*-guanosine triphosphate (GTP) (active state). In cancer cells, unlike in normal cells, *ras*-GTP is slowly hydrolyzed to the inactive state (i.e, *ras*-GDP), and this results in unregulated cell growth. It is estimated that 20% of all human tumors are linked to malfunctions of the *ras* oncogene; the largest malfunctions are found in pancreatic (90%), colon (50%), and lung (40%) cancers (1). In recent years, enormous effort has been focused on understanding the prenyl transfer event of *ras* protein, a process that is required for *ras* to bind to cellular membranes and send signals for cellular proliferation. Farnesylation of *ras* with farnesyl pyrophosphate occurs in the presence of farnesyl protein transferase (FPT). Farnesylated *ras* undergoes further processing by the cleavage of the fourth amino acid from the carboxyl terminus and methylation of the newly formed carboxylic acid with methyl transferase to give the methyl ester. Thus, transformed farnesylated *ras* is responsible for sending the signal to the cell for proliferation. Several potent inhibitors of FPT have been discovered (2), including Sch 66336, which is undergoing clinical trial against various forms of cancer. Several clinical trials are under way to determine the clinical efficacy of FPT inhibitors in humans.

Traditionally, noncovalent complexes can be detected by gel permeation chromatography and centrifugation. However, these methods provide little information regarding the molecular weight and binding stoichiometry of the complex. Other techniques such as X-ray crystallography and nuclear magnetic resonance (NMR) spectroscopy have been used successfully where an adequate supply of the protein is available (NMR) and the protein is crystalline (X-ray). In recent years, mass spectrometry (MS) has been developed to provide an alternative approach for the detection of noncovalent complexes, and in this chapter we summarize some of these results.

Introduction of the electrospray ionization technique (ESI) has revolutionized the applicability of MS for the study of proteins of high molecular weights either by themselves or when they are bound to ligands (3). In ESI, gas-phase highly charged ions are generated by spraying the sample solution through a needle under a strong electric field. The ionization occurs by protonation or other gas-phase ion–molecule reactions, resulting in the formation of multiply charged ions of the protein sample. The above process is mild and allows large polar molecules to be desolvated and ionized in an intact form into the gas phase with very little fragmentation. This unique characteristic of ESI may allow the preservation of protein conformations and noncovalent associations in the gas phase, thus making ESI a suitable technique for the detection of the noncovalent complexes. In addition, recent developments of MS instrumentation allow a sufficiently high mass range to be used for observing noncovalent complexes that bear some excess charges (4). It should be noted that another ionization technique [matrix-assisted laser desorption ionization (MALDI) (5–7)] has also been shown to be applicable to the detection of noncovalent complexes (8), often using only the "first shot" MALDI signals (9). However, ESI-MS is the preferred technique for the detection of noncovalent complexes, primarily because it allows more facile introduction of samples to the instrument from an aqueous medium.

The first studies of noncovalent complexes by ESI-MS were reported by Ganem et al. (10) on the binding between FKBP, an immunosuppressive protein, rapamycin, and FK506 (an immunosuppressive agent). This work was closely followed by Katta and Chait's study (11) on the heme–globin complex in myoglobin and Baca and Kent's work (12) on the ternary complex of dimeric HIV-1 protease and a substrate-based inhibitor. Subsequently, numerous investigations of noncovalent complexes have been reported using ESI-MS (13–20). The advantages of using ESI-MS in detecting noncovalent complexes are that ESI-MS provides stoichiometric information from much less material in far shorter time and structural insights into specific noncovalent associations in solution. In the following sections, several representative examples of noncovalent complexes are discussed, including protein–ligand interactions, drug–protein–ligand complexes, protein–metal complexes, protein–protein associations, protein–nucleic acid interactions, and drug–nucleic acid complexes. Experimental considerations and future prospects are also presented.

II. PROTEIN–LIGAND INTERACTIONS

One of the important noncovalent complexes in cancer research is the association of the *ras* protein with either GDP or GTP. Understanding these noncovalent interactions not only provides insight into the mechanism of enzyme interactions but also forms a basis for developing a strategy for the discovery of inhibitors for the relevant disease processes.

The *ras*-GDP has a dissociation constant (K_D) of 20.3 pM. Figure 1 illustrates an ESI-MS spectrum of *ras*-GDP complex (0.5 μg/μL) in 2 mM ammonium acetate and 0.5% acetic acid (1:1) solution, pH 2.5 (15). A broad distribution of multiply charged ions was observed, with the charge state ranging from +9 to +17. The deconvoluted mass spectrum yielded an average

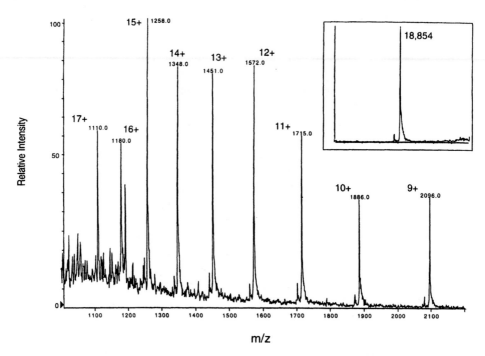

Figure 1 Electrospray ionization mass spectrum of *ras*-GDP obtained at pH 2.5 in 2 mM ammonium acetate buffer using a 20 μM protein solution. Insert shows the deconvoluted mass spectrum of the apo-*ras*, giving a molecular mass of 18,854 Da. (Reproduced with permission from Ref. 15.)

molecular mass of 18,854 Da, as shown in the insert in Fig. 1. This is in good agreement with the theoretical molecular mass of 18,853.3 Da for the free apo-*ras* protein. Apparently, the *ras*-GDP complex was dissociated under the experimental conditions at lower pH values, and only the denatured *ras* protein could be detected. When acetonitrile/water (50:50) was used as the solvent with pH values in the range of 2–2.4, the distribution of multiply charged ions was shifted to lower values (*m/z* 858), indicating a complete denaturation or unfolding of the *ras* protein. The shifting of the charge state distributions was reported to indicate the folding and unfolding of proteins by ESI-MS (14). The difference in the distribution of charge states is related to the accessibility of ionizable groups during the ESI process. An unfolded protein with basic sites is more accessible and more likely to accumulate charges than when it is in the more compact native state.

A. Effect of pH

When the pH was adjusted to 5.2 (in 2 mM ammonium acetate), a different set of multiply charged ions was observed (Fig. 2) (15). The deconvolution of the ESI-MS spectrum yielded an average molecular mass of 19,295.6 Da (insert), which is in good agreement with the expected molecular mass (19,296 Da) of the *ras*-GDP complex. This observation was further confirmed by a series of adduct ions due to the successive addition of sodium cations to the *ras*-GDP complex.

When the pH value was lowered to 4.7 (in 2 mM ammonium acetate), a different ESI-MS spectrum was generated (Fig. 3) that has a broader distribution

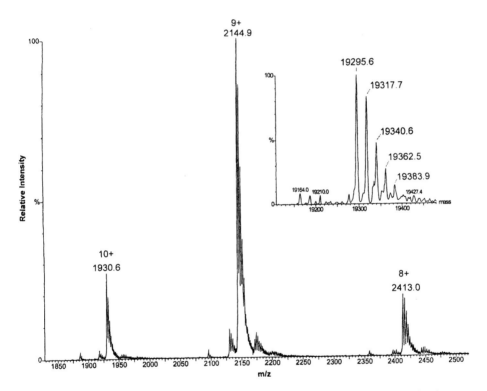

Figure 2 Electrospray ionization mass spectrum of the *ras*-GDP complex obtained at pH 5.2 in 2 mM ammonium acetate buffer using a 20 μM protein solution. Insert shows the deconvoluted mass spectrum of the *ras*-GDP complex, giving a measured molecular mass of 19,295.6 Da. Additional peaks in the spectrum correspond to sodium ion adducts. (Reproduced with permission from Ref. 15.)

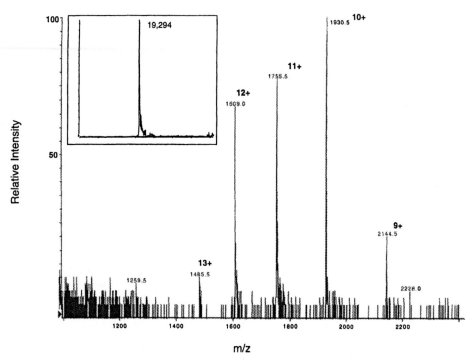

Figure 3 Electrospray ionization mass spectrum of the *ras*-GDP complex at pH 4.7 in 2 mM ammonium acetate buffer. Insert shows the deconvoluted mass spectrum of the *ras*-GDP complex, giving a molecular mass of 19,294 Da. (Reproduced with permission from Ref. 15.)

of multiply charged ions centered at the +11 charge state (15). This reflects a small unfolding of the *ras*-GDP complex but without any loss of the GDP nucleotide. The deconvoluted spectrum provided an average molecular mass of 19,294 Da, which is consistent with the theoretical molecular mass of the *ras*-GDP complex.

When the ESI-MS experiment was performed at pH 3, the ESI-MS spectrum exhibited two sets of ion distributions (B9–B11 and A10–A22) (Fig. 4) (15). The deconvoluted spectrum of these two sets of ions shows that these are species of two molecular masses, 19,294.4 Da and 18,851.4 Da, respectively (insert in Fig. 4), which are the intact *ras*-GDP complex and the free apo-*ras* protein. It is important to note that the charge distributions of the A series ions are much broader and have shifted to much lower *m/z* values, indicating

Detection of Noncovalent Complexes

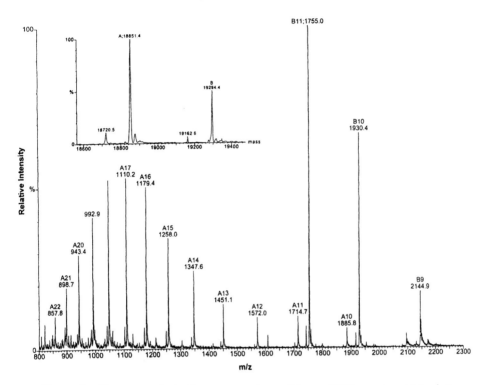

Figure 4 Electrospray ionization mass spectrum of the *ras*-GDP complex obtained at pH 3 in a 2 mM ammonium acetate buffer. Insert shows the deconvoluted mass spectrum of the *ras*-GDP complex, giving a molecular mass of 19,294.4 Da for the complex and a molecular mass of 18,851.4 Da for apo-*ras*. (Reproduced with permission from Ref. 15.)

substantial unfolding of the free apo-*ras* protein. The pH effects on the formation of *ras*-GDP complex by ESI-MS are consistent with the data obtained from solution-state NMR (21). At a lower pH value (pH 2–2.4), the noncovalent complex can be completely destroyed as illustrated in the case of *ras*-GDP complex (15).

B. Effect of Solvent

In another experiment to study the effect of organic solvents on the formation of *ras*-GDP complex, methanol was added to the *ras*-GDP solution at pH 5.8 (in 2 mM ammonium acetate) to give a 20% methanol solution. As expected, substantial dissociation of *ras*-GDP complex occurred. This is reflected by the observation of two species having molecular masses of 19,294 Da and 18,851 Da as can be seen in the deconvoluted spectrum (Fig. 5) (15). The addition of methanol or another organic solvent (e.g., acetonitrile) disrupts the hydrophobic interactions and causes the unfolding of the protein. As this unfolding process continues, more basic sites are available for protonation, giving rise to a broader charge state distribution. The narrower charge state distribution (high mass region) implies that the protein is in a folded conformation, maintaining the intact complex structure in solution. When the percentage of methanol was raised to 50% in the *ras*-GDP solution, this noncovalent complex was no longer detectable.

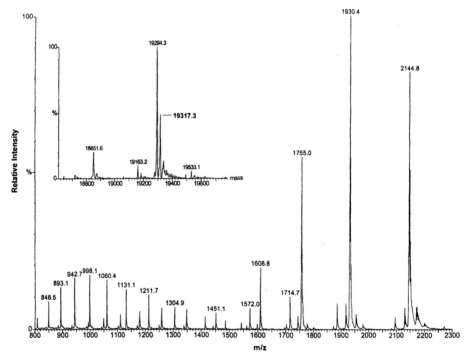

Figure 5 Electrospray ionization mass spectrum of the *ras*-GDP complex in 20% methanol and 2 mM ammonium acetate buffer at pH 5.8. Insert shows the deconvoluted spectrum, giving molecular masses of 18,851.6 Da (apo-*ras*), 19,294.3 Da (*ras*-GDP), and 19,317.3 Da (*ras*-GDP-Na). (Reproduced with permission from Ref. 15.)

C. Effect of ESI Cone Voltage

In a typical ESI experiment, the orifice voltage (or cone voltage) is kept at low values to maintain the stability of the complex. A higher orifice voltage increases the internal energy imparted to the complex and destabilizes the complex. As expected, the effect of the orifice voltage on the abundance of the complex depends on the binding energy of the noncovalent complex. In the case of the *ras*-GDP complex, the binding is relatively strong; the complex does not begin to dissociate until an orifice voltage of 120 V was reached.

Another consideration in detection of noncovalent complexes is the specificity of the noncovalent interaction. It is not unusual to observe a nonspecific gas-phase complex by ESI-MS, especially at elevated protein concentrations. One characteristic feature of the nonspecific gas-phase complex is that a relatively large amount of material is needed to observe it. The observation of a specific noncovalent association requires much less material. Therefore, dilution experiments should be performed to differentiate specific noncovalent interactions from nonspecific gas-phase complexes. In addition, changes of pH values and organic solvents should also have significant impacts on the observed ESI charge distributions of specific noncovalent complexes.

III. DRUG–PROTEIN–LIGAND COMPLEX

A. *ras*-GDP-Ligand

The observation of the *ras*-GDP complex by ESI-MS forms the basis for the screening of potential inhibitors that bind and inactivate the *ras* protein by preventing the exchange of GTP for GDP. For example, Sch 54341 has been found to form a noncovalent ternary complex, *ras*-GDP-Sch 54341 (see Fig. 6 for structure), in 1:1 stoichiometry. A deconvoluted mass spectrum (Fig. 6) shows the formation of the *ras*-GDP-Sch 54341 ternary complex with an average molecular mass of 19,570 Da (22). In the vitro activity (IC_{50}) of Sch 54341 was found to be 50 μM, which is consistent with the formation of a noncovalent complex. A second potential inhibitor, Sch 54292 (IC_{50} 1 μM; see Fig. 7 for structure), was also found to form a noncovalent complex with *ras*-GDP (Fig. 7) (22). The molecular mass at 19,816 Da is that of the noncovalent ternary complex *ras*-GDP-Sch54292. Although the bonding of several of these noncovalent complexes is relatively weak, this screening strategy has been successfully applied to screen over 50 potential inhibitors of *ras*-GDP. This approach is extremely useful for the study of weakly bound inhibitors in an efficient way, as it was also shown in the screening of aldose reductase inhibitors (23). This ESI-based

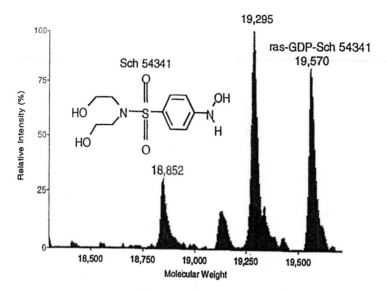

Figure 6 Deconvoluted ESI mass spectrum of *ras*-GDP-Sch 54341 noncovalent complex. (Reproduced with permission from Ref. 22.)

methodology can potentially lead to the identification of more potent therapeutical compounds based on their structure–activity relationships. The other important finding in these studies is that the small-molecule inhibitors do not appear to bind to GDP; instead, they bind to the *ras* protein.

B. Determination of Binding Sites

An approach involving site-specific chemical modifications of an amino acid can be applied to determine drug–protein or protein–ligand binding sites. In the drug–protein complex, the amino acids near the binding pocket as well as those in the protein interior will be hindered from undergoing chemical modifications, whereas the exposed amino acids will readily undergo modifications. The drug-binding site of the protein can be established by determining which amino acids are hindered from specific chemical modifications as a consequence of complex formation (24). In the case of the denatured protein, however, these amino acids should be readily available for chemical modification.

The specific binding site of Sch 54292 in its ternary complex with *ras* and GDP was determined by employing the approach of chemical derivatization and LC/MS analysis of the derivatized products (22). It is known that

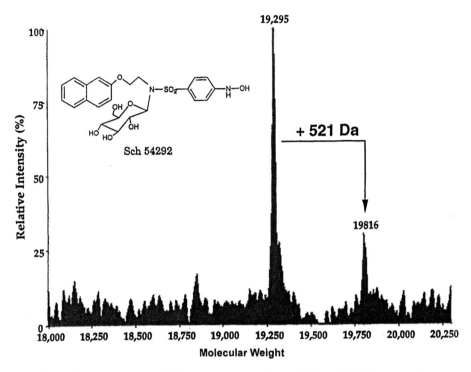

Figure 7 Deconvoluted ESI mass spectrum of *ras*-GDP-Sch 54292 noncovalent complex. (Reproduced with permission from Ref. 22.)

succinic anhydride reacts specifically with the ε-amino group of exposed lysine (Lys) residues on the surface of a protein under appropriate conditions. The hindered Lys residues will remain unmodified and can be identified by enzymatic digestion of the succinylated protein followed by LC/MS analysis of the generated peptide fragments. The degree of chemical modification can also be assigned on the basis of molecular mass determination of peptide fragments by ESI-MS. In the case of *ras* protein, there are a total of nine Lys groups available for succinylation, as indicated in the deconvoluted ESI spectrum (Fig. 8A). In contrast, the *ras*-GDP complex exhibited six Lys groups undergoing succinylation (Fig. 8B).

The smaller number of Lys groups available for succinylation in the *ras*-GDP may be due to steric hindrance in the complex caused by GDP. It was reasoned that the Lys group involved in Sch 54292 binding or adjacent to its binding site in the *ras*-GDP-Sch 54292 complex may be further hindered by

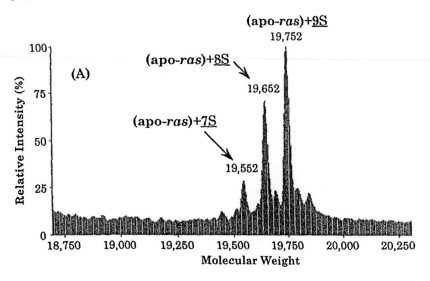

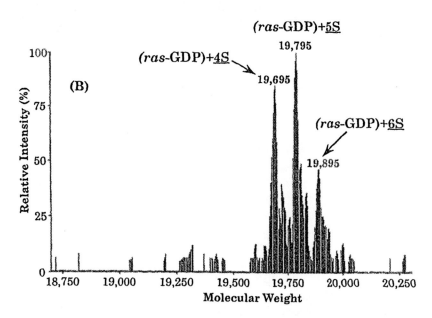

Figure 8 (A) The deconvoluted ESI spectrum of apo-*ras* protein that had undergone succinylation. (B) The deconvoluted ESI mass spectrum of *ras*-GDP complex that had undergone succinylation.

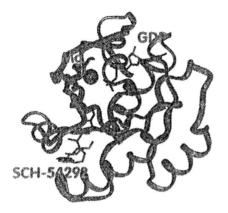

Figure 9 A three-dimensional structure of *ras*-GDP-Sch 54292 obtained from NMR and molecular modeling studies. (Reproduced with permission from Ref. 21.)

Sch 54292 and thus protected from succinylation. The succinate labeling patterns for apo-*ras*, *ras*-GDP, and *ras*-GDP-Sch 54292 were compared to indicate which Lys residues were involved in Sch 54292 binding. In each of the experiments, derivatization of the exposed Lys residues was performed by using a tenfold molar excess of succinic anhydride in 25 mM Tris buffer (pH 7.4) containing 5 mM $MgCl_2$. Following digestion of the derivatized proteins with endoproteinase Lys-C and peptide mapping by LC/MS analysis of the resulting peptide fragments, the identities of unmodified Lys residues were determined. The experimental data indicated that Sch 54292 was bound to *ras*-GDP near the region of Lys-101 without displacing the nucleotide. Figure 9 illustrates a three-dimensional structure of *ras*-GDP-Sch 54292 complex obtained from NMR and molecular modeling studies (21), which is consistent with the binding site of Sch 54292 in the complex, as determined by chemical derivatization and LC/ESI-MS mapping.

IV. PROTEIN–METAL COMPLEXES

A. Calmodulin/Ca

Protein–metal complexes involve both ionic interactions and hydrophobic interactions. Examples include noncovalent association of Ca^{2+}-bound calmodulin with a target protein melittin (25,26) and HCV–metal complex (15). ESI-MS has been found to be very effective in detecting these complexes.

Calmodulin is a small, compact, highly conserved cytoplasmic Ca^{2+}-binding protein. It is present in all eukaryotic cells studied, and calmodulin-like proteins have been identified in prokaryotic cells. The interactions of calmodulin and its target enzymes have been studied extensively. It was found that calcium binding to calmodulin is cooperative with respect to the two globular domains of calmodulin, and that Ca^{2+}-loaded calmodulin undergoes a conformational change that exposes its hydrophobic surface and allows the binding of the target enzymes. There are a variety of downstream targets for Ca^{2+}-bound calmodulin, including activating ion channels and receptor proteins. Both hydrophobic and ionic interactions are responsible for the association of the Ca^{2+}-loaded calmodulin with enzymes.

A study of small hydrophobic peptides such as melittin with Ca^{2+}-calmodulin complex should provide some details about their interactions. Gross and coworkers (25) investigated the complex formation of Ca^{2+}-calmodulin with melittin by negative ion ESI-MS. They found that no binding of the calmodulin to the melittin occurred in the absence of calcium. When calcium-saturated calmodulin was loaded with melittin, the noncovalent complex Ca^{2+}-calmodulin–melittin was observed (Fig. 10). The deconvoluted spectrum shows that the complex is formed with no less than four Ca^{2+} ions and the binding stoichiometry of melittin to calmodulin is 1:1. It is known that the calmodulin specifically binds to at least four Ca^{2+} ions. With the binding of four or more Ca^{2+} ions, the calmodulin molecules have a smaller negative charge and undergo conformational changes that expose a new surface to both hydrophobic and ionic interactions in the binding to melittin. This study clearly indicates the nature of the Ca^{2+}-induced noncovalent association of calmodulin with the small hydrophobic peptides. These complexes are being explored by H/D exchange and ESI-MS (see discussions in Chapter 10 on H/D exchange in solution by ESI-MS).

B. HCV/Zn

Another example involving a protein–metal complex is the HCV-Zn complex. HCV is a major etiological hepatitis agent. The HCV protease domain (HCV NS3) contains residues 1–181 of the N-terminus portion of the HCV protein. It is tetrahedrally coordinated by three cysteines and one histidine to a zinc ion, which stabilizes the tertiary structure. This HCV protease-Zn complex was observed by ESI-MS (27). Figure 11 illustrates the deconvoluted spectra for HCV protease at pH 3 and pH 7 (15). The intact complex was formed with an average molecular weight of 21,141 Da at pH 7 (Fig. 11B),

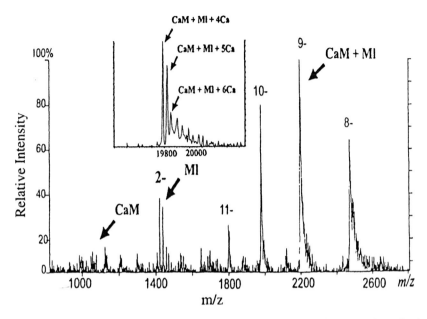

Figure 10 Electrospray ionization mass spectrum of noncovalent complex of calcium-loaded calmodulin with melittin. The insert shows a deconvoluted spectrum. CaM, Ml, and [CaM + MI + xCa] refer to calmodulin, melittin, and the complex, respectively. (Reproduced with permission from Ref. 25.)

whereas the free protease domain (MW 21,076 Da) was observed at pH 3 (Fig. 11A). The binding in the HCV-Zn complex is stronger than that in the ras-GDP because increasing the orifice voltage to 160 V does not affect its stability. This may reflect the nature of the HCV–Zn interaction, which is ionic rather than noncovalent.

V. PROTEIN–PROTEIN ASSOCIATION

One of the most important applications of ESI-MS is the study of protein–protein interactions such as the binding of immunoglobulin E, vascular endothelial growth factor, interleukin-2, or interleukin-5 to their respective receptors. The interleukins remain attractive drug targets in therapeutic areas of allergies, cancer, and asthma. The associated protein–protein interfaces often have some small regions that are critical to binding. The targeting of these small regions by

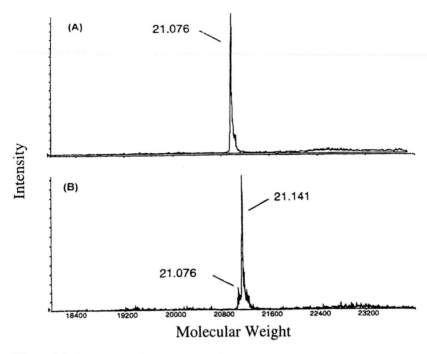

Figure 11 Deconvoluted mass spectra of HCV protease domain obtained at (A) pH 3 and (B) pH 7. (Reproduced with permission from Ref. 15.)

small-molecule inhibitors may be capable of disrupting undesirable protein–protein interactions.

A. Leucine Zipper Peptides

One of early studies on noncovalent dimers of leucine zipper peptides illustrated the applicability of ESI-MS (28). The leucine zipper motif has been found in the C-termini of several DNA-binding transcriptional activator proteins. The dimerization of the leucine zipper motif is necessary for DNA binding and plays an important role in DNA replication, recombination, strand scission, and transcription (28). McLafferty and coworkers (28) detected the dimers of the leucine zipper peptides GCN4-p1 and N16V by size-exclusion liquid chromatography/ESI-MS. They found that N16V has a greater tendency than GCN4-p1 to form dimers. This is in agreement with the behavior of these peptides in aqueous solution demonstrated by other methods (28).

B. γ-IFN

Mature human γ-IFN is a protein consisting of 143 amino acids with an excess of basic residues. Its homodimer (molecular mass 33,815 Da) is the active form of this protein. With carefully controlled experimental conditions for ESI, the γ-IFN dimer can be preserved in the gas phase. Figure 12A displays the ESI mass spectrum for γ-IFN at pH 9 (29). An average molecular mass of 33,819 Da was obtained from the experiment, which indicated the formation of the γ-IFN dimer. Note that the *m/z* values for the ESI signals of the γ-IFN monomer overlap with those corresponding to even-charge states of the dimer. Thus, the presence of the dimer is based on the detection of the odd-charge state ions, as shown in Fig. 12.

Dilution experiments are performed to eliminate the possibility of forming nonspecific protein aggregates in the gas phase. In this case, a solution containing an equimolar mixture of γ-IFN and cytochrome c (35 μM) was analyzed by ESI-MS, and only γ-IFN dimer was observed. As discussed in previous sections, the orifice voltage has a significant effect on the stability of noncovalent complexes, depending on the binding energy of the complex. The effect of the orifice voltage on γ-IFN dimer is shown in Fig. 12. It can be seen that the weakly bound γ-IFN dimer starts to dissociate at an orifice voltage of 60 V and completely disappears when the orifice potential reaches 100 V.

C. Large Protein Complexes

There are many challenges in detecting protein–protein complexes and preserving these complexes in the gas phase by ESI-MS. The first is the charge repulsion between neighboring protein subunits. Another is the loss of hydrophobic interactions as solvent is taken away from the protein surface during the electrospray process. The final obstacle is that the complex has to survive collisions with gas molecules during the transfer from the high pressure ESI source to the vacuum of the mass spectrometer. With many protein–protein complexes at very high molecular masses, appropriate MS instrumentation with a high mass range is necessary to detect them. Careful manipulations of the pressure differentials throughout the mass spectrometer are critical to maintaining the integrity of the complex (30). Green et al. (31) detected approximately 200 kDa globin dodecamer subassemblies in hexagonal bilayer hemoglobins (Hb's) on a quadrupole time-of-flight (QTOF) mass spectrometer with a mass range of 14,000 Da. In that study, the ESI-MS of the subassemblies obtained by gel filtration of partially dissociated

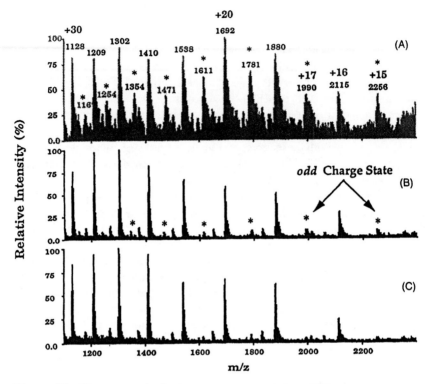

Figure 12 Electrospray ionization mass spectra of the γ-IFN dimer noncovalent complex obtained from aqueous solution (pH ≈ 9) at an orifice voltage (OR) of (A) 50 V, (B) 70 V, (C) 100 V. The odd-charge states are denoted with an asterisk. (Reproduced with permission from Ref. 29.)

L. terrestris and *Arenicola marina* Hb's showed the presence of noncovalent complexes of monomer (M) and disulfide-bonded trimer (T) subunits with masses in the 213.3–215.4 and 204.6–205.6 kDa ranges, respectively (31). They found that the observed mass of the *L. terrestris* subassembly decreased linearly with an increase in the declustering voltage, ranging from approximately 215.4 kDa at 60 V to approximately 213.3 kDa at 200 V. In contrast, the mass of the *Arenicola marina* complex decreased linearly as the potential was increased from 60 V to 120 V and reached an asymptote at approximately 204.6 kDa (180–200 V). The decrease in mass is probably due to the progressive removal of complexed water and alkali metal cations. The ESI-MS at an acidic pH showed both subassemblies to consist of only M and T subunits, having the composition M_3T_3 as demonstrated by the experimental molecular weights (31). Because there are three isoforms of M and four isoforms of T in *L. terrestris* and two isoforms of M and five isoforms of T in *Arenicola marina*, the masses of M_3T_3 are not unique. Based on a random assembly model, the known ESI-MS masses and relative abundances of the M and T subunit isoforms, the expected mass of randomly assembled subassemblies was calculated to be 213.436 kDa for *Lumbricus* Hb and 204.342 kDa for *Arenicola* Hb, which is in good agreement with the experimental values (31).

In another study, Robinson and coworkers (32) employed a nanoflow electrospray with a QTOF mass spectrometer to study the intact GroEL complex. The GroEL proteins belong to the chaperonins that help their substrate proteins fold to native-like states on the chaperone-assisted protein folding pathway. These helper proteins safeguard newly synthesized proteins from unproductive interactions that may lead to aggregation. The detection by ESI of the intact GroEL chaperonin complex may shed light on the mechanism of action of this class of proteins.

The theoretical molecular mass of the GroEL multiprotein complex is 800,758 Da. The QTOF mass spectrometer has an unlimited mass range in theory and is quite suitable for detecting such a complex. Figure 13 shows a nanoflow ESI-MS spectrum obtained from a solution of GroEL at pH 5.0 (32). A series of peaks are centered at $m/z \approx 10,000$, and the deconvolution of the spectrum gives rise to a molecular mass of $803,742 \pm 616$ Da for the intact GroEL complex. The number of subunits in the complex is determined to be 14 (32). It is interesting to note that the GroEL$_{14\text{-mer}}$ is quite stable in the gas phase. It does not dissociate upon an increase of translational energy in the ESI source region of the mass spectrometer. However, an increase in collision energy in the source region leads to a change in the charge state distributions, as illustrated in Fig. 14. The stepwise increases in the collision energy generate an intermediate charge state series centered at $m/z \approx 6000$. With a further increase in collision energy, no detectable ions of $m/z \approx 10,000$ remain, and the dominant species is centered at $m/z \approx 6000$. The molecular mass of this species is $401,556 \pm 202$ Da, in good agreement with the theoretical value for half the mass of the intact complex. At the highest collision energy, a well-resolved charge state series centered at $m/z \approx 2000$ is observed, corresponding to $57,287 \pm 1$ Da in mass. This mass is approximately one-seventh of the mass of the intermediate, illustrating the 14-subunit stoichiometry of the intact complex. Clearly, the intact GroEL$_{14\text{-mer}}$ dissociates into a GroEL heptamer intermediate and a GroEL monomer, with an increase in collision energy. The results demonstrate the remarkable

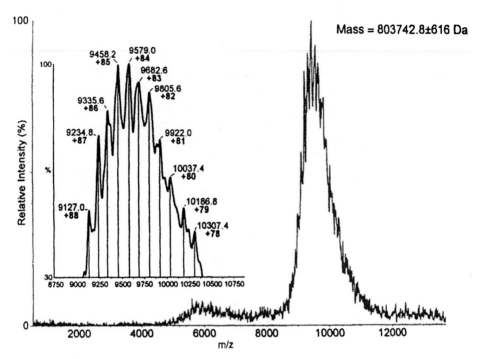

Figure 13 Nanoflow ESI mass spectrum obtained from a solution of GroEL at pH 5.0 at a protein concentration of 1 μM. The insert shows the expanded mass region. (Reproduced with permission from Ref. 32.)

ability of ESI-MS to detect macromolecular assemblies with molecular masses close to 1,000,000 Da.

More recently, Robinson and coworkers (33) reported the detection of the protein shell—or capsid—of the bacteriophage MS2, a virus that infects bacterial cells. The intact capsid consists of 180 copies of the coat protein. The researchers used the same methodology to determine the protein complex with collisional dampening and also to obtain information about the arrangement of the capsid's protein subunits without the cooling step (33). Their data suggest that the protein coat is constructed from hexameric building blocks, which is consistent with the trimer of dimers observed in the X-ray crystal structure (33). This methodology is potentially useful for establishing connectivities in macromolecular complexes; information on such complexes will be essential in the postgenomic era for describing the function and interactions of individual proteins.

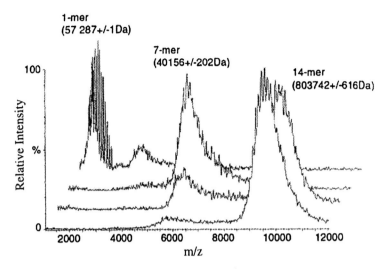

Figure 14 Electrospray ionization TOF mass spectrum of the collision-induced dissociation of the GroEL$_{14\text{-mer}}$ showing the formation of the heptameric intermediate and a well-resolved monomer. (Reproduced with permission from Ref. 32.)

VI. PROTEIN–NUCLEIC ACID INTERACTIONS

The ionic interactions between proteins and nucleic acids (DNA and RNA) have been probed by ESI-MS. Early work by Smith and coworkers (34) demonstrated the detection of noncovalent complexes between a single-stranded DNA-binding protein (gene V) and a variety of poly(dT) and poly(dA) oligonucleotide sequences by ESI-MS. A 16-mer oligonucleotide gave predominantly a 4:1 (protein monomer/oligonucleotide) complex. This binding stoichiometry was also measured independently by using size-exclusion chromatography, and the result was consistent with the ESI-MS data (34). In another study, Smith and coworkers (35) detected a transcription factor PU.1 protein–DNA complex. They showed that PU.1 formed a complex only with the wild-type DNA sequence from a pool of related DNA sequences (wild-type and mutant). This provided strong evidence that this protein–DNA complex observed in the gas phase was held together by a sequence-specific interaction.

Standing and coworkers (36) used an ESI-TOF mass spectrometer to study noncovalent interactions between *trp* apo repressor (TrpR), a DNA-binding protein that is involved in the regulation of tryptophan biosynthesis, and its specific operator DNA. They found that the *trp* repressor homodimer bound to a 21 base pair DNA sequence, whereas the repressor remained monomeric in the absence of this sequence (36). Recent work by Smith and coworkers (37) also indicated that stable gas-phase complexes can be observed between a bacteriophage T4 regA protein, a unique translational regulator, and RNA with target sequences.

In general, the interactions between proteins and nucleic acids are very sequence-specific. Specific ions of noncovalent protein and nucleotide complexes can be selectively dissociated by changing the solution conditions and by increasing the desolvation potential (38). This method of characterizing protein–nucleic acid interactions complements protein chemical approaches and other methods of structural determination.

VII. DRUG–NUCLEIC ACID COMPLEXES

With recent developments in genomics, more attention has been given to design novel drugs that bind to targeted oligonucleotides. Most bioanalytical techniques such as gel electrophoresis do not provide efficient means to detect the binding of small molecules to large oligonucleotide targets. ESI-MS has the advantage of superior mass accuracy and sensitivity and may be an ideal tool for determining drug-binding stoichiometry, binding affinities, and the nature of binding (39,40).

A. Antibiotic Binding to HIV RNA

In one study, Loo and coworkers (41) investigated the aminoglycoside antibiotic binding to HIV-1 RNA of the transactivation responsive element (TAR). Replication of the HIV requires the complexation of the viral transactivator protein (Tat) to TAR that is located at the 5′ end of mRNA. The Tat protein contains a C-terminal arginine-rich region that interacts specifically with a three-nucleotide pyrimidine bulge in TAR. Because the Tat–TAR complexation is critical for HIV replication, drugs that interfere with the interaction of Tat protein and TAR RNA may be potential inhibitors of HIV replication.

Members of the aminoglycoside antibiotics are known to bind to a wide variety of RNA molecules. Neomycin and streptomycin inhibit the formation of the Tat protein–TAR RNA complex. Loo and coworkers found that neomycin has a maximum binding stoichiometry of 3 to TAR RNA and of 2 to the Tat–TAR complex (41). A high affinity binding site of neomycin was found to be near the three-nucleotide bulge region of TAR RNA (41), which is consistent with previous solution-phase footprinting measurements. The ESI-MS competition binding experiments also indicated that neomycin has a higher affinity than streptomycin toward TAR RNA. The stability of TAR RNA–aminoglycoside complexes in the gas phase was found to be much higher than that of a noncovalent complex between TAR RNA and a small-molecule inhibitor (2,4,5,6-tetraaminoquinazoline) of TAR RNA, although this small-molecule inhibitor has a solution binding affinity similar to that of the aminoglycosides (41). Such differences in stability may be due to the different interactions in the noncovalent complexes. The small-molecule inhibitor interacts with TAR RNA via hydrophobic interactions, whereas the aminoglycosides bind to RNAs through electrostatic forces (41). The electrostatic forces are generally strengthened in a solvent-free environment, and therefore complexes with the electrostatic forces are very stable in the gas phase. This study clearly demonstrates the utility of ESI-MS in probing drug–RNA interactions.

B. Drug–DNA Interactions

The noncovalent interaction of small organic molecules with duplex DNA forms the molecular basis of many antitumor, antiviral, and antibiotic drugs. The DNA-binding drugs interact with duplex DNA in two principal ways: groove binding and intercalation. Most small molecules bind to the minor groove of B-DNA because of the stronger van der Waals contacts in this region. Those minor-groove ligands that are positively charged at physiological pH often prefer A and T sites simply because there is a negative electrostatic potential in the minor groove of the AT-rich region, whereas many intercalators prefer the G and C regions of DNA.

Gross and coworkers (42) used ESI-MS to rapidly determine the noncovalent binding of various drugs with DNA and to assess their relative affinities and stoichiometries. They used a set of self-complementary oligodeoxynucleotides that differ in length (6-mer to 12-mer), motif (GC-rich or AT-rich), and sequence. Two groups of drugs were tested to bind to DNA. Group I drugs (distamycin, Hoechst 33258, Hoechst 33342, berenil, and actinomycin D) are classic minor-groove binders and intercalators. Group II drugs (porphyrin H_2TMpyP-4, metalloporphyrin CuTMpyP-4, FeTMpyP-4, MnTMpyP-4, and [Ru(II) 12S4dppz]Cl$_2$) bind to DNA via mixed modes (groove binders and intercalators). The data confirmed the binding stoichiometry and showed preferred binding of minor-groove binders (distamycin, Hoechst 33258, Hoechst 33342, and berenil) to AT-rich oligomers and preferred interaction of the intercalator actinomycin D with GC-rich oligomers (42).

For group II drugs, the H_2TMpyP-4 and CuTMpyP-4 bind to DNA through mixed modes, whereas FeTMpyP-4 and MnTMpyP-4 interact with DNA by minor-groove binding. The competitive binding experiments indicated that the binding affinities with duplex 5'-CGCAAATTTGCG-3' in group

I decrease from Hoechst 33342, Hoechst 33258, distamycin to berenil, whereas the order of binding affinities for group II with duplex 5'ATATAT-3' is $H_2TMpyP-4 \approx CuTMpyP-4 > FeTMpyP-4 \approx MnTMpyP-4$ (42).

This work illustrates the strength of ESI-MS for the detection of noncovalent complexes of drugs and DNA with small amounts of compounds (less than 1 nmol per analysis was used in the experiments). More important, the binding mode can be readily established by examining the selectivities of test duplexes to the candidate drugs.

VIII. CONCLUSIONS

Detection of noncovalent complexes by ESI-MS has evolved into an important field in biological mass spectrometry. The superior sensitivity and rapid throughput afforded by ESI-MS offer advantages over other traditional methods in detecting noncovalent complexes. The stoichiometry, binding affinity, and nature of binding in noncovalent complexes can be readily obtained by ESI-MS. The significant improvement in MS instrumentation, such as ESI-TOF, is enhancing the experimental capability of measuring macromolecular assemblies. With well-refined ESI conditions and carefully designed experiments, more classes of noncovalent complexes can be explored by ESI-MS.

Future directions in the studies of noncovalent complexes by ESI-MS will include faster screening or identification of specific compounds that interact with targeted proteins or nucleic acids in pharmaceutical research. Continued and expanded studies of macromolecular assemblies by ESI-MS will contribute to a better understanding of protein–protein and protein–nucleic acid interactions, which in turn will provide answers to many biological processes such as receptor interactions and polynucleotide recognition.

REFERENCES

1. M Barbacid. Ras oncogenes: Their role in neoplasia. Eur J Clin Invest 20:225–235, 1990.
2. AG Taveras, J Deskus, J Chao, CJ Vaccaro, FG Njoroge, B Vibulbhan, P Pinto, S Remiszewski, JD Rosario, RJ Doll, C Alvarez, T Lalwani, AK Mallams, RR Rossman, A Afonso, VM Girijavallabhan, AK Ganguly, B Pramanik, L Heimark, WR Bishop, L Wang, P Kirschmeier, L James, D Carr, R Patton,

MS Bryant, AA Nomeir, M Liu. Identification of pharmacokinetically stable 3, 10-dibromo-8-chlorobenzocycloheptapyridine farnesyl protein transferase inhibitors with potent enzyme and cellular activities. J Med Chem 42:2651–2661, 1999.
3. JB Fenn, M Mann, CK Meng, SF Wong, CM Whitehouse. Electrospray ionization for mass spectrometry of large biomolecules. Science 246:64–71, 1989.
4. RG Cooks, G Chen, P Wong, H Wollnik. Mass spectrometers. In: GL Trigg, Ed. Encyclopedia of Applied Physics. New York: VCH, 1997, Vol 19, pp 289–330.
5. M Karas, U Bahr. Matrix-assisted laser desorption-ionization (MALDI) mass spectrometry: Principles and applications. In: RM Caprioli, A Malorni, G Sindona, eds. Selected Topics in Mass Spectrometry in the Biomolecular Sciences. Amsterdam, The Netherlands: Kluwer Academic, 1997, pp 33–53.
6. F Hillenkamp, M Karas, RC Beavis, BT Chait. Matrix-assisted laser desorption/ionization mass spectrometry of biopolyners. Anal Chem 63:1193A–1203A, 1991.
7. S Berkenkamp, F Kirpekar, F Hillenkamp. Infrared MALDI mass spectrometry of large nucleic acids. Science 281:260–262, 1998.
8. TB Farmer, RM Caprioli. Determination of protein-protein interactions by matrix-assisted laser desorption/ionization mass spectrometry. J Mass Spectrom 33:697–704, 1998.
9. T Vogl, J Roth, C Sorg, F Hillenkamp, K Strupat. Calcium-induced noncovalent linked tetramers of MRP8 and MRP14 detected by ultraviolet matrix-assisted laser desorption/ionization mass spectrometry. J Am Soc Mass Spectrom 10:1124–1130, 1999.
10. B Ganem, YT Li, JD Henion. Detection of non-covalent receptor-ligand complexes by mass spectrometry. J Am Chem Soc 113:6294–6296, 1991.
11. V Katta, BT Chait. Observation of the heme-globin complex in native myoglobin by electrospray-ionization mass spectrometry. J Am Chem Soc 113:8534–8535, 1991.
12. M Baca, SBH Kent. Direct observation of a ternary complex between the dimeric enzyme HIV-1 protease and a substrate-based inhibitor. J Am Chem Soc 114:3992–3993, 1992.
13. JA Loo. Studying of non-covalent complexes by electrospray ionzation mass spectrometry. Mass Spectrom Rev 16:1–23, 1997.
14. RL Winston, MC Fitzgerald. Mass spectrometry as a readout of protein structure and function. Mass Spectrom Rev 16:165–179, 1997.
15. BN Pramanik, PL Bartner, UA Mirza, YH Liu, AK Ganguly. Electrospray ionization mass spectrometry for the study of non-covalent complexes: An emerging technology. J Mass Spectrom 33:911–920, 1998.
16. AM Last, CV Robinson. Protein folding and interactions revealed by mass spectrometry. Curr Opin Chem Biol 3:564–570, 1999.
17. N Xu, L Pasa-Tolic, RD Smith, S Ni, BD Thrall. Electrospray ionization-mass spectrometry study of the interaction of cisplatin-adducted oligonucleotides

with human XPA minimal binding domain protein. Anal Biochem 272:26–33, 1999.
18. KP Madhusudanan, SB Katti, W Haq, PK Misra. Antisense peptide interactions studied by electrospray ionization mass spectrometry. J Mass Spectrom 35:237–241, 2000.
19. M Bartok, PT Szabo, T Bartok, G Szollosi. Identification of ethyl pyruvate and dihydrocinchonidine adducts by electrospray ionization mass spectrometry. Rapid Commun Mass Spectrom 14:509–514, 2000.
20. Y Wang, M Schubert, A Ingendoh, J Franzen. Analysis of non-covalent protein complexes up to 290 kDa using electrospray ionization and ion trap mass spectrometry. Rapid Commun Mass Spectrom 14:12–17, 2000.
21. AK Ganguly, YS Wang, BN Pramanik, RJ Doll, ME Snow, AG Taveras, S Remiszewski, D Cesarz, JD Rosario, B Vibulbhan, JE Brown, P Kirschmeier, EC Huang, L Heimark, A Tsarbopoulos, VM Girijavallabhan, RM Aust, EL Brown, DM Delisle, SA Fuhrman, TF Hendrickson, CR Kissinger, RA Love, WA Sisson, JE Villafranca, SE Webber. Interaction of a novel GDP exchange inhibitor with the ras protein. Biochemistry 37:15631–15637, 1998.
22. AK Ganguly, BN Pramanik, EC Huang, S Liberles, L Heimark, YH Liu, A Tsarbopoulos, RJ Doll, AG Taveras, S Remiszewski, ME Snow, YS Wang, B Vibulbhan, D Cesarz, JE Brown, J del Rosario, L James, P Kirschmeier, V Girijavallabhan. Detection and structural characterization of ras oncoprotein-inhibitors complexes by electrospray mass spectrometry. Bioorg Med Chem 5:817–820, 1997.
23. N Potier, P Barth, D Tritsch, JF Biellmann, AA Van Dorsselaer. Study of non-covalent enzyme-inhibitor complexes of aldose reductase by electrospray mass spectrometry. Eur J Biochem 243:274–282, 1997.
24. RF Steiner, S Albaugh, C Fenselau, C Murphy, M Vestling. A mass spectrometry method for mapping the interface topography of interacting proteins, illustrated by the melittin-calmodulin system. Anal Biochem 196:120–125, 1991.
25. OV Nemirovskiy, R Ramanathan, ML Gross. Investigation of calcium-induced, noncovalent association of calmodulin with melittin by electrospray ionization mass spectrometry. J Am Soc Mass Spectrom 8:809–812, 1997.
26. TD Veenstra, AJ Tomlinson, L Benson, R Kumar, S Naylor. Low temperature aqueous electrospray ionization mass spectrometry of noncovalent complexes. J Am Soc Mass Spectrom 9:580–584, 1998.
27. UA Mirza, YH Liu, L Ramanathan, L Heimark, BN Pramanik, B Malcolm, P Weber, AK Ganguly. Characterization of hepatitis C virus by mass spectrometry: detection of protein-Zn complex. Presented at the 46th ASMS Conf on Mass Spectrometry and Allied Topics, Orlando, FL, 1998.
28. YT Li, YL Hsieh, JD Henion, MW Senko, FW McLafferty, B Ganem. Mass spectrometric studies on noncovalent dimers of leucine zipper peptides. J Am Chem Soc 115:8409–8413, 1993.
29. EC Huang, BN Pramanik, A Tsarbopoulos, P Reichert, AK Ganguly, PP Trotta, TL Nagabhushan. Application of electrospray mass spectrometry in probing

protein-protein and protein-ligand non-covalent interactions. J Am Soc Mass Spectrom 4:624–626, 1993.
30. AN Krutchinsky, IV Chernushevich, VL Spicer, W Ens, KG Standing. Collisional damping interface for an electrospray ionization time-of-flight mass spectrometer. J Am Soc Mass Spectrom 9:569–579, 1998.
31. BN Green, RS Bordoli, LG Hanin, FH Lallier, A Toulmond, SN Vinogradov. Electrospray ionization mass spectrometric determination of the molecular mass of the approximately 200-kDa globin dodecamer subassemblies in hexagonal bilayer hemoglobins. J Biol Chem 274:28206–28212, 1999.
32. AA Rostom, CV Robinson. Detection of the intact GroEL chaperonin assembly by mass spectrometry. J Am Chem Soc 121:4718–4719, 1999.
33. MA Tito, K Tars, K Valegard, J Hajdu, C V Robinson. Electrospray time of flight mass spectrometry of the intact MS2 virus capsid. J Am Chem Soc 122:3550, 2000.
34. X Cheng, AC Harms, PN Goudreau, TC Terwilliger, RD Smith. Direct measurment of oligonucleotide binding stoichiometry of gene V proteins by mass spectrometry. Proc Natl Acad Sci USA 93:7022–7027, 1996.
35. X Cheng, PE Morin, AC Harms, JE Bruce, Y Ben-David, RD Smith. Mass spectrometric characterization of sequence-specific complexes of DNA and transcription factor OU.1 DNA binding domain. Anal Biochem 239:35–40, 1996.
36. N Potier, LJ Donald, I Chernushevich, A Ayed, W Ens, CH Arrowsmith, KG Standing, HW Duckworth. Study of a noncovalent trp repressor: DNA operator complex by electrospray ionization time-of-flight mass spectrometry. Protein Sci 7:1388–1395, 1998.
37. C Liu, LP Tolic, SA Hofstadler, AC Harms, RD Smith, C Kang, N Sinha. Probing RegA/RNA interactions using electrospray ionization-Fourier transform ion cyclotron resonance-mass spectrometry. Anal Biochem 262:67–76, 1998.
38. M Przybylski, J Kast, MO Glocker, E Durr, HR Bosshard, S Nock, M Sprinzl. Mass Spectrometric approaches to molecular characterization of protein-nucleic acid interactions. Toxicol Lett 82–83:567–575, 1995.
39. A Kapur, JL Beck, MM Sheil. Observation of daunomycin and nogalamycin complexes with duplex DNA using electrospray ionization mass spectrometry. Rapid Commun Mass Spectrom 13:2489–2497, 1999.
40. V Gabelica, E De Pauw, F Rosu. Interaction between antitumor drugs and a double-stranded oligonucleotide studied by electrospray ionization mass spectrometry. J Mass Spectrom 34:1328–1337, 1999.
41. KA Sannes-Lowery, H-Y Mei, JA Loo. Studying aminoglycoside antibiotic binding to HIV-1 TAR RNA by electrospray ionization mass spectrometry. Int J Mass Spectrom 193:115–122, 1999.
42. KX Wan, T Shibue, ML Gross. Non-covalent complexes between DNA-binding drugs and double-stranded oligodeoxynucleotides: A study by ESI ion-trap mass spectrometry. J Am Chem Soc 122:300–307, 2000.

10
Application of Hydrogen Exchange and Electrospray Ionization Mass Spectrometry in Studies of Protein Structure and Dynamics

Yinsheng Wang and Michael L. Gross
Washington University, St. Louis, Missouri

I. INTRODUCTION

Development of electrospray ionization (ESI) (1) and matrix-assisted laser desorption ionization (MALDI) (2) makes mass spectrometry (MS) a sensitive technique for sequencing peptides and proteins. But can mass spectrometry play a role in determining the higher order structure and function of proteins, for which there is fundamental interest in their structure and conformational dynamics in solution? Additional and related questions arise. How does a mutation change a protein's conformation? How does a protein fold to its stable conformation? What are the folding intermediates and their populations, and what is the energy profile leading to those intermediates and from there to the folded conformation? Can these questions be answered by mass spectrometric measurements?

Proteins function through interactions with themselves, with small ligands, or with other proteins and DNA. A set of second-generation questions focus on which part of a protein is involved in the interactions and how the conformation of the protein changes in response to those interactions.

Recent advances in combinatorial approaches for constructing native proteins and in recombinant technologies for generating protein and peptide drugs also provide incentive for the development of efficient and dependable screening methods aimed at understanding protein structure and interaction. Furthermore, advances in proteomics will require methods not only to identify proteins but also to provide insight into their conformations and interactions. Efficiency in this pursuit will require methods to give broad, low resolution pictures of a protein's conformation and interactions before it is submitted to detailed characterization by biochemical and biophysical methods such as X-ray crystallography and nuclear magnetic resonance (NMR).

Protection from hydrogen exchange at amide linkages is a characteristic feature of a folded protein's structure (3,4), and this is the property that can be probed by hydrogen–deuterium (H/D) exchange and mass spectrometry. Although H/D exchange is usually followed with NMR, the rates and extent of H/D exchange can also be monitored by using mass spectrometry (MS), as has been pioneered principally by Smith's group at the University of Nebraska. Because the sensitivity of modern-day mass spectrometry is high, its combination with hydrogen exchange offers the opportunity to rapidly explore high order structure and interactions for proteins whose concentrations in solution are 1000 times or more dilute than those that can be probed by NMR.

Because the study of hydrogen exchange of proteins by MS has been reviewed elsewhere (5–7), we focus in this chapter on those applications of ESI-MS and hydrogen exchange that were described mainly in articles published since the last review appeared. Although MALDI MS has also been applied in hydrogen exchange studies of proteins (8–12), a review of this approach is not consistent with the aims of a monograph devoted to ESI-MS.

II. METHODS AND EXPERIMENTAL DESIGN

A. Background for Hydrogen Exchange of Proteins

Exchangeable hydrogens on a protein can be divided into two groups based on their exchange rates. Hydrogens on the amino terminus or on certain amino acid side chains (on lysine, aspartic acid, serine, etc.) exchange very rapidly, whereas those on the amide linkages exchange slowly. The rate of exchange of an amide hydrogen in random-chain polypeptides can be calculated if one considers the steric and inductive effects of neighboring amino acid side chains (13,14). Different side chains can affect exchange rates by over an order of magnitude. Furthermore, the exchange can be either acid- or base-catalyzed:

$$k_{ex} = k_H[H^+] + k_{OH}[OH^-] \tag{1}$$

where k_H and k_{OH} are the rate constants for the acid- and base-catalyzed reactions, respectively. Under conditions where the exchange rate is small, water also catalyzes the exchange reaction (14). Studies of polyalanines showed that the exchange rate is slowest when the pH is approximately 3. When the pH is greater than 3, the reaction is base-catalyzed, and the rate will increase tenfold with an increase of one pH unit (14). The rate is also temperature-dependent and increases by two- to threefold for every 10 °C temperature increase. These effects show that it is essential to control carefully the pH and temperature of the exchange reaction. Furthermore, they can be exploited to "quench" the exchange, exchange back the deuterons on active side chains with protons, and admit the quenched solution into a mass spectrometer to determine the molecular weight and the extent of exchange of the protein.

For a structured protein, there are two mechanisms for hydrogen exchange. According to the first mechanism, amide hydrogens in a folded protein exchange directly without the protein unfolding:

$$F(H) \rightleftharpoons F(D) \tag{2}$$

For the second mechanism, the amide hydrogens do not exchange until part of the protein unfolds (local unfolding) or the whole protein unfolds (global unfolding):

$$F(H) \underset{k_{-1}}{\overset{k_1}{\rightleftharpoons}} U(H) \overset{k_2}{\longrightarrow} U(D) \underset{k_1}{\overset{k_{-1}}{\rightleftharpoons}} F(D) \tag{3}$$

In Eqs. (2) and (3), $F(.)$ and $U(.)$ designate the folded and unfolded forms, respectively, and H and D represent the unexchanged and exchanged forms, respectively. The rate constants k_1, k_{-1}, and k_2 are for the reactions corresponding to unfolding, refolding, and H/D exchange, respectively.

The amide hydrogen atoms that are in the core of a folded protein or that are involved in intramolecular hydrogen bonding are slower to exchange than those that are exposed to solvent. The hydrogen-exchange rate of an amide hydrogen in a structured protein can be 10^8 times slower than that of an amide located in a random-chain polypeptide (7). As a result, amide hydrogen atoms on a folded protein can act as a continuous sensor extending along the entire protein backbone, providing a basis for using hydrogen exchange as a tool for studying protein structure and dynamics.

If the unfolded state is stabilized by urea or other denaturing conditions, $k_2 \gg k_{-1}$, the rate of refolding (k_{-1}) may be determined by measuring the exchange rate (k_{meas}):

$$k_{meas} = k_{-1} \qquad (4)$$

The kinetic measurement is made by incubating a protein in the presence of denaturant and D_2O followed by quenching at various time points and by measuring the extent of deuterium incorporation by ESI-MS. The protein exposure time in D_2O depends on whether continuous or pulse labeling is used (see Sect. III.B.2 for details). The exchange mechanism here is designated as EX1. On the other hand, if refolding rates are much faster than exchange rates, the measured rate of exchange is second order and termed EX2:

$$k_{meas} = K_{unf} k_2 \qquad (5)$$

where k_{meas}, k_2, and K_{unf} are the measured rate constant for exchange, the rate constant for random chain polypeptides, and the equilibrium constant for the folding/unfolding reaction, respectively (15).

For the EX2 mechanism, Eq. (6) connects measurable hydrogen exchange rates with structural stability, which is represented by K_{unf} and therefore with the structural stabilization free energy ($\Delta G°$), as shown in the equation (16)

$$\Delta G° = -RT \ln K_{unf} = -RT \ln(k_{meas}/k_2) \qquad (6)$$

When protein folding occurs by a two-state mechanism (no intermediates), the free energy changes for unfolding at different concentrations of denaturant may be determined. Those free energy changes are given the equation

$$\Delta G°(Den) = \Delta G°(0) - m[Den] \qquad (7)$$

where $\Delta G°$ (Den) is the structural stabilization energy at a certain denaturant concentration (17). If the equation applies, then the structural stabilization free energy at zero denaturant concentration, $\Delta G°(0)$, is the intercept of a plot of $\Delta G°$(Den) vs. denaturant concentration, [Den]. The slope m of the curve may be viewed as a measure of the ability of a denaturant to unfold a protein.

B. Experimental Methods for Measuring H/D Exchange by MS

The most straightforward measurement for mass spectrometry is that of the molecular weight. The difference between the molecular weight of the exchanged protein and that of the native protein is the number of acidic hydrogens that have undergone exchange. The extent of exchange in the protein offers a global view of the solvent accessibility of the protein. A schematic overview of the method is shown in Fig. 1. The exchange-in is usually done at room temperature by diluting a stock solution of protein with a buffered D_2O solution at a pH of approximately 7. After incubation for a certain time, adding a buffer at pH 2.5–3.0 and decreasing the temperature to 0°C quench the reaction. Under the quench conditions, the half-life for hydrogen exchange at a peptide amide linkage is increased to 10–40 min (14,18). Pepsin digestion may then be used if the protein is to be fragmented, as discussed in the next section. Before analysis by ESI-MS, the sample is often desalted by HPLC with a 5 min wash with 5–90% CH_3CN in 0.05% trifluoroacetic acid.

C. Protein Fragmentation Coupled with Mass Spectrometry

The approach to obtain higher resolution data at the peptide level of a protein is to fragment the labeled protein into its constituent peptides and measure the extent of exchange for each peptide (19). A detailed description of this approach is given in the review by Smith et al. (7). This approach was originally developed by Rosa and Richards (20) and modified by Englander et al. (18), who used acid proteases to fragment proteins labeled with tritium. The approach was later adapted for measurement of hydrogen–deuterium (H/D) exchange by MS (19). Referring to Fig. 1, we see that following the quench the protein is digested by pepsin, and the peptides that result from the proteolysis are separated by HPLC and analyzed by mass spectrometry. Although pepsin is a nonspecific protease, its cleavage sites are reproducible for a given protein if the digestion conditions are controlled to be nearly identical from experiment to experiment.

The important information obtained by the protein fragmentation-MS approach is the determination of the deuterium content of a particular portion of the protein after the entire protein in D_2O has been incubated for a certain time and the portion of interest has been produced by enzymatic digestion. To determine the extent of exchange in each individual peptide, the peptides generated by pepsin digestion must be separated and identified, and this is achieved by HPLC

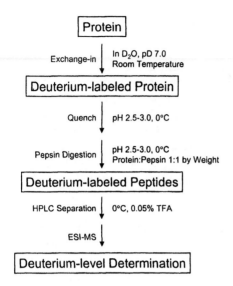

Figure 1 Schematic diagram of the protein fragmentation–mass spectrometry approach.

and tandem mass spectrometry (MS/MS) (21) or by carboxy-terminal sequencing (22) if a low resolution mass spectrometer is used. Separation can be avoided by using a simple strategy of measuring the molecular masses of peptides of ^{13}C, ^{15}N doubly depleted protein and a mass spectrometer capable of exact-mass measurements (e.g., Fourier transform ion cyclotron resonance, FTICR) (23–25). The sequence coverage varies from 80% to 100%, depending on the size of the protein and whether a high resolving power mass spectrometer is used (23,26–29).

Deuterium loss during pepsin digestion and HPLC separation can be adjusted by including in the study reference samples that are 0% or 100% deuterated. These peptides are submitted to the same protocol as the peptides from the exchanged protein. Zhang and Smith (19) showed the detailed procedure and the evaluation of the adjustment procedures.

D. Determination of the Rate Constant for Deuterium Incorporation

Two methods are commonly used for the determination of a rate constant for deuterium incorporation. One is to use a nonlinear least squares fitting routine for the time-dependent exchange data, a procedure that was first implemented by Zhang and Smith (19):

$$[N - D] = \sum_{i=1}^{N} e^{-k_i t} \tag{8}$$

The term D is the deuterium level of a peptide with N amide linkages following incubation of the intact protein in D_2O for time t, k_i is the pseudo first-order rate constant for isotopic exchange at an individual amide linkage. The analysis can lead to the determination of the distribution of exchange rate constants of amide hydrogen atoms in individual peptide segments.

A second approach is to employ a maximum entropy method (MEM) analysis to calculate the most probable rate-constant distributions for hydrogen exchange. An algorithm for applying MEM to analyze progress curves of exchange-in was devised by Zhang et al. (30), a method that maximizes the entropy (S):

$$S = -\sum f_k \left[\ln\left(\frac{f_k}{A}\right) - 1 \right] \tag{9}$$

The term f_k is the probability for obtaining rate constant k, and A is the most probable value of f_k. Entropy maximization is subject to the constraint

$$X^2 = \sum_t \frac{(D_t^{cal} - D_t^{exp})^2}{\sigma_t^2} \tag{10}$$

where D_t^{exp} is the observed deuterium incorporation at time t, D_t^{cal} is the deuterium incorporation at time t calculated from the sum of exponentials, and σ_t is the standard deviation of the extent of deuteration at time t. Because MEM inherently incorporates experimental errors, it provides a more unbiased result, thereby avoiding over- or underinterpretation of the data (30). A comparison of the two approaches showed that they gave similar results (Fig. 2) (28).

III. APPLICATIONS

In this section we discuss some representative applications of hydrogen exchange and MS in studies of protein–ligand interactions, protein folding and dynamics, the screening of recombinant protein drugs, and the screening of proteins from recombinant libraries. In that way, we present an overview of current applications and discuss advances aimed at determining the extent of exchange at single-amide resolution. The attribution of an advance to a certain section in the organization is somewhat arbitrary because some studies involve more than one application.

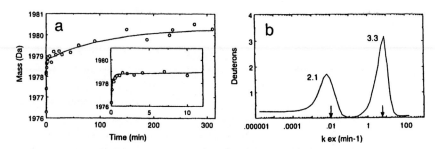

Figure 2 Results from the nonlinear least squares (NLSQ) and maximum entropy method (MEM) analyses for peptide 14, which contains residues 111–126. (a) Progress curves for in-exchange, showing best fits from NLSQ. (b) MEM analysis showing distribution of rate constants, with corresponding values of the number of amides and rate constants from NLSQ indicated by numbers and arrows, respectively. Note that the areas under the MEM peaks approximate the number of deuterons in each rate class determined by NLSQ. (Adapted, with permission, from Fig. 4 of Ref. 28.)

A. Protein–Ligand and Protein–Metal Ion Interaction

Many proteins function through their interaction with ligands. These ligands can be metal ions, small substrates, or peptides. Important information such as the conformational change upon ligand binding and the location of the ligand-binding site can be inferred from hydrogen exchange studies using MS.

1. Small-Ion Interaction with Proteins

Hydrogen–deuterium exchange combined with mass spectrometry can be used to follow the conformational changes induced by the binding of a metal ion to a protein. We discuss briefly examples in which three Ca^{2+}-binding proteins have been studied. Calmodulin (CaM), as one example, binds with Ca^{2+}. The Ca^{2+}, which acts as a messenger, causes the conformation of calmodulin to change to one that can interact with target enzymes. Although the crystal structure of Ca^{2+}-loaded calmodulin is known, the corresponding structure of the apo-calmodulin has not been solved. Therefore, it was of interest and of some importance to measure the extent of hydrogen exchange as the apoprotein interacts with Ca^{2+} and assumes a new conformation.

To follow this process, Gross and coworkers (31) implemented a titration with Ca^{2+} to monitor deuterium uptake as the Ca^{2+} concentration increases. The titration reveals that the extent of exchange decreases as the Ca^{2+} concentration increases, and the amount of incorporated deuterium reaches a minimum when the CaM-4 Ca^{2+} is at least 85% of all forms of the Ca^{2+}–calmodulin complex. Titration experiments of this type not only provide a quantitative assessment of the stoichiometry of protein–ligand complexes but also reveal, in a global way, protein folding as a function of metal ion binding.

The studies of the refolding of bovine lactalbumin (BLA), a Ca^{2+}-binding protein involved in lactose transport (32), showed that the rate of refolding is faster by more than two orders of magnitude in the presence of Ca^{2+} than in its absence (33). The extent of hydrogen exchange protection observed in partially structured species in the presence of Ca^{2+} is consistently higher than in the corresponding species in the absence of Ca^{2+}. This suggests that the molten globule state, which forms rapidly after initiation of folding in the presence of Ca^{2+}, contains a somewhat more persistent structure than the one that forms in the absence of Ca^{2+} (33).

Another Ca^{2+}-binding protein studied by H/D exchange is human cardiac troponin C, which plays a role in cardiac muscle contraction. Using solution-phase H/D exchange coupled with pepsin digestion, HPLC and ESI-Fourier transform ion cyclotron resonance (FTICR) MS, Marshall and coworkers (25) detected small conformational changes induced by Ca^{2+} binding. The changes are so small that they were difficult to detect by NMR (34–36).

Using a similar approach, Marshall and coworkers (23) studied another protein, *Yersinia* protein tyrosine phosphatase (PTPase), and its binding with vanadate, which is a competitive inhibitor for PTPase. Vanadate inhibits by acting as a transition state analog for the PTPase-catalyzed reaction because when bound it adopts a pentavalent geometry. Binding to vanadate causes the rate of H/D exchange to decrease throughout the whole protein. The presence of vanadate especially protects H/D exchange in residues that are in direct contact with the oxyanion and in residues the stabilize the active site structure.

Two studies (23,25) clearly demonstrate the advantages of FTICR MS for measurement of the rates of H/D exchange. Exploiting the high mass accuracy offered by FTICR and using a $^{13}C,^{15}N$ isotopically doubly depleted protein, Marshall and coworkers (23,25) assigned most proteolytic fragments from the protein by peptide masses alone. The spectrometer allows the number of deuterons incorporated to be measured with an accuracy of 0.5 Da, or within ~3% (23). Because FTICR MS has high mass resolving power, isotopic peaks from all the proteolytic fragments are well resolved, and this separation of peaks facilitates assigning the charge state and distinguishing between barely separated or overlapped isotopic distributions from different proteolytic fragments (23).

2. Protein– and Cofactor–Substrate Interaction

Recent studies show that both the substrate-binding region and the conformational changes induced by substrate binding can be identified by H/D exchange and mass spectrometry. Two proteins, diaminopimelate dehydrogenase and dihydropicolinate reductase, are involved in L-lysine biosynthesis and have NAD(P)H as a cofactor. Hydrogen–deuterium exchange combined with pepsin digestion, HPLC, and ESI-MS identified several peptides whose amide hydrogen exchange rate decreased upon substrate and cofactor binding. Some of those peptides are from regions that probably bind substrate and cofactor. Others are located at the interdomain hinge region and may be exchangeable in the "open," catalytically inactive conformation but cannot be exchanged in the "closed," catalytically active conformation after substrate binding and domain closure occur (26,27). The results are consistent with the crystal structures of the enzyme–cofactor complexes (37,38).

3. Protein–Peptide Interaction

Src-homolog-2 (SH2) domains are small, approximately 100 amino acid protein modules that are present in a number of signal transduction proteins (39,40). H/D exchange combined with pepsin digestion, HPLC, and MS showed that binding to a 12-residue, high affinity peptide decreased hydrogen exchange along much of the SH2 backbone. The result indicated that the peptide binding stabilizes motions within the SH2 domain (29).

Hydrogen–deuterium exchange, in combination with protein fragmentation and MS, was also applied to study the interaction of insulin-like growth factor I (IGF-I) and IGF-I binding protein 1. In this study, the focus was on the substrate and the locations of the hydrophobic core of IGF-1. Two regions that are involved in binding to IGF-I binding protein 1 were identified. The outcome is consistent with the interpretation from other approaches (41). The use of a short column (5 cm in length) for desalting and immobilizing the pepsin onto a stainless steel column is worth noting, because it brings automation and reproducibility to the protein fragmentation method.

The various regions of a protein involved in interaction and changes in conformation can be identified by H/D exchange using MS. But without prior knowledge of the interaction, it is difficult to distinguish the two and thereby identify the substrate-binding region.

B. Protein Folding and Dynamics

The folding of a protein to its functional state from information encoded in the amino acid sequence is key to the conversion of genetic information into biological activity. In recent years, MS studies of H/D exchange have provided valuable information about the conformations of native protein, and studies of the refolding of proteins in vitro after chemical denaturation have begun to give insight into the nature of this complex process. In this section we discuss the application of H/D exchange and mass spectrometry to studies comparing the natural conformations of wild-type and mutant proteins and to studies of the folding dynamics of proteins.

1. Natural Conformation

The FK506-binding protein (FKBP) is a 107 amino acid protein that exhibits peptidyl-prolyl *cis-trans*-isomerase activity in vitro. The rates for H/D exchange of individual peptic peptides of recombinant (C22A) human FKBP were obtained by applying the maximum entropy method to the hydrogen-exchange kinetic curve and were compared with those obtained by NMR. The rate constant distributions determined by the two methods agree well, and the methods are complementary. MS is more sensitive, but NMR is more capable of determining exchange rates for individual amides (30).

Hydrogen-deuterium exchange and ESI-MS were also successfully applied to differentiate the folding of wild-type and mutant proteins. Transthyretin (TTR) is a plasma protein involved in the binding of thyroid hormones and vitamin A. Wild-type TTR forms an amyloid in a disease known as senile systematic amyloidosis (SSA) (42), and variant TTR molecules aggregate to form fibrils in familial amyloidotic polyneuropathy (43). H/D exchange measurements by nanoflow ESI-MS produce data that allow the comparison of the solution structure of Val30Met variant with that of wild-type TTR at a pH of 4.5 and a protein concentration at 0.09 μM (0.005 mg/mL) (44). The concentration is much too low for contemporary NMR to be applied. At these dilute concentrations, both proteins are monomeric and populate amyloidogenic intermediates (45,46). Of 118 backbone amide hydrogen groups, 55 in the wild type were protected from exchange, whereas only 20 in the Val30Met variant were protected, suggesting partial unfolding of the β-sheet structure in Val30Met. These results provide not only new insights into the correlation between tetramer stability and amyloidogenicity but also support for a possible route to fibril formation and transient unfolding of the TTR monomer (44).

To establish and validate the mass spectrometric method, many early MS studies were directed to proteins whose structures had already been determined by X-ray crystallography or NMR. Backbone conformation can sometimes be approximated by sequence alignment to proteins with a similar function and for which a 3-D structure is available. As an example, H/D exchange and MS were used to obtain information about the solution structures of wild-type and constitutively active mutants of MAP kinase kinase 1, for which the backbone conformation was obtained by aligning to cAMP-dependent kinase (47).

Compared with the wild-type protein, three peptides from the mutant protein showed decreased exchange, whereas nine showed increased exchange. The decreases are consistent with electrostriction or reduced solvent access owing to domain closure or the formation of new hydrogen or salt bonds around the catalytic cleft and within the activation lip.

2. Unfolding Dynamics

An unfolded protein shows several distinct features in its ESI mass spectrum. One is charge state distribution; the unfolded state accommodates more charges than the folded state (5). For a protein that is partially folded, the folded portion will display lower charge states, whereas the unfolded portion will display higher charge states, reflecting these differences in a bimodal charge state distribution. For example, cellular retinoic acid–binding protein I (CRABP I) in aqueous solution and in 10% ethanol gives a bimodal distribution of charge states, including a narrow (+11 to +14) distribution (48,49). Thermal denaturation studies show that a transition of charge states to a bimodal distribution occurred in the same temperature range as the transition of the H/D exchange mechanism of *E. coli* thioredoxin from EX2 to EX1 (50). This result is consistent with those from CD measurements in the near-UV.

The mass spectral peak width for a protein molecular ion produced by ESI is another indicator for the folding state of a protein. Narrower peak widths at half-height show a smaller distribution of exchanged species, and this usually means that there is a smaller distribution of conformers. Therefore, a locally unfolded protein can be differentiated from a globally unfolded protein on the basis of peak width; the partially unfolded species gives a narrower peak width (51).

Another, more important feature of an ESI mass spectrum that reflects a protein conformation is the mass profile. For samples with homologous structures, the extent of H/D exchange in each segment is the same for all molecules, and as a result a single envelope of isotopic peaks representing the random distribution of deuterium will be observed for the segment. For samples that are heterogeneous with respect to structure, however, one single segment may be folded in some molecules but remain unfolded in others. As a result, bimodal isotope patterns (i.e., two isotopic envelopes) will be seen. An example (Fig. 3) is provided by the mass profiles of a doubly charged peptide at different refolding times (52). The deuterium levels in the unfolded populations of the peptide are the same as those of the 100% deuterated reference sample (Fig. 3f), showing that exchange of amide peptides is complete in 10 s.

Rabbit muscle aldolase is a homotetramer in which each subunit adopts an α/β barrel structure, where β strands are connected with α-helices (53).

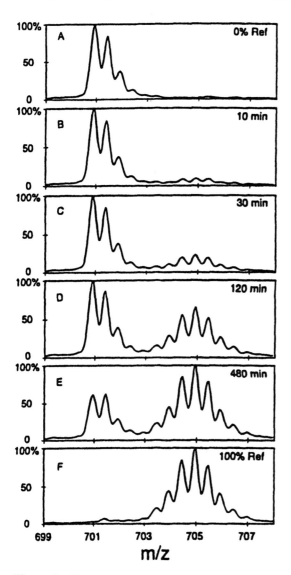

Figure 3 Electrospray ionization mass spectra of doubly charged peptide comprising residues 257–269 of rabbit muscle aldolase, which was incubated in 3 M urea–H_2O for 10–480 min, then pulse labeled for 10 s in 3 M urea–D_2O (panels B–E). Mass spectra of the same segment of aldolase that contained no deuterium or was completely exchanged in D_2O are presented in panels A and F, respectively. (Taken, with permission, from Fig. 1 of Ref. 52.)

Smith and coworkers (52,54–56) conducted systematic studies of the refolding of this relatively large protein (158,000 Da), in which the unfolding is not yet amenable to study by NMR. Some peptides of the aldolase had undergone either full or no exchange (i.e., no population of partial exchange was seen). This observation is direct evidence that the regions of all aldolase molecules are either completely folded or completely unfolded, and no intermediate states exist (57), a clear indication that folding of the protein is cooperative.

The rate constants for unfolding of all the peptides in aldolase were established after pepsin digestion by three-component data fitting, and the peptides were grouped by their unfolding rate constants: Peptides with similar unfolding rates were classified as having single, unfolding domains. Three unfolding domains of rabbit muscle aldolase, in 3 M urea, were identified, and the rate constants were determined to be 0.10, 0.036, and 0.0064 min^{-1}, respectively (52).

The analysis of those unfolding domains showed that multiple noncontiguous segments of the backbone may constitute an unfolding domain and different regions of a helix may belong to different unfolding domains. Three peptides (with residues 204–206, 166–170, and 251–255), which are in three helices in the subunit-binding surfaces, showed folding lifetimes that are longer than 50 h in 3.5 M urea. The result is consistent with size-exclusion chromatography studies, which show that aldolase exists as a tetramer even in the presence of 4 M urea (52,56).

The unfolding domains of aldolase can also be identified from studies of the intact protein. On the basis of the analysis of the $[M + 47 H]^{47+}$ ion, four species with distinct isotopic envelopes were identified in the refolding. The average mass determined from the peak envelopes representing individual folding intermediates and the 92% recovery of deuterium showed that each unfolding domain consisted of approximately 107 residues (56).

With the bimodal mass profiles in hand, the fractions of the folded molecule and the unfolding intermediates can be calculated by using the peak areas of the envelopes corresponding to the folded conformer, unfolding intermediates, and unfolded conformer.

Not only can the folding intermediates and their populations be determined, but also a refolding model can be proposed from the H/D exchange results obtained by ESI-MS/MS. The folding of a protein can be either cooperative, in which two or three domains in the same molecule unfold and refold simultaneously, or sequential. Although the structures for the intermediates in unfolding and refolding inherent to those two models are identical, kinetics of the two models will lead to significantly different populations of folding intermediates. Those different populations can be determined by measuring the peak areas corresponding to the intermediates in the ESI mass spectrum.

Therefore, the folding model can be identified. Deng et al. (56) applied this method for the refolding of aldolase in urea and determined that the folding occurs in a sequential manner.

The study of protein unfolding under different denaturing conditions can also provide useful information. Again using aldolase as a model system, Deng and Smith (54) studied unfolding at different concentrations of urea. The results show that there is little change in deuterium levels of the unfolding intermediates; therefore, the concentration of urea does not have much effect on the number of residues composing each of the unfolding domains. Major changes, however, occur in the fractional populations of those different structural forms (54). Deng and Smith (54) further used those different populations to identify new solvent-accessible surface areas that are exposed upon unfolding of the protein. A measure of the new surface area is given by the *m* value or slope of the plot made in accordance with Eq (7) (17).

Smith and coworkers (55) compared continuous and pulsed labeling for the study of the unfolding of aldolase. The procedure they used for pulse labeling differs from the one for continuous labeling in that the protein is incubated in H_2O–urea instead of D_2O–urea for various times. Then, in pulse labeling, the solution is exposed to D_2O–urea (20-fold dilution) for 10 s to allow the protein to exchange before quenching, whereas in continuous labeling it is quenched immediately. During the incubation in the presence of urea, the population of unfolded forms increases until it reaches equilibrium. The equilibrium between folding and unfolding occurs in the presence of D_2O in continuous labeling but in the presence of H_2O in pulse labeling. A rapid-mixing device is used for pulse labeling (58).

The exposure time, which depends on pH, temperature, and the amino acid sequence, can be estimated by using parameters derived from studies of model peptides (14). The minimum exposure time and the choice of labeling methods depend on the time needed to complete the exchange of all amide hydrogen atoms in the unfolded structure (14). The incubation time, however, should be sufficiently short so that only some of the amide hydrogens in the folded structure exchange completely during the exposure time (55). Smith and coworkers (55) identified only three or four constituent peptides in rabbit muscle aldolase that showed well-resolved bimodal isotopic distributions by continuous labeling. When they used pulsed labeling, they identified more than 30 peptides that showed a bimodal distribution. These results show that amide hydrogens of a significant number of peptides in the folded structure exchange completely during exposure in continuous labeling.

C. Screening Proteins That Are Recombinant Drugs

New protein- and peptide-based drugs continue to emerge, making necessary the development of rapid and sensitive methods for checking consistency between and within batches of those drugs. These tests should ensure product quality. Moreover, if they are physical-chemical in nature, they have the additional advantage of replacing animal tests. The combination of H/D exchange and ESI-MS is a tool for monitoring four insulins used for treating insulin-dependent diabetes (59). The extent of hydrogen exchange for less than 2 µg of sample is sufficient for discriminating among the different insulins. After 60 min, bovine, porcine, r-human, and LysPro insulins (the first two are produced naturally, whereas the latter two are recombinant human proteins) had exchanged 25, 28, 30, and 38 amide protons, respectively (59). Furthermore, the H/D exchange data can distinguish active and denatured forms of the protein.

D. Screening Proteins from Combinatorial Libraries

Finding native-like proteins in libraries of de novo proteins, which were designed by using a "binary code strategy" (60), is difficult and tedious, especially when the fraction of non-native structures that compose the library is large. Protection from hydrogen exchange is a characteristic property of native-like structures, and measuring the extent of protection may be useful for screening the large number of samples that can be generated by combinatorial means before they are submitted to arduous separation and purification (61). In a group of four proteins—M60, MF, M13 and M86—which are derived from an initial collection of 29 proteins (60), Hecht and coworkers (61) found that M60 was more resistant to hydrogen exchange than the other proteins in the group. This result is consistent with other evidence that M60 possesses properties of native-like structures (e.g., cooperative thermal denaturation and chemical shift dispersion in NMR, as characterized previously) (62), but the results can be obtained much more rapidly by MS than by the NMR method.

After the utility of the method was demonstrated, Rosenbaum et al. (61) further applied it to a new batch of de novo proteins—G73, L52, K14, I13, and D2—which had been characterized minimally. The data showed that only G73 showed significant protection from hydrogen exchange. Therefore, the authors purified G73 and obtained its one-dimensional NMR spectrum. NMR showed the expected dispersion in chemical shifts for this protein, confirming its native-like structure (61). The MS approach to obtaining H/D exchange data has the advantages of rapid processing, small sample requirement, low solution concentrations (which minimize aggregation), and tolerance to impurities. There is, however, a limitation to this approach: It is difficult to distinguish a native-like structure from an oligomerized structure by this method only. But this limitation applies to any screening method of native-like properties, and hydrogen-exchange MS is less likely to suffer because oligomerization can be minimized by conducting the study at low concentrations of protein, which is not often possible when some other physical-chemical methods are used.

Electrospray ionization MS and H/D exchange were also applied to probe the interactions of aromatic side chains of amino acid residues with porphyrins in specially designed hemoproteins (63). The success highlights the ability of this approach to reveal structural differences between complex molecules that might otherwise be difficult to measure in a rapid manner.

E. Single-Amide Resolution

Nuclear magnetic resonance had been the only practical method for determining deuterium levels at individual peptide linkages. Collision-induced dissociation (CID) of deuterated peptides extricated from an exchanged protein can also, in principle, give data for determining deuterium levels at individual peptide linkages. There is concern, however, that results may be misleading because intramolecular proton scrambling can occur during the fragmentation of peptides (64–66). A recent study by Deng et al. (67) showed that scrambling may not be a serious problem under the right conditions and for the correct choice of fragment ions. Deng et al. used cytochrome c as a model, because the exchange rates of all but the most rapidly exchanging amide protons in cytochrome c are known from NMR measurements (68). Product ion spectra of the doubly charged peptide constituents showed that mainly y and b ions are produced, and the masses of the b ions showed the deuterium levels of individual amide linkages. The levels are consistent with the amide exchange rates obtained by NMR (67). The deuterium levels for individual amides determined using the masses of the y ions are often consistent with those based on the masses of b ions, but several discrepancies exist, suggesting that some proton scrambling occurs for the formation of y ions during CID (67).

The possibility of using CID was further confirmed for "in-source fragmentation" in an ESI-FTICR mass spectrometer (48,49). By increasing the skimmer potential from 80 to 300 V, Eyles et al. (48,49) caused fragmentation of the multiply charged CRABP I in the nozzle–skimmer region of the electrospray source of an FTICR mass spectrometer. Cleavage of 44 of a total of 156 peptide linkages occurred as a result of the CID process, resulting in a series of b and y ions (48,49). In this study, only protiated fragments were seen in the

N-terminal His-tag region (b_{21}^{3+} ion in Fig. 4); this suggests that hydrogen scrambling does not occur to any significant extent. If exchange were to occur, some deuterium incorporation would be expected in the peptide segment containing the His tag.

These studies show that the extent of isotope exchange at individual amide linkages can be measured by using ESI and tandem MS. This capability opens the door for mass spectrometry to determine highly localized structural changes in proteins. In-source fragmentation and its combination with the high mass resolving power and high mass accuracy offered by FTICR-MS is one approach, and the combination can provide accurate, site-specific information

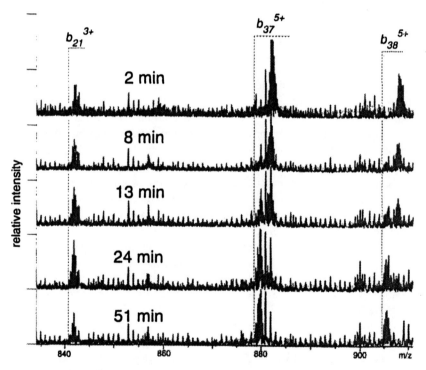

Figure 4 Time-dependent mass spectra showing the fragment ion peaks from a CAD spectrum of multiply charged CRABP I. The spectra are from in-source collisional activation. Charge state envelopes of intact protein ions range from +11 to +14. The dashed lines on the left of each isotopic cluster indicate the position of the fully charged fragment ion peak (monoisotopic mass). (Taken, with permission, from Fig. 5 of Ref. 49.)

on the solvent accessibility of various structural elements within a protein as a function of time. Therefore, one can characterize conformational stability in a site-specific fashion. A limitation, however, is revealed by a study of CRABP I protein. A large segment of the protein (Leu43–Leu121) remained intact when submitted to in-source fragmentation. As a result, the site-specific deuterium incorporation in this large segment could not be measured (48,49). The fragmentation efficiency using hexapole trapping (69) can be enhanced by taking advantage of the long ion accumulation time in the device, providing a potential to cover nearly the entire structure (48). If this extensive fragmentation is general, sequence coverage by this approach will be very useful.

Fragment ions can be immediately separated and analyzed in the mass spectrometer when dissociation occurs directly in the ESI interface region. Therefore, the time needed for sample workup (i.e., pepsin digestion and HPLC separation) is minimized, and back-exchange is reduced by the in-source fragmentation approach.

IV. CONCLUSIONS AND FUTURE DIRECTIONS

Protection from hydrogen exchange is a signature of a structured protein. Hydrogen–deuterium exchange by MS has established its potential for the study of structural and conformational dynamics of proteins. Wild-type and mutant proteins can have dramatically different H/D exchange behavior, which may be a significant measure of their function. By using denaturing reagents, intermediates in protein refolding can be identified, and their populations can be determined, providing energetics of protein refolding. By using H/D exchange and MS, the domains for ligand binding can be located and conformational changes induced by ligand binding can be identified. Hydrogen–deuterium exchange by MS has also been applied as a rapid screening method for tailored proteins produced by combinatorial methods as well as for protein drugs produced by recombinant technologies.

Proteomics is emerging rapidly, and MS will play an important role in protein identification. We expect that the combination of H/D exchange and MS will be an added tool to give information about the structure or folding of newly identified proteins and guide the use of theory in predicting their folding behavior. One approach is to compare the H/D exchange behavior, obtained rapidly by MS, of a new protein with those of homologs having known structures.

Pepsin digestion of the deuterium-incorporated protein followed by MS improves the resolution of the method, enabling the determination of deuterium incorporation at the peptide level. Data from CID in the MS/MS mode or from in-source fragmentation should further improve the resolution to the level of individual amide linkages. This advance changes the notion that only NMR can achieve the determination of exchange at the amino acid level.

ACKNOWLEDGMENT

This work was supported by the National Centers of Research Resources of the NIH (grant P41RR00954).

REFERENCES

1. JB Fenn, M Mann, CK Meng, SF Wong, CM Whitehouse. Science 246:64–71, 1989.
2. M Karas, D Bachmann, U Bahr, F Hillenkamp. Int J Mass Spectrom Ion Process 78:53–68, 1987.
3. TE Creighton. Proteins: Structures and Molecular Properties. 2nd ed. New York: Freeman, 1993.
4. SF Betz, DP Raleigh, WF DeGrado. Curr Opin Struct Biol 3:601–610, 1993.
5. A Miranker, CV Robinson, SE Radford, CM Dobson. FASEB J 10:93–101, 1996.
6. DL Smith, K Dharmasiri. NATO ASI Ser, Ser C 510:45–58, 1998.
7. DL Smith, Y Deng, Z Zhang. J Mass Spectrom 32:135–146, 1997.
8. J Buijs, CC Vera, E Ayala, E Steensma, P Haakansson, S Oscarsson. Anal Chem 71:3219–3225, 1999.
9. ID Figueroa, DH Russell. J Am Soc Mass Spectrom 10:719–731, 1999.
10. JG Mandell, AM Falick, EA Komives. Anal Chem 70:3987–3995, 1998.
11. JG Mandell, AM Falick, EA Komives. Mass Spectrom Biol Med 91–109, 2000.
12. J Villanueva, F Canals, V Villegas, E Querol, FX Aviles. FEBS Lett 472:27–33, 2000.
13. GP Connelly, Y Bai, MF Jeng, SW Englander. Proteins 17:87–92, 1993.
14. Y Bai, JS Milne, L Mayne, SW Englander. Proteins 17:75–86, 1993.
15. S Neilson. Adv Prot Chem 17:75, 1966.
16. SW Englander, L Mayne. Annu Rev Biophys Biomol Struct 21:243–265, 1992.
17. CN Pace. Methods Enzymol 131:266–280, 1986.
18. JJ Englander, JR Rogero, SW Englander. Anal Biochem 147:234–244, 1985.
19. Z Zhang, DL Smith. Protein Sci 2:522–531, 1993.
20. JJ Rosa, FM Richards. J Mol Biol 133:399–416, 1979.
21. K Biemann. Methods Enzymol 193:455–479, 1990.
22. RM Caprioli, T Fan. Anal Biochem 154:596–603, 1986.

23. F Wang, W Li, MR Emmett, CL Hendrickson, AG Marshall, Y-L Zhang, L Wu, Z-Y Zhang. Biochemistry 37:15289–15299, 1998.
24. F Wang, MA Freitas, AG Marshall, BD Sykes. Int J Mass Spectrom 192:319–325, 1999.
25. F Wang, W Li, MR Emmett, AG Marshall, D Corson, BD Sykes. J Am Soc Mass Spectrom 10:703–710, 1999.
26. F Wang, JS Blanchard, X-j Tang. Biochemistry 36:3755–3759, 1997.
27. F Wang, G Scapin, JS Blanchard, RH Angeletti. Protein Sci 7:293–299, 1998.
28. KA Resing, AN Hoofnagle, NG Ahn. J Am Soc Mass Spectrom 10:685–702, 1999.
29. JR Engen, WH Gmeiner, TE Smithgall, DL Smith. Biochemistry 38:8926–8935, 1999.
30. Z Zhang, W Li, IM Logan, M Li, AG Marshall. Protein Sci 6:2203–2217, 1997.
31. O Nemirovskiy, DE Giblin, ML Gross. J Am Soc Mass Spectrom 10:711–718, 1999.
32. K Kuwajima. Proteins 6:87–103, 1989.
33. V Forge, RT Wijesinha, J Balbach, K Brew, CV Robinson, C Redfield, CM Dobson. J Mol Biol 288:673–688, 1999.
34. MX Li, SM Gagne, L Spyracopoulos, CP Kloks, G Audette, M Chandra, RJ Solaro, LB Smillie, BD Sykes. Biochemistry 36:12519–12525, 1997.
35. L Spyracopoulos, MX Li, SK Sia, SM Gagne, M Chandra, RJ Solaro, BD Sykes. Biochemistry 36:12138–12146, 1997.
36. SK Sia, MX Li, L Spyracopoulos, SM Gagne, W Liu, JA Putkey, BD Sykes. J Biol Chem 272:18216–18221, 1997.
37. G Scapin, SG Reddy, R Zheng, JS Blanchard. Biochemistry 36:15081–15088, 1997.
38. G Scapin, SG Reddy, JS Blanchard. Biochemistry 35:13540–13551, 1996.
39. GB Cohen, R Ren, D Baltimore. Cell 80:237–248, 1995.
40. T Pawson, J Schlessinger. Curr Biol 3:434–442, 1993.
41. H Ehring. Anal Biochem 267:252–259, 1999.
42. A Gustavsson, H Jahr, R Tobiassen, DR Jacobson, K Sletten, P Westermark. Lab Invest 73:703–708, 1995.
43. M Savaiva. Human Mutat 5:191–196, 1995.
44. EJ Nettleton, M Sunde, Z Lai, JW Kelly, CM Dobson, CV Robinson. J Mol Biol 281:553–564, 1998.
45. HA Lashuel, Z Lai, JW Kelly. Biochemistry 37:17851–17864, 1998.
46. Z Lai, W Colon, JW Kelly. Biochemistry 35:6470–6482, 1996.
47. KA Resing, NG Ahn. Biochemistry 37:463–475, 1998.
48. SJ Eyles, JP Speir, GH Kruppa, LM Gierasch, IA Kaltashov. J Am Chem Soc 122:495–500, 2000.
49. SJ Eyles, T Dresch, LM Gierasch, IA Kaltashov. J Mass Spectrom 34:1289–1295, 1999.
50. CS Maier, MI Schimerlik, ML Deinzer. Biochemistry 38:1136–1143, 1999.
51. EW Chung, EJ Nettleton, CJ Morgan, M Grob, A Miranker, SE Radford, CM Dobson, CV Robinson. Protein Sci 6:1316–1324, 1997.

52. Y Deng, DL Smith. Biochemistry 37:6256–6262, 1998.
53. N Blom, J Sygusch. Nat Struct Biol 4:36–39, 1997.
54. Y Deng, DL Smith. Anal Biochem 276:150–160, 1999.
55. Y Deng, Z Zhang, DL Smith. J Am Soc Mass Spectrom 10:675–684, 1999.
56. Y Deng, DL Smith. J Mol Biol 294:247–258, 1999.
57. Y Bai, TR Sosnick, L Mayne, SW Englander. Science 269:192–197, 1995.
58. K Dharmasiri, DL Smith. Anal Chem 68:2340–2344, 1996.
59. R Ramanathan, ML Gross, WL Zielinski, TP Layloff. Anal Chem 69:5142–5145, 1997.
60. S Kamtekar, JM Schiffer, H Xiong, JM Babik, MH Hecht. Science 262:1680–1685, 1993.
61. DM Rosenbaum, S Roy, MH Hecht. J Am Chem Soc 121:9509–9513, 1999.
62. S Roy, G Ratnaswamy, JA Boice, F Fairman, G McLendon, MH Hecht. J Am Chem Soc 119:5302–5306, 1997.
63. D Liu, DA Williamson, ML Kennedy, TD Williams, MM Morton, DR Benson. J Am Chem Soc 121:11798–11812, 1999.
64. FW McLafferty, Z Guan, U Haupts, TD Wood, NL Kelleher. J Am Chem Soc 120:4732–4740, 1998.
65. RS Johnson, D Krylov, KA Walsh. J Mass Spectrom 30:386–387, 1995.
66. AG Harrison, T Yalcin. Int J Mass Spectrom Ion Process 165:339–347, 1997.
67. Y Deng, H Pan, DL Smith. J Am Chem Soc 121:1966–1967, 1999.
68. JS Milne, L Mayne, H Roder, AJ Wand, SW Englander. Protein Sci 7:739–745, 1998.
69. K Sannes-Lowery, RH Griffey, GH Kruppa, JP Speir, SA Hofstadler. Rapid Commun Mass Spectrom 12:1957–1961, 1998.

11
Microelectrospray Analysis Combined with Microdialysis Sampling of Neuropeptides and Drugs

Per E. Andrén
Uppsala University, Uppsala, Sweden

Richard M. Caprioli
Vanderbilt University, Nashville, Tennessee

I. INTRODUCTION

Micro-electrospray ionization (micro-ESI) (1,2) is a miniaturized low flow rate (submicroliter per minute), high ionization efficiency adaptation of the basic electrospray ionization technique. Similar devices have also been described, most notably, nanospray ionization (3) and picospray ionization (4). Micro-ESI is most suitable for on-line analysis, for example, for on-line infusion of samples from a holding loop of an injection valve in a pump-driven system or from the output of a narrow-bore or capillary bore column from a micro liquid chromatograph or an electrophoresis capillary. Flow rates are typically between 50 nL/min and 1 µL/min from a capillary needle having a nozzle orifice diameter of approximately 5–10 µm. "Nanospray" refers to a method of low flow rate (from about 20 nL/min to several hundred nanoliters per minute) achieved for a liquid sample that is placed in a capillary terminating in a narrow orifice (about 2–3 µm in diameter), where the spray condition is achieved without the aid of a pump. Picospray ionization is a still lower flow rate version for which capillary nozzle orifice diameters are 1–2 μm and picoliter per minute flow rates can be achieved.

A few of the many arrangements and designs of the devices that have been constructed over the years are shown in Fig. 1. A micro-ESI device is depicted in Fig. 1a that can easily be used with on-line separation processes, preferably ones that are themselves low flow rate systems. This design incorporates a provision for including adsorbent material (e.g., LC packing) that provides a highly efficient means for desalting, preconcentration, and some fractionation of analytes in a sample (1). Low flow rates are usually achieved with the aid of a syringe pump. High voltage is applied upstream of the nozzle, eliminating the need to coat the nozzle with an electrically conductive layer. Finally, by spraying directly from the tip of the micro-LC column or needle, dead volume connections and transfer lines are eliminated.

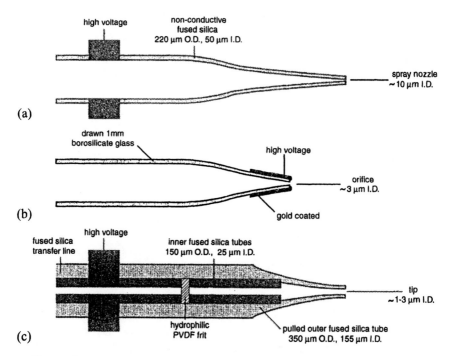

Figure 1 Three basic designs of low flow rate, high efficiency electrospray devices. (a) Microelectrospray; (b) nanoelectrospray; (c) a microelectrospray design incorporating a frit.

A nanospray device, depicted in Fig. 1b, contains a capillary needle loaded with a liquid sample. The low flow rate of the sample spray is obtained through the applied voltage, eliminating the need for a mechanical pumping system. This design is capable of providing a steady low flow rate of liquid for hours from just a few microliters of sample. The micro-ESI device shown in Fig. 1c incorporates a frit to filter particulate matter before the liquid passes through the restricter nozzle, minimizing clogging of the nozzle (5). Designs using glass frits, membrane material, solid packing, etc., to filter the solvent/sample flow have also been reported.

A number of variations and improvements of the basic designs described above have appeared over the years, some unique to particular instruments. Although there are some differences among the designs and performances of these devices, for the purposes of this chapter we use the term "micro-ESI" to describe miniaturized electrospray ionization devices with which very low flow rates and high ionization efficiency can be achieved.

II. INSTRUMENTATION

A typical micro-ESI instrumental setup for continuous infusion of a sample from a low volume loop of an injection valve is shown in Fig. 2. The pump of choice is the syringe pump because of the steady, continuous delivery of solvent that can be achieved. All tubing is narrow-bore (less than 75 μm diameter) fused silica capillary tubing, and connections are low or zero dead-volume fittings. Even so, great care must be taken to use the correct ferrules and to make connections that do not create dead volumes. High voltage is applied to the solution upstream of the nozzle in this design, so solvents must be electrically conductive. This can be achieved simply by adding 0.1–1.0% acetic acid, or other electrolyte, to the solvent if necessary. If preconcentration and desalting are desirable, the needle can be packed with adsorbent packing that can bind the analyte. At the desired time, the analyte is desorbed from the packing material by a solvent change and is subsequently spray-ionized for analysis.

A. Capillary Spray Needles

To reduce the volume of liquid flowing into the mass spectrometry source, small inner diameter capillary needles are used, terminating in the spray nozzle. Initially, these needles were constructed by hand, but today they are commercially available. Nevertheless, their construction will be briefly

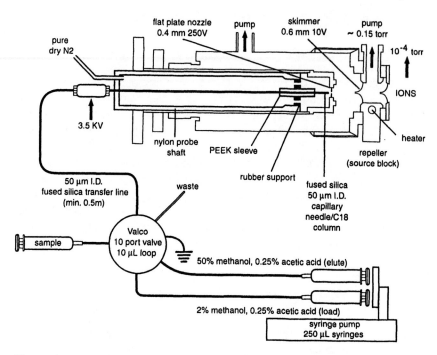

Figure 2 A typical micro-ESI instrumental setup. (Reproduced with permission from Ref. 1.)

described (1). Fused silica capillary tubing, typically 50 μm inner diameter and 200 μm outer diameter, is used as the starting material. For larger i.d. tubing it may be desirable to etch away the excess fused silica at the nozzle. The polyimide coating is removed from the tip (over a length of approximately 0.5 cm), and the tip is then lowered into a 49% solution of hydrofluoric acid (HF), with distilled water pumped through at a rate of 500 nL/min. The tip is then carefully ground by hand using a ceramic cutting stone to obtain a nozzle with a smooth end. The opposite end of the needle is attached to a stainless steel zero dead-volume fitting, and the high voltage is applied to the spray needle through the solvent at this union. Depending upon experimental conditions (e.g., diameter and length of the capillary), the needle voltage varies from 1.0 to 3.5 kV. With these spray needles it is possible to achieve very high sensitivity and high signal stability, making them suitable for multihour use. The spray needles are robust, but because of the narrow orifice they are vulnerable to clogging if solvents and liquid samples are not carefully filtered.

B. Achieving High Sensitivity Measurements

The development of micro-ESI for the high sensitivity detection of neuropeptides was achieved by using needles packed with C18 particles (1,2), similar to those used for LC columns, as shown in Fig. 3. This arrangement allows peptides to be desalted and preconcentrated in a single step, as described below in more detail. In addition, it is important to place the tip of the spray needle as close as possible to the nozzle orifice entrance into the mass spectrometer so as to obtain maximum transfer efficiency of the sample. For analyzers such as quadrupole mass filters, scanning narrow ranges (10–20 m/z units) further improves the signal-to-noise ratio. In all, this methodology increases the sensitivity considerably so as to allow analysis of neuropeptides in the low attomole per microliter range.

Neurotensin gives an abundant signal by electrospraying aqueous solutions. Figure 4 shows the product ion mass spectrum of the precursor molecular ion $[M + 3H]^{3+}$ (m/z 558.7) of neurotensin and a doubly charged product ion (m/z 578.9) produced from the continuous infusion of 32 amol/µL in 50% methanol containing 0.25% acetic acid flowing at a rate of about 800 nL/min (2). The spectrum was obtained from an average of 16 scans of 0.5 s each, consuming less than 3.2 amol of peptide. Of course, the ultimate sensitivity achievable in a given experiment depends on the nature of the peptide of interest, and not all peptides will give the same results as neurotensin. For example, when analyzing microdialysate samples containing physiological levels of salt, greater sample manipulation is necessary to remove the salt, causing some losses. The detection limit becomes poorer and drops to approximately 50–100 amol (total amount injected).

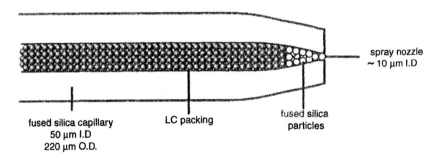

Figure 3 The fused silica capillary spray needle of the micro-ESI source. The tip of the needle contains LC packing to achieve preconcentration and desalting of the sample.

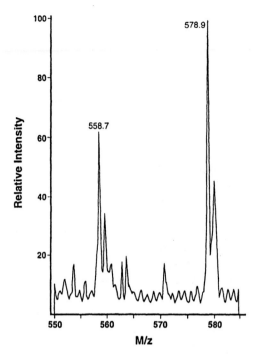

Figure 4 Product ion mass spectrum of the [M+3H]$^{3+}$ ion of neurotensin (m/z 558.7) and the doubly charged product ion at m/z 578.9 produced by the continuous infusion of a solution containing 32 amol/µL (32 pM). The spectrum was obtained from an average of sixteen 0.5 s scans.

C. On-Line Prepurification of Samples

Desalting of samples is of utmost importance for determination of low levels of peptides, for example, in measuring neuropeptides in the brains of small animals at endogenous levels. This is best achieved through the use of an on-line packed capillary column for concentrating and desalting the samples. The end of the capillary column can also serve as the spray needle in the mass spectrometer, eliminating dead volume in the system. Several investigators reported procedures for packing fused silica capillaries (1,6,7). Using a capillary with a needle nozzle (prepared as described in Sect. II.A), a frit can be constructed by tapping in about a 100 µm length of 10 µm spherical silica particles and sintering them into place by using a microtorch. A slurry of methanol and 10 µm porous spherical C18 particles is then loaded into the fused silica column via a stainless steel reservoir by using a syringe pump; a 5 cm length of column is packed. The length of the packing, of course, depends on the type and amount of analyte in the sample to be analyzed. As an example of a typical analysis, peptide samples (10 μL) are loaded on a capillary C18 column by a microliter syringe at a flow rate of 1 μL/min. The mobile phase contains 2% methanol–0.25% acetic acid in distilled water. After 15 min of washing at 1 μL/min, neuropeptides are eluted directly into the mass spectrometer by using 50% methanol–0.25% acetic acid in distilled water at 0.3 μL/min. This integrated sample processing and micro-ESI source allow relatively large volumes of dilute solutions to be analyzed.

D. In Vivo Microdialysis with Micro-ESI and Tandem MS

Microdialysis for in vivo analysis has gained wide recognition as a sampling technique for the measurement of many neurochemicals (8,9). The technique is based on the principle that the microdialysis probe imitates the function of a capillary blood vessel. A hollow microfiber with a semipermeable membrane at the tip of the microdialysis probe is implanted within the extracellular space of a tissue of interest in a living animal and perfused with fluid. The composition of the microdialysis perfusate is similar to that of the extracellular fluid owing to diffusion through the membrane. Microdialysis allows direct sampling of the extracellular fluid in anesthetized or freely moving animals, and it can be used in nearly every tissue or organ, including blood. Furthermore, it is possible to sample continuously for hours or days from a single animal without dehydrating the animal. The microdialysis membrane has a molecular weight cutoff such that large molecules like enzymes and other proteins are precluded from the dialysate, thus providing relatively clean samples. Among the disadvantages of microdialysis sampling are the relatively low analyte concentrations obtained and the concomitant high salt content.

The sensitivity and specificity of analysis of small volume microdialysate samples containing low concentrations of analyte are enhanced by coupling this micro-ESI source with a tandem quadrupole, ion-trap, or quadrupole time-of-flight (QTOF) mass spectrometer. The tandem MS (MS/MS) analysis mode adds specificity to the analysis and furthermore greatly reduces the contribution of chemical background noise to the signal. Several applications of the microdialysis sampling technique combined with micro-ESI-MS/MS to the analysis of neuropeptides in the extracellular fluid of rat brain as well as other neuroactive compounds are described in the following pages.

III. DETERMINATION OF EXTRACELLULAR RELEASE OF ENDOGENOUS NEUROTENSIN

Neurotensin (NT), pGlu-Leu-Tyr-Glu-Asn-Lys-Pro-Arg-Arg-Pro-Tyr-Ile-Leu, is a tridecapeptide that is localized in neurons and terminals throughout the brain. It has a variety of biological activities as a central neurotransmitter or neuromodulator. NT is closely associated with central nervous system (CNS) dopamine neurons and interacts with dopamine at physiological, anatomical, and behavioral levels. Neurotensin is colocalized with dopaminergic neurons in the hypothalamus and midbrain (10) and exerts hypothermic and analgesic effects when injected into the CNS. From a clinical point of view, studies with NT have led to implications of its involvement in schizophrenia, Parkinson's disease, Huntington's chorea, and Alzheimer's disease (11–14). Several of the NT fragments also retain biological activity (15,16).

Two brain areas were chosen for the analysis using in vivo microdialysis of basal and KCl-induced stimulation of endogenous levels of NT: the hypothalamus with a reported high tissue concentration of NT and the globus pallidus/ventral pallidum with a lower tissue concentration of NT (17,18). For quantitative experiments, synthetic NT in standards made up in artificial CSF and endogenous NT in microdialysis samples were analyzed by using on-line packed C18 capillaries and micro-ESI-MS/MS. Neuropeptides were eluted from the capillary C18 columns with 50% methanol–0.25% acetic acid. The NT peak in the mass chromatogram was approximately 15 s wide. To obtain statistically reliable measurements from the mass spectra, 20 scans were averaged over the apex of the peak in the mass chromatogram. Standard curves produced after injection of 38.6, 155, 618, and 1230 amol of NT (in 10 µL artificial CSF) showed the linearity and interassay variability of the micro-ESI-MS/MS system (Fig. 5). The standard curves had an average r^2 linear coefficient of 0.993 and a standard error of the mean (SEM) of <17% ($n = 5$). The lowest amount used in generating the standard curve, 38.6 amol, had a signal-to-noise (S/N) ratio of 1.3 ± 0.2 (mean ± SEM), and the highest (1330 amol), 16.7 ± 1.1.

Basal levels of NT were established in both the hypothalamus and globus pallidus/ventral pallidum. The levels were slightly higher in the hypothalamus (approximately 100 amol per 10 µL of dialysate) than in the globus pallidus/ventral pallidum region (approximately 74 amol per 10 µL of dialysate). Peptidase inhibitors were not included in the perfusate or given systemically, because such agents can raise the level of the neuropeptides in the synaptic cleft and otherwise affect the regulatory mechanisms. The basal amounts of NT appear, however, to be close to the limit of detection of the mass spectrometric

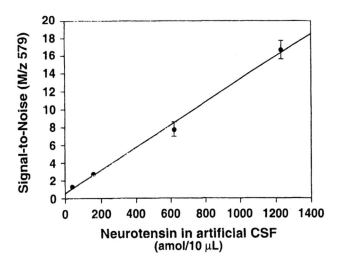

Figure 5 Standard curve for neurotensin. Standards were dissolved in artificial cerebrospinal fluid, and 10 μL samples were loaded onto the C18 packed micro-ESI needle with 2% methanol–0.25% acetic acid through an injection valve. Samples were eluted into the mass spectrometer with 50% methanol–0.25% acetic acid.

system. The signal-to-noise (S/N) ratio accompanying the determination of basal levels of NT in the hypothalamus was approximately 2.5 for the mass spectrometric analysis.

Release of NT in the hypothalamus was stimulated by direct depolarization by presenting a (1.0 μL) bolus of artificial CSF supplemented with 100 mM KCl to the microdialysis probe. This perfusion increased the recovered NT to 548 ± 90 amol/10 μL in the hypothalamus (544% increase compared to the baseline). In the globus pallidus/ventral pallidum region, KCl stimulation elevated the NT release to 499 ± 99 amol/10 μL during a 30 min sample period (674% increase compared to the baseline) (Fig. 6). It should be pointed out that there was a large variation in the absolute increase produced by KCl as evidenced by the standard error of the mean (SEM) in Fig. 6. The mass spectrum in the region of the m/z 579 ion during the stimulation of KCl in the hypothalamus is given in Fig. 7.

Although it is very difficult to accurately extrapolate in vitro recoveries to in vivo extracellular concentrations, a rough estimate of the basal extracellular concentration measured in the present experiments can be made. The approximate concentration of NT in the dialysates collected under

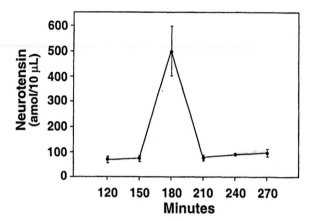

Figure 6 Time course of basal and KCl-induced release of neurotensin in globus pallidus/ventral pallidum. The increase in baseline levels was evoked by 1.0 μL of 100 mM KCl included in the perfusion medium during a 30 min collection period of the microdialysis sampling. Values represent mean ± SEM of absolute amounts of neurotensin in a 10 μL sample ($n = 8$).

basal conditions were approximately 70–100 amol/10 μL. Using an average relative recovery of 24% for NT at the flow rate of 0.4 μL/min gives an extracellular concentration of 30–42 amol/μL. Again, this is only a rough estimate; extrapolation of microdialysate concentrations obtained in the present study to the absolute concentration of NT in the extracellular fluid in vivo is complex and not straightforward. This is mainly the result of the limitations of the microdialysate technique, such as the efficiency of transfer of analytes across the membrane, and also to the effects of varying tissue density (19,20). In addition, the calculation relies on the assumption that relative recoveries determined in vitro can be extrapolated to the in vivo environment. Benveniste (21) suggested that such calculations should be multiplied by a factor of approximately 10 to account for the complexity and volume fraction differences between brain tissue and solution. Previously reported levels of basal NT in brain regions from microdialysates measured by radioimmunoassay (RIA) are somewhat higher than those found in the present study (22–26). However, the other studies used microdialysis flow rates higher than the 0.4 μL/min used here, so a direct comparison is difficult. Nevertheless, this example illustrates the utility of using mass spectrometry and in vivo microdialysis to monitor extracellular levels of endogenous NT

Microdialysis of Neuropeptides and Drugs

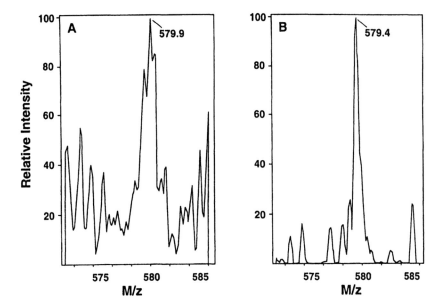

Figure 7 Mass spectra of (A) baseline and (B) KCl-stimulated release of endogenous neurotensin from the hypothalamus of a freely moving rat. The spectrum from the baseline shows a significant m/z 579 peak over background. The KCl-stimulated release of neurotensin shows that the signal-to-noise ratio of the m/z 579 ion has increased severalfold.

from two brain areas, hypothalamus and the globus pallidus/ventral pallidum, in the rat. Reports of successful in vivo microdialysis measurements of NT in the rat brain were previously published, but these studies used analysis by RIA (22,27).

Microdialysis is widely used for monitoring a number of neuroactive substances in the brain extracellular environment. The inviting simplicity of the technique has made it useful to many neuroscientists, but usually the limiting factor is the analytical characterization following sample collection. Any substance that is able to diffuse across the membrane is capable of being monitored, provided that a sufficiently sensitive assay exists for its measurement. For neuropeptides, depending on size and hydrophobicity, recovery through the membrane can be an additional problem. Generally, recoveries do not exceed 10–15% for most peptides. Another factor, restriction of diffusion through the brain microenvironment, is also an important issue and can further reduce the recovery of the substance of interest.

IV. STUDIES OF NEUROTENSIN METABOLISM

Previous studies of NT metabolism that employed in vivo microdialysis and mass spectrometry showed the presence of an intermediate metabolite (28), NT1-12, which had not been observed from in vitro studies. To study the formation of this peptide metabolite as well as other metabolic fragments, microdialysis and nanoflow capillary LC/micro-ESI-MS and MALDI (matrix-assisted laser desorption ionization) MS were employed. Experiments were carried out by administering low concentrations of exogenous NT into the extracellular fluid through a microdialysis probe implanted in the striatum of an anesthetized rat. The identities of the peptide metabolic fragments collected through this probe and estimates of their levels in dialysate were obtained by using mass spectrometry in combination with on-line LC. The in vivo metabolism of NT, the involvement of striatal carboxypeptidase A (CPA) activity in the formation of the NT fragment NT1-12, and the involvement of other enzymes for the further degradation of this metabolic intermediate in vivo were studied.

During infusion of NT into the rat striatum, dialysate was collected in 30 min intervals and analyzed by capillary LC/micro-ESI-MS. The metabolic fragments identified in the striatum were the N-terminal fragments NT(1–12), NT(1–11), NT(1–10), and NT(1–8), the C-terminal fragment NT(9–13), and internal fragments NT(3–8) and NT(7–11). Figure 8 shows the summed spectra for components eluting from the capillary column for one such experiment. Less than unit mass resolving power was used to acquire mass spectra to maximize sensitivity, and therefore ions due to naturally abundant isotopes were not resolved. A low dose experiment was also performed with the infusion of 100 fmol/μL NT into the striatum to determine whether the metabolic fragment pattern was dose-dependent. The same metabolic distribution of NT fragments was found as for the higher infused concentrations (1 or 10 pmol/μL) of NT.

For verification and analysis of the sample by another MS ionization method, the MALDI mass spectra of in vivo microdialysates were obtained. The results showed the same terminal fragments of NT that were observed using micro-ESI-MS analysis.

When specific enzyme inhibitors were infused together with NT into the striatum through the microdialysis probe, the formation of the NT fragment NT(1–12) could be studied. The results of the experiments, summarized in Fig. 9, show that a CPA enzyme (or enzyme having similar activity) is present in the striatum of rat brain. The formation of NT(1–12) decreased by 70% when 100 μM benzylsuccinic acid, a potent CPA inhibitor, was infused into the brain. The experiments also showed that the conversion of NT to NT(1–12) is

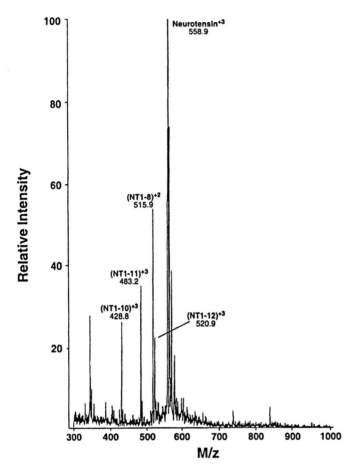

Figure 8 Micro-ESI mass spectrum of in vivo microdialysate containing artificial cerebrospinal fluid obtained from the metabolism of exogenous neurotensin in the rat brain. The metabolism in the striatum was studied by perfusing 1.0 µM neurotensin through the microdialysis probe. The probe was used both to deliver the peptide into the brain and to collect the processed products at a flow rate of 300 nL/min. Samples were taken every 30 min.

only one of many alternative pathways of NT metabolism in vivo. The inhibition of the major central nervous system neuropeptide-metabolizing enzyme, metalloendopeptidase 24.15, using the specific inhibitor CPP (*N*-[1-(*R*,*S*)-carboxy-3-phenylpropyl]-Ala-Ala-Phe-*p*-aminobenzoate) increased the NT(1–12) formation by 300%. However, inhibitors of neutral endopeptidase

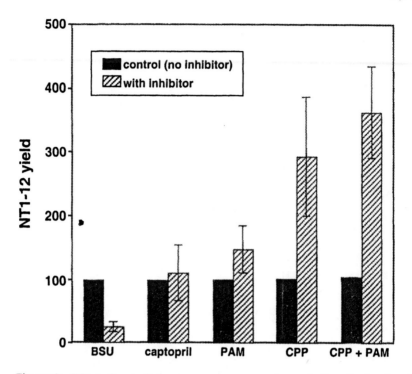

Figure 9 Effect of proteolytic enzymes on neurotensin metabolism. Brain microdialysis (striatum) of 100 fmol/μL neurotensin in the absence or presence of benzylsuccinic acid (BSU, 100 μM), captopril (40 μM), phosphoramidon (PAM, 40 μM), CPP (40 μM), CPP + PAM (100 μM and 10 μM). Data are mean ± SD value ($n = 3-5$).

24.11 and angiotensin-converting enzyme had significantly less effect. A combined inhibition with both CPP and phosphoramidon (PAM) made NT(1–12) the major product of NT metabolism and increased its formation by 360%.

The present example demonstrates that NT is metabolized in vivo by proteolytic enzymes in the extracellular fluid within the rat striatum. It is proposed that CPA activity is likely responsible for the formation of the intermediate NT fragment NT(1–12). Several brain peptidases were reported to hydrolyze NT at various peptide bonds (see Fig. 10), but it is not known which of these enzymes participate in the degradation of NT in vivo. Extracellular degradation was suggested as the process of inactivation, and numerous purified peptidases that are capable of cleaving NT were found in the brain and cerebrospinal fluid, including the angiotensin-converting enzyme proline endopeptidase, metalloendopeptidase 24.15 and 24.16, and neutral endopeptidase 24.11 (29,30).

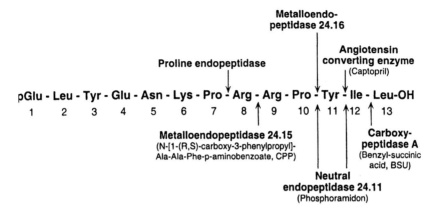

Figure 10 Summary of enzymes that inhibit on NT1–12 yeild. Benzylsuccinic acid (BSU) is an inhibitor of carboxypeptidase A; phosphoramidon is an inhibitor of neutral endopeptidase 24.15; N-[1-(R,S)-carboxy-3-phenylpropyl]-Ala-Ala-Phe-p-aminobenzoate (CPP) is a metalloendopeptidase 24.11 inhibitor; and captopril is an angiotensin-converting enzyme inhibitor.

The principle of "fixed time assay" was used for the characterization of the enzymes involved in NT(1–12) fragment formation and degradation. If it is assumed that the initial rate of the reaction is constant, then the amount of NT(1–12) fragments formed during the first 30 min of NT perfusion should be proportional to the enzyme activity present. The coinfusion of a specific enzyme inhibitor with NT on the same animal should cause a change in the concentration of NT(1–12) if that specific enzyme is involved in the formation or degradation of NT(1–12). This principle has been applied in the in vitro microslice study of neurotensin metabolism (30), although in the present microdialysis experiments the reactions were carried out in vivo, with much lower substrate concentrations and a faster removal of the products from the system.

The physiological significance of the CPA-involved NT(1–12) formation is unclear, but because the C-terminal sequence is required for NT activity, one can hypothesize that CPA cleavage provides an alternative mechanism for the downregulation of NT-stimulated physiological responses, thus giving flexibility to fine-tune neurological functions.

In vivo microdialysis and mass spectrometry is a powerful combination for the study of metabolic processes of complex reaction systems in vitro. The in vivo microdialysis mass spectrometric techniques have several advantages over the microslice/HPLC techniques for the study of neuropeptide metabolism.

First, the experiments are performed in a live animal. Second, the metabolites are isolated from the enzymes at the site of collection. Third, mass spectrometry gives molecular specificity and ultrahigh sensitivity.

V. MICROELECTROSPRAY AND IN VIVO MICRODIALYSIS OF DRUGS

Microdialysis and micro-ESI-MS/MS were employed for the detection and quantitation of the drug CGP-36742 in rat brain after the drug was administered intravenously (IV) or by mouth (PO). The microdialysates containing CGP-36742 were analyzed directly by micro-ESI-MS/MS without prior sample purification. This drug is a GABA-B antagonist with high affinity and specificity for GABA-B receptors (31). Previous attempts to determine CGP-36742 in the brain by using radiolabeled CGP-36742 were unsuccessful.

$$H_2NCH_2CH_2CH_2\overset{\overset{\displaystyle O}{\|}}{\underset{\underset{\displaystyle OH}{|}}{P}}CH_2CH_2CH_2CH_3$$

Chemical Structure of CGP-36742 (3-aminopropyl-N-butylphosphinic acid)

The ability of a novel neuroactive compound to traverse the blood-brain barrier is an important parameter in drug discovery, and the dynamics of the turnover of the drug is a key issue in its therapeutic efficacy. Thus, in vivo studies of a novel compound and its brain bioavailability are important parts of drug design. Often, because of the low tissue concentrations of these compounds after systemic administration, many novel compounds are radiolabeled and administered to an animal model. The animal is later euthanized, and its tissues are assayed for levels of radioactivity, a putative marker for the presence of the drug. The radiolabeling of the compound greatly enhances detection sensitivity. However, this technique has several disadvantages. The structural identity of the radioactive analytes remains unknown, because only total radioactivity is measured. Pharmacokinetic studies measuring the time course for the appearance and subsequent disappearance of a drug require large numbers of experimental animals, each animal representing one time point. Finally, the use of radioactivity can be hazardous to laboratory personnel, and the expense of both the purchase and the disposal of radioisotopes can be prohibitively costly.

The kinetics of metabolism of CGP-36742 were successfully measured by micro-ESI-MS/MS. The microdialysis samples in the present example were analyzed without prior sample purification. The dialysates were thawed on ice and diluted 1:4. Dilution was necessary to reduce the concentration of the endogenous salts that were in the dialysate sample and to add solvent components necessary for maintaining a stable ESI spray. Desalting with nanoflow capillary liquid chromatography was not possible, because the compound was not suitably retained on the reverse-phase or other resin. All analyses were performed by tandem MS (MS/MS) for two reasons: In the MS mode, the background ion of m/z 182 significantly interfered with measurement of the drug molecular ion (of m/z 180.2), and MS/MS provided much greater specificity for the intact precursor compound and lowered the background noise significantly.

The time course for the appearance and disappearance of CGP-36742 after IV administration was very similar in the frontal cortex and in the ventricle. Observations from Fig. 11 indicate an approximate half-life of the drug in the brain of approximately 30 min. Peak drug levels in the ventricle were significantly higher (>fivefold) than in the frontal cortex. The increased levels of CGP-36742 in the third ventricle compared to the frontal cortex may be due to a combination of several factors. The ventricle is a more dynamic region in that the microdialysis probe is surrounded by a pool of CSF and not cells as in the frontal cortex. This absence of tissue tortuosity would greatly enhance the recovery of the drug across the microdialysis membrane (32). The drug levels in the third ventricle may also be higher owing to greater transfer of the drug to the CSF. There are seven small regions that line the ventricle and lack a tight brain-blood barrier. These areas, the circumventricular organs, are passively permeable to hydrophilic solutes such as the CGP-36742 drug (33).

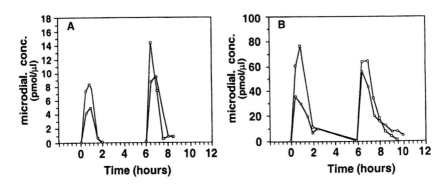

Figure 11 Kinetic analysis of CGP-36742 in microdialysates of (A) rat frontal cortex ($n = 2$) and (B) third ventricle ($n = 2$) after IV injections of 100 mg/kg at 0 and 6 h.

In vivo microdialysis in combination with micro-ESI-MS/MS provides highly specific monitoring of systemically administered pharmacological agents in target tissues. This work demonstrates that microdialysis combined with micro-ESI-MS/MS can be applied effectively for a study of the complete time course of an unmodified drug in rat brain in a single animal.

VI. CONCLUSIONS

The micro-ESI source provides a very high sensitivity, low flow rate adaptation of the basic ESI design. It allows higher liquid mass transfer and ionization efficiency because the spray plume is narrow, the droplet diameters small, and the entrance orifice close to the mass spectrometer. When coupled to reverse-phase packing in any one of several ways, the preconcentration, desalting, and fractionation advantages give rise to an extraordinarily high sensitivity, permitting the determination of peptides in the low attomole per microliter concentration range in many cases.

REFERENCES

1. MR Emmett, RM Caprioli. Micro-electrospray mass-spectrometry—Ultra-high-sensitivity analysis of peptides and proteins. J Am Soc Mass Spectrom 5:605–613, 1994.
2. PE Andren, MR Emmett, RM Caprioli. Micro-electrospray: Zeptomole/attomole per microliter sensitivity for peptides. J Am Soc Mass Spectrom 5:867–869, 1994.
3. M Wilm, M Mann. Analytical properties of the nanoelectrospray ion source. Anal Chem 68:1–8, 1996.
4. GA Valaskovic, NL Kelleher, DP Little, DJ Aaserud, FW McLafferty. Attomole-sensitivity electrospray source for large-molecule mass-spectrometry. Anal Chem 67:3802–3805, 1995.
5. MT Davis, DC Stahl, SA Hefta, TD Lee. A microscale electrospray interface for online, capillary liquid-chromatography tandem mass-spectrometry of complex peptide mixtures. Anal Chem 67:4549–4556, 1995.
6. MA Moseley, LJ Deterding, KB Tomer, JW Jorgenson. Nanoscale packed-capillary liquid chromatography coupled with mass spectrometry using a coaxial continuous-flow fast atom bombardment interface. Anal Chem 63:1467–1473, 1991.
7. RT Kennedy, MD Oates, BR Cooper, B Nickerson, JW Jorgenson. Microcolumn separations and the analysis of single cells. Science 246:57–63, 1989.
8. U Ungerstedt, C Pycock. Functional correlates of dopamine neurotransmission. Bull Schweiz Akad Med Wiss 30:44–55, 1974.

9. U Ungerstedt, A Hallstrom. In vivo microdialysis: A new approach to the analysis of neurotransmitters in the brain. Life Sci 41:861–864, 1987.
10. T Hokfelt, BJ Everitt, E Theodorsson-Norheim, M Goldstein. Occurrence of neurotensinlike immunoreactivity in subpopulations of hypothalamic, mesencephalic, and medullary catecholamine neurons. J Comp Neurol 222:543–559, 1984.
11. A Fernandez, P Jenner, CD Marsden, ML De Ceballos. Characterization of neurotensin-like immunoreactivity in human basal ganglia: Increased neurotensin levels in substantia nigra in Parkinson's disease. Peptides 16:339–346, 1995.
12. CB Nemeroff, WW Youngblood, PJ Manberg, AJ Prange Jr, JS Kizer. Regional brain concentrations of neuropeptides in Huntington's chorea and schizophrenia. Science 221:972–975, 1983.
13. E Widerlov, LH Lindstrom, G Besev, PJ Manberg, CB Nemeroff, GR Breese, JS Kizer, AJ Prange Jr. Subnormal CSF levels of neurotensin in a subgroup of schizophrenic patients: Normalization after neuroleptic treatment. Am J Psychiatr 139:1122–1126, 1982.
14. IN Ferrier, AJ Cross, JA Johnson, GW Roberts, TJ Crow, JA Corsellis, YC Lee, D O'Shaughnessy, TE Adrian, GP McGregor et al. Neuropeptides in Alzheimer type dementia. J Neurol Sci 62:159–170, 1983.
15. DE Hernandez, CM Richardson, CB Nemeroff, RC Orlando, S St-Pierre, F Rioux, AJ Prange Jr. Evidence for biological activity of two N-terminal fragments of neurotensin, neurotensin1–8 and neurotensin1–10. Brain Res 301:153–156, 1984.
16. RD Pinnock. Neurotensin depolarizes substantia nigra dopamine neurones. Brain Res 338:151–154, 1985.
17. RM Kobayashi, M Brown, W Vale. Regional distribution of neurotensin and somatostatin in rat brain. Brain Res 126:584–588, 1977.
18. PC Emson, M Goedert, P Horsfield, F Rioux, S St Pierre. The regional distribution and chromatographic characterisation of neurotensin-like immunoreactivity in the rat central nervous system. J Neurochem 38:992–999, 1982.
19. H Benveniste, AJ Hansen, NS Ottosen. Determination of brain interstitial concentrations by microdialysis. J Neurochem 52:1741–1750, 1989.
20. PF Morrison, PM Bungay, JK Hsiao, BA Ball, IN Mefford, RL Dedrick. Quantitative microdialysis: Analysis of transients and application to pharmacokinetics in brain. J Neurochem 57:103–119, 1991.
21. H Benveniste. Brain microdialysis. J Neurochem 52:1667–1679, 1989.
22. AJ Bean, MJ During, RH Roth. Stimulation-induced release of coexistent transmitters in the prefrontal cortex: An in vivo microdialysis study of dopamine and neurotensin release. J Neurochem 53:655–657, 1989.
23. AJ Bean, MJ During, AY Deutch, RH Roth. Effects of dopamine depletion on striatal neurotensin: Biochemical and immunohistochemical studies. J Neurosci 9:4430–4438, 1989.
24. W Huang, GR Hanson. Differential effect of haloperidol on release of neurotensin in extrapyramidal and limbic systems. Eur J Pharmacol 332:15–21, 1997.

25. JD Wagstaff, JW Gibb, GR Hanson. Dopamine D2-receptors regulate neurotensin release from nucleus accumbens and striatum as measured by in vivo microdialysis. Brain Res 721:196–203, 1996.
26. JD Wagstaff, JW Gibb, GR Hanson. Microdialysis assessment of methamphetamine-induced changes in extracellular neurotensin in the striatum and nucleus accumbens. Pharmacol Exp Ther 278:547–554, 1996.
27. NT Maidment, BJ Siddall, VR Rudolph, E Erdelyi, CJ Evans. Dual determination of extracellular cholecystokinin and neurotensin fragments in rat forebrain: Microdialysis combined with a sequential multiple antigen radioimmunoassay. Neuroscience 45:81–93, 1991.
28. H Zhang, PE Andrén, RM Caprioli. Micro-preparation procedure for high-sensitivity matrix-assisted laser desorption ionization mass spectrometry. J Mass Spectrom 30:1768–1771, 1995.
29. TP Davis, A Culling Berglund. Neuroleptic drug-treatment alters in vitro central neurotensin metabolism. Psychoneuroendocrinology 12:253–260, 1987.
30. TP Davis, TJ Gillespie, PNM Konings. Specificity of neurotensin metabolism by regional rat-brain slices. J Neurochem 58:608–617, 1992.
31. G Bonanno, M Raiteri. Multiple GABA(B) receptors. Trends Pharm Sci 14:259–261, 1993.
32. H Benveniste, PC Huttemeier. Microdialysis: Theory and application. Prog Neurobiol 35:195–215, 1990.
33. RD Broadwell, WA Banks. Cell biological perspective for the transcytosis of peptides and proteins through the mammalian blood-brain fluid barriers. In: WM Pardridge, ed. The Blood-Brain Barrier: Cellular and Molecular Biology. New York: Raven Press, 1993, pp xix, 476.

Index

Acetic acid, 26, 155
Actinonin, 170
ADC, analog-to-digital converter, 46
ADME, 211
Ammonium acetate, 29, 155
Ammonium hydroxide, 155
Antibiotics, 164
    aminoglycosides, 164
    β-lactams, 164
    cephalosporins, 164
    dicloxacillin, 164
    everninomicin, 173
    gentamicin B, 169
    macrolides, 164
    penicillins, 164
    quinolones, 164
    sarafloxacin, 164
    tylosin, 164

Averaging algorithm, 51

B-ions, 266
Bis(ethylhexyl) phthalate, 28
Blackbody infrared
    dissociation, 41
Breakdown curves, 21

CAD, collision-activated dissociation,
    271, 272, 273
Cesium iodide, 48
Charged residue mechanism,
    17, 18, 19
Charge exchange reactions, 254
Charge state distribution, multiple
    charged ions, 7, 8, 17, 21, 31, 43,
    51, 53, 153, 154, 285
Charge-stripping reactions, 33

Chemical derivatization, 160, 212, 223, 370
Cholesterol, 160
CID, collision-induced dissociation, 6, 13, 19, 29, 32, 36, 41, 43, 47, 150, 216, 223, 303
Combinatorial chemistry, 149, 159, 187–206
 affinity selection, 196
 chemiluminescent nitrogen detector, 194
 ELSD, Evaporative light scattering detection, 196
 parallel synthesis, 187
 pulsed ultrafiltration, 196
 solid-phase synthesis, 187, 188
Complex structures
 actinonin, 172
 βB3, 273
 CCR5 antagonist, 156, 157
 celestone phosphate, 159
 cerebroside, 179
 cholesterol, 161
 dicloxacillin, 166
 gentamicin B, 171
 metoprolol, 216
 mometasone furoate, 163
 N-linked oligosaccharide, 294
 NT1-12, 424
 pentapeptide, 154
 quinine, 38
 ras-GDP-Sch 54292, 373
 reserpine, 20
 sarafloxacin, 165
 Sch 54341, 370
 Sch 54292, 371
 Sch 27899, 173, 178
 tylosin A, 168
Component detection algorithm, 58
Corona discharge, 6, 11, 33, 160

Deconvolution algorithm, 54
Desolvation, 6, 7, 11, 13, 21, 31
Dimethyl formamide, 27
Dimethyl sulfoxide, 26
Direct liquid introduction, 2
Double-focusing, 177

Drug metabolism, 211
 biotransformation, 226
 duty cycle, 219, 232
 pharmacokinetics, 211, 220
 quantitative analysis, 219
 SIM, selected ion monitoring, 25, 35, 37, 232
 SRM, selected reaction monitoring, 25, 35, 36, 38, 213
 toxicokinetics, 220
Dynamic range, 17, 313

Edman sequencing, 269, 293
Electron capture dissociation, 303
Electrospray
 ion spray, 8, 9, 21, 24
 microspray, 23
 nanospray, 21, 23, 24, 34, 53, 105–141
 picospray, 411
 z-spray, 23
Exact-mass measurement, 177, 179, 203, 394

FIA, flow injection analysis, 151, 194
FK506, 363
Formic acid, 155
FTICR, fourier transform ion cyclotron resonance, 3, 9, 39, 40, 42, 48, 198, 254, 315
FWHM, full width at half-maximum, 50

GC/MS, 34, 42, 46, 190, 211, 212, 219
GDP, 361, 363
Glycoproteins, 283
 CHO IL-5, 288
 CHO IL-4, 286, 290, 296, 297, 299, 302
 IL-5, 298
 IL-4, 286, 291
 site heterogeneity, 286
GTP, 362, 363

H/D exchange, 43, 390
 protein folding, 398
 unfolding dynamics, 400

Index

In-source CID, 19, 35
Intercalators, 383
Ion evaporation, 17, 18, 19, 115
Ion-ion reactions, 43
Ionization, 1
   APCI, 2, 3, 6, 7, 9, 10, 11, 12, 14, 29, 33, 34, 160, 190, 211
   API, 3, 7, 10, 12, 29, 211
   CI, 1, 2, 3, 7, 160, 168
   DCI, 6
   EI, 1, 2, 3, 42, 160, 168
   electrohydrodynamic ionization, 6, 15
   FAB, 1, 3, 12, 29, 168, 251, 295, 296
   FD, Field desorption, 4, 12
   LSIMS, 4
   MALDI, 4, 12, 34, 42, 126, 257, 283, 295, 302, 363, 389, 422
   PDMS, 4
   SIMS, 4, 6, 12
   thermospray, 3, 7, 190
Ion-molecule reactions, 41, 43
Isopropanol, 27
ITMS, 9, 213, 214

LC/MS/MS, 132, 162, 265

MAGIC, 3
Magnetic sector, 36, 252
Maximum entropy method, 56
Microdialysis, 28, 255, 411
Minor groove binders, 383
Mometasone furoate, 162
MS/MS, 19, 35, 36, 44, 46, 107, 119, 126, 128, 135, 150, 156, 219, 227, 271
MS^n, 164, 223, 273, 312, 323
Multiplicative correlation algorithm, 56
Multisprayer, 238, 241

Noncovalent interactions, 361
   binding affinity, 384
   gamma-interferon, 361, 377
   GroEL complex, 379
   HCV-Zn, 374

[Noncovalent interactions]
   ras-GDP, Ras-guanosine diphosphate, 361, 362, 363, 364, 367, 371
   ras-GTP, Ras-guanosine triphosphate, 362
   stoichiometry, 262, 384

PAHs, polynuclear aromatic hydrocarbons, 16
Partial correlation method, 56
Perfluoroalkyl triazines, 49
Pharmaceuticals, 149
   degradation products, 161
   high throughput analysis, 159
   impurities, 161
   molecular weight determination, 151
Photodissociation, 41, 150
   infrared multiphoton laser-induced dissociation, 324
   IRMPD, 337
Pneumatic nebulization, 21, 22, 107
Polyethylene glycol, 27
Polypropylene glycol, 27
Porphyrin, 16
Proteins, peptides:
   bovine fetuin, 301
   bovine insulin, 3
   bovine lactalbumin, 397
   bovine pancreatic trypsin inhibitor, 259
   bovine serum albumin, 252
   calmodulin, 373, 374, 396
   cytochrome c, 405
   FKBP, 399
   FPT, farnesyl protein transferase, 362
   gramicidin S, 31
   insulin, 251, 252
   lentil lectin, 254
   leucine enkephalin, 31
   leucine zipper peptides, 376
   lysozyme, 31, 37, 40
   melittin, 273, 373, 374
   myoglobin, 31, 37, 40, 256
   neurotensin, 415, 418
   ovalbumin, 294
   rabbit muscle aldolase, 400

ribonuclease B, 294
ubiquitin, 53, 273
ultramark 1621, 49
Proteomics, 256, 307–354, 390, 407
  accurate mass tag, 315, 327
  enzymatic digestion, 260
  gene expression, 308
  ICAT, isotope coded affinity tags, 333, 351, 352, 353
  pepsin digestion, 393, 398, 402, 407
  phosphorylation, 258, 284, 293, 333
  post-translational modifications, 322, 326, 332, 333
  proteolyic digestion, 119, 295
  Sequest, 265
  trypsin digestion, 268
Proton-bound dimers, 151

QTOF, 47, 198, 201, 223, 254, 268, 290, 379, 417

Radio-immunoassay, 420
Rayleigh stability limit, 17
Reconstructed ion chromatogram, 176
Reflectron, 44
Resolving power, 37, 40, 50, 52, 176, 179, 216, 254, 258, 273, 311, 394

Separation techniques
  capillary electrochromatography, 2
  CE, capillary electrophoresis, 2, 9, 18, 24, 25, 34, 107, 116, 218, 263, 266, 288
  CIEF, 316, 317, 320, 342, 346, 354
  electrophoretic mobility, 268
  gel electrophoresis, 106, 135, 256, 265
  gel permeation chromatography, 2
  IEF, isoelectric focusing, 258
  ion-exchange chromatography, 28

[Separation techniques]
  normal phase liquid chromatography, 34
  polyacrylamide gel electrophoresis, 257, 309
  SDS-PAGE, 129, 288
  size exclusion chromatography, 197, 269
  supercritical fluid chromatography, 2, 196
  thin-layer chromatography, 189
  2D-PAGE, 309, 310, 315, 316, 317, 320, 327, 341
SID, surface-induced dissociation, 47
Sodium dodecyl sulfate, 28, 310
Solid-phase extraction, 220, 230
SORI, sustained off-resonance irradiation, 41, 323, 324
Space-charge, 40, 42, 43
Sphingolipids, 177
  cerebrosides, 177, 179, 180
Structure-activity relationship, 257
SWIFT, stored waveform inverse fourier transform, 323, 338

Tandem-in-space, 41
Tandem-in-time, 41
Tandem mass spectrometry, 35, 38, 107, 150, 271
Taylor cone, 107, 109, 113
TDC, time-to-digital converter, 45, 47
Tetrahydrofuran, 27
Time-of-flight, 6, 25, 39, 44, 176, 190, 215, 252, 289
Trifluoroacetic acid, 263
Triple-quadrupole, 36, 212, 299

Unit resolution, 50

Y-ions, 266

CPSIA information can be obtained at www.ICGtesting.com
Printed in the USA
LVOW08*0411240414

382988LV00004B/26/P